METHODS IN MOLECULAR BIOLOGY™

Series Editor
John M. Walker
School of Life Sciences
University of Hertfordshire
Hatfield, Hertfordshire, AL10 9AB, UK

For further volumes:
http://www.springer.com/series/7651

Data Production and Analysis in Population Genomics

Methods and Protocols

Edited by

François Pompanon and Aurélie Bonin

Université Grenoble I, Laboratoire d'Ecologie Alpine,
Grenoble, France

 Humana Press

Editors
François Pompanon
Université Grenoble 1
Laboratoire d'Ecologie Alpine
CNRS-UMR5553, 2233 rue de la Piscine
38041 Grenoble, Cedex 09, France

Aurélie Bonin
Université Grenoble 1
Laboratoire d'Ecologie Alpine
CNRS-UMR5553, 2233 rue de la Piscine
38041 Grenoble, Cedex 09, France

ISSN 1064-3745   ISSN 1940-6029 (electronic)
ISBN 978-1-61779-869-6   ISBN 978-1-61779-870-2 (eBook)
DOI 10.1007/978-1-61779-870-2
Springer New York Heidelberg Dordrecht London

Library of Congress Control Number: 2012937649

Printed on acid-free paper

Humana Press is a brand of Springer
Springer is part of Springer Science+Business Media (www.springer.com)

Preface

Population genomics is a recently emerged discipline, which aims at understanding how evolutionary processes influence genetic variation across genomes. Its underlying principle involves genome-wide screening of several hundred or thousand loci across many individuals from several populations, in order to disentangle locus-specific (e.g., mutation, recombination, selection) from genome-wide (e.g., genetic drift, gene flow, inbreeding) effects. With such wealth of data, classical population genetics statistics, such as population differentiation or linkage disequilibrium, can then be studied as variables varying along the genome to understand which evolutionary forces drive its evolutionary dynamics.

In the past, characterizing polymorphism at the genome scale has been greatly facilitated by technological advances like capillary electrophoresis, which allows the simultaneous genotyping of numerous loci, such as Amplified Fragment Length Polymorphism (AFLP) markers, or DNA chips for Single Nucleotide Polymorphism (SNP) typing. Today, in the era of cheaper next-generation sequencing (NGS), population genomics takes on a new dimension and is meant to soar. Because it is not as daunting anymore to obtain whole genome data for any species of interest, population genomics is now conceivable in a wide range of fields, from medicine and pharmacology to ecology and evolutionary biology. However, because of the lack of reference genome and of enough a priori data on the polymorphism, population genomics analyses of populations will still involve higher constraints for researchers working on nonmodel organisms, as regards the choice of the genotyping/sequencing technique or that of the analysis methods. This observation guided our selection of chapters, which purposely put emphasis on various protocols and methods that are applicable to species where genomics resources are still scarce.

Setting up a population genomics study implies facing three main challenges. First, one has to devise a sampling and/or experimental design suitable to address the biological question of interest. Then, one has to implement the best genotyping or sequencing method to obtain the required data, given the time and cost constraints as well as the other genetic resources already available. Finally, one has to make the most of the (generally huge) dataset produced and use appropriate analysis methods to reach a biologically relevant conclusion.

The first challenge is addressed in Part I of this book, which is entitled "Sampling and Experimental Design." Chapter 1 is primarily concerned with the optimization of sampling and explains how basing a sampling design on model-based stratification using the climatic and/or biological spaces may be more efficient than basing it on the geographic space. Chapter 2 deals with the largely overlooked issue of sample tagging in NGS and presents bioinformatic tools for creating sets of tag and managing multiplexes of samples. Chapter 3 covers all aspects of SNP discovery in nonmodel organisms using Roche transcriptome sequencing, from tissue choice to publication of the work. Chapter 4 introduces ISIF, a program that helps design a successful AFLP experiment and choose the best restriction enzymes and combinations of selective bases based on a priori information from a close reference genome.

Part II, "Producing Data" brings together lab protocols aiming at generating high-quality genotyping or sequencing data. Chapter 5 focuses on Diversity Arrays Technology (DArT), a high-throughput genotyping technique that has already proven its worth for a wide range of nonmodel species, and Chapter 6 reports two protocols allowing the isolation and sequencing of AFLP fragments of interest for subsequent analysis.

When NGS is a reasonable option, many labs decide to outsource the sequencing task to a platform/company that can afford to acquire and maintain next-generation sequencers. Moreover, the fast evolution of NGS techniques is likely to make the current protocols quickly obsolete. Therefore, we decided not to emphasize such protocols unless they address a very specific problem. For example, Chapter 7 deals with whole-genome sequencing of ancient (or degraded) DNA, Chapter 8 describes transcriptome sequencing using the 454 platform of Roche, and Chapter 9 is dedicated to the practical implementation of paired-end Illumina sequencing of RAD (Restriction-site Associated DNA) fragments to reveal SNP markers in nonmodel organisms.

Part III entitled "Analyzing Data" compiles new statistical methods and bioinformatic tools to meet the last challenge pertaining to data management and analysis in population genomics. Chapter 10 presents RawGeno, a free program designed to automatically score AFLP profiles and identify reliable markers. Chapter 11 gives an overview of the most efficient methods allowing haplotype reconstruction. Chapter 12 tackles the issue of allele versus paralog determination in 454 transcriptomic data, and proposes a series of bioinformatic scripts to identify true allelic clusters in new datasets. Chapter 13 addresses the common problem of multiple-testing related to the typically huge numbers of loci analyzed in population genomics datasets, which can lead to the detection of many false positives. Chapter 14 outlines the analytical specificities of RAD-seq and similar data and introduces a new computational pipeline called Stacks for the analysis of RAD-seq data in organisms with and without a reference genome. Chapter 15 describes METAPOP, a computer application that provides an analysis of gene and allelic diversity in subdivided populations from molecular genotype or coancestry data.

One of the central goals of population genomics is to identify loci underlying phenotypic variation or adaptation in natural populations and this is reflected in the last three chapters of this book. Chapter 16 presents the R-DetSel software package, which implements a coalescent-based method to detect markers that deviate from neutral expectation in pairwise comparisons of diverging populations. Chapter 17 shows how allele distribution models can be exploited to detect loci displaying signatures of selection and illustrates this approach in cichlid fish using the software MATSAM. Finally, Chapter 18 presents a new method allowing the quantification of the genetic component of phenotypic variance in wild populations using phenotypic trait values and multilocus genetic data available simultaneously for a sample of individuals from the same population.

Whether presenting a specific protocol or an overview of several methods, each chapter aims at providing guidelines to help choose and implement the best experimental or analytical strategy for a given purpose. In this respect, the Notes section is particularly valuable since it gathers practical information and tips rarely highlighted in scientific articles. The methods and protocols described in these chapters were selected because they are likely to be of interest to a wide readership and we hope that that they will contribute to the success of many population genomics studies in the future.

Grenoble, France   *François Pompanon*
                                                                                                      Aurélie Bonin

Contents

PART III   ANALYZING DATA

Contributors

CÉCILE H. ALBERT • *Laboratoire d'Ecologie Alpine, UMR CNRS 5553, Université Joseph Fourier, Grenoble, France; Department of Biology, McGill University, Montreal, QC, Canada*

DAVID E. ALQUEZAR-PLANAS • *Centre for GeoGenetics, Natural History Museum of Denmark, Copenhagen, Denmark*

NADIR ALVAREZ • *Department of Ecology and Evolution, University of Lausanne, Lausanne, Switzerland*

NILS ARRIGO • *Laboratory of Evolutionary Botany, Institute of Biology, University of Neuchâtel, Neuchâtel, Switzerland; Department of Ecology and Evolution, University of Lausanne, Lausanne, Switzerland*

MALGORZATA ASCHENBRENNER-KILIAN • *Diversity Arrays Technology Pty Ltd, Yarralumla, Canberra, ACT, Australia*

ETIENNE BEZAULT • *Department of Aquatic Ecology & Evolution, Institute of Ecology and Evolution (IEE), University of Bern, Bern, Switzerland; Centre of Ecology, Evolution and Biogeochemistry, Swiss Federal Institute for Aquatic Science and Technology (EAWAG), Kastanienbaum, Switzerland; Department of Biology, Reed College Portland, OR, USA*

CHRISTELLE BLASSIAU • *Laboratoire Stress Abiotiques et Différenciation des Végétaux Cultivés, UMR INRA/USTL 1281, Villeneuve d'Ascq, France*

HÉLÈNE BLOIS • *Diversity Arrays Technology Pty Ltd, Yarralumla, Canberra, ACT, Australia*

AURÉLIE BONIN • *Laboratoire d'Ecologie Alpine, UMR CNRS 5553, Université Joseph Fourier, Grenoble, France*

MATTHIEU BOUAZIZ • *Department of Biostatistics, Pharnext, Paris, France; Statistics and Genome laboratory, UMR CNRS 8071, USC INRA, University of Evry, Val d'Essonne, France*

ARMANDO CABALLERO • *Departamento de Bioquímica, Genética e Inmunología, Facultad de Biología, Universidad de Vigo, Vigo, Spain*

VANESSA CAIG • *Diversity Arrays Technology Pty Ltd, Yarralumla, Canberra, ACT, Australia*

JASON CARLING • *Diversity Arrays Technology Pty Ltd, Yarralumla, Canberra, ACT, Australia*

JULIAN CATCHEN • *Center for Ecology and Evolutionary Biology, University of Oregon, Eugene, OR, USA*

CYRIL CAYLA • *Diversity Arrays Technology Pty Ltd, Yarralumla, Canberra, ACT, Australia*

JEONG-HYEON CHOI • *The Center for Genomics and Bioinformatics, Indiana University, Bloomington, IN, USA*

ERIC COISSAC • *Laboratoire d'Ecologie Alpine, UMR CNRS 5553, Université Joseph Fourier, Grenoble, France*

WILLIAM A. CRESKO • *Center for Ecology and Evolutionary Biology, University of Oregon, Eugene, OR, USA*

OLIVIER DELANEAU • *Chaire de Bioinformatique, Conservatoire National des Arts et Métiers, Paris, France*

LAURENCE DESPRÉS • *Laboratoire d'Ecologie Alpine, UMR CNRS 5553, Université Joseph Fourier, Grenoble, France*

KATRINA M. DLUGOSCH • *Department of Ecology & Evolutionary Biology, University of Arizona, Tucson, AZ, USA*

PAUL D. ETTER • *Institute of Molecular Biology, University of Oregon, Eugene, OR, USA*

MARGARET EVERS • *Diversity Arrays Technology Pty Ltd, Yarralumla, Canberra, ACT, Australia*

SARAH L. FORDYCE • *Centre for GeoGenetics, Natural History Museum of Denmark, Copenhagen, Denmark*

MICKAËL GUEDJ • *Department of Biostatistics, Pharnext, Paris, France*

KATARZYNA HELLER-USZYNSKA • *Diversity Arrays Technology Pty Ltd, Yarralumla, Canberra, ACT, Australia*

CARLOS M. HERRERA • *Estación Biológica de Doñana, Consejo Superior de Investigaciones Científicas (CSIC), Isla de La Cartuja, Sevilla, Spain*

PAUL A. HOHENLOHE • *Center for Ecology and Evolutionary Biology, University of Oregon, Eugene, OR, USA*

PUTHICK HOK • *Diversity Arrays Technology Pty Ltd, Yarralumla, Canberra, ACT, Australia*

ROLF HOLDEREGGER • *WSL Swiss Federal Research Institute, Birmensdorf, Switzerland*

COLLEEN HOPPER • *Diversity Arrays Technology Pty Ltd, Yarralumla, Canberra, ACT, Australia*

ERIC HUTTNER • *Diversity Arrays Technology Pty Ltd, Yarralumla, Canberra, ACT, Australia*

DAMIAN JACCOUD • *Diversity Arrays Technology Pty Ltd, Yarralumla, Canberra, ACT, Australia*

MARINE JEANMOUGIN • *Department of Biostatistics, Pharnext, Paris, France; Statistics and Genome laboratory, UMR CNRS 8071, USC INRA, University of Evry, Val d'Essonne, France*

ERIC JOHNSON • *Institute of Molecular Biology, University of Oregon, Eugene, OR, USA*

STÉPHANE JOOST • *Laboratory of Geographic Information Systems (LASIG), Institute of Environmental Engineering (IIE), Ecole Polytechnique Fédérale de Lausanne (EPFL), Lausanne, Switzerland*

MICHAEL KALBERMATTEN • *Laboratory of Geographic Information Systems (LASIG), Institute of Environmental Engineering (IIE), Ecole Polytechnique Fédérale de Lausanne (EPFL), Lausanne, Switzerland*

NOLAN C. KANE • *Botany Department, University of British Columbia, Vancouver, BC, Canada*

ANDRZEJ KILIAN • *Diversity Arrays Technology Pty Ltd, Yarralumla, Canberra, ACT, Australia*

ZHAO LAI • *The Center for Genomics and Bioinformatics, Indiana University, Bloomington, IN, USA*

STÉPHANIE MANEL • *Laboratoire Population Environnement Développement, UMR 151 UP/IRD, Université Aix-Marseille, Marseille, France; Laboratoire d'Ecologie Alpine, UMR CNRS 5553, Université Joseph Fourier, Grenoble, France*

CLAIRE-LISE MEYER • *Laboratoire Génétique et Evolution des Populations Végétales, FRE CNRS 3268, Université Lille 1, Cité Scientifique, Villeneuve d'Ascq, France*

MARGOT PARIS • *Institute of Integrative Biology, ETH Zurich, Zurich, Switzerland*

KAIMAN PENG • *Diversity Arrays Technology Pty Ltd, Yarralumla, Canberra, ACT, Australia*

ANDRÉS PÉREZ-FIGUEROA • *Facultad de Biología, Departamento de Bioquímica, Genética e Inmunología Facultad de Biología, Universidad de Vigo, Vigo, Spain*

FRANÇOIS POMPANON • *Laboratoire d'Ecologie Alpine, UMR CNRS 5553, Université Joseph Fourier, Grenoble, France*

LOREN H. RIESEBERG • *The Center for Genomics and Bioinformatics, Indiana University, Bloomington, IN, USA; Botany Department, University of British Columbia, Vancouver, BC, Canada*

SILVIA T. RODRÍGUEZ-RAMILO • *Departamento de Bioquímica, Genética e Inmunología Facultad de Biología, Universidad de Vigo, Vigo, Spain*

OLE SEEHAUSEN • *Department of Aquatic Ecology & Evolution, Institute of Ecology and Evolution (IEE), University of Bern, Bern, Switzerland; Centre of Ecology, Evolution and Biogeochemistry, Swiss Federal Institute for Aquatic Science and Technology (EAWAG), Kastanienbaum, Switzerland*

PIERRE TABERLET • *Laboratoire d'Ecologie Alpine, UMR CNRS 5553, Université Joseph Fourier, Grenoble, France*

GRZEGORZ USZYNSKI • *Diversity Arrays Technology Pty Ltd, Yarralumla, Canberra, ACT, Australia*

RENAUD VITALIS • *CNRS, INRA, UMR CBGP (INRA–IRD–CIRAD–Montpellier SupAgro), Montferrier-sur-Lez Cedex, France*

XINGUO WANG • *David H. Murdock Research Institute, Kannapolis, NC, USA*

PETER WENZL • *Diversity Arrays Technology Pty Ltd, Yarralumla, Canberra, ACT, Australia*

CHRISTOPHER W. WHEAT • *Department of Biological and Environmental Sciences, University of Helsinki, Helsinki, Finland; Centre for Ecology and Conservation, School of Biosciences, University of Exeter, Penryn, Cornwall, UK*

LING XIA • *Diversity Arrays Technology Pty Ltd, Yarralumla, Canberra, ACT, Australia*

NIGEL G. YOCCOZ • *Fishers and Economics, Department of Arctic and Marine Biology, Faculty of Biosciences, University of Tromsø, Tromsø, Norway*

JEAN-FRANÇOIS ZAGURY • *Conservatoire National des Arts et Metiers, Paris, France*

YI ZOU • *The Center for Genomics and Bioinformatics, Indiana University, Bloomington, IN, USA*

Part I

Sampling and Experimental Design

Chapter 1

Sampling in Landscape Genomics

Stéphanie Manel, Cécile H. Albert, and Nigel G. Yoccoz

Abstract

Landscape genomics, based on the sampling of individuals genotyped for a large number of markers, may lead to the identification of regions of the genome correlated to selection pressures caused by the environment. In this chapter, we discuss sampling strategies to be used in a landscape genomics approach. We suggest that designs based on model-based stratification using the climatic and/or biological spaces are in general more efficient than designs based on the geographic space. More work is needed to identify designs that allow disentangling environmental selection pressures versus other processes such as range expansions or hierarchical population structure.

Key words: Landscape genomics, Sampling, Population genetics, Genetic adaptive diversity

1. Introduction

Various methods allow the identification of genomic regions under the influence of natural or artificial selection, such as quantitative trait loci (QTL) analysis, association mapping and quantitative genetics studies (1). However, for non-model species, for which most of the genome is unknown, the task can be laborious. A landscape genomics approach, based on correlation between locus frequencies and environmental variables used to detect regions of the genome under selection, is a promising approach (2). Landscape genomics has the remarkable characteristic of not requiring phenotypic data on the adaptive trait(s) of interest, but focuses on the genotypes and assumed selective pressures.

The landscape genomics approach is based on two kinds of survey: a survey of many genetic loci (typically several hundred or more) scattered throughout the genome and a survey of individuals. Both should be designed in order to efficiently and unbiasedly

François Pompanon and Aurélie Bonin (eds.), *Data Production and Analysis in Population Genomics: Methods and Protocols*, Methods in Molecular Biology, vol. 888, DOI 10.1007/978-1-61779-870-2_1, © Springer Science+Business Media New York 2012

discover genomic regions under selection, either directly or more likely through physical linkage (3). Sampling design is therefore a fundamental aspect of landscape genomics, but it has not been so thoroughly investigated as the analytical methods needed to understand the correlative patterns. Compared to the traditional genetic approach focused on differentiation between populations (e.g., Fst), landscape genomics analyses allele frequencies or proportions at each observed location and at one locus/gene and aims at understanding the kinds of association (linear, quadratic, etc.) observed depending on the level of a spatial hierarchy (e.g., individual, population, metapopulation; Fig. 1). The ultimate objective of a landscape genomics study is to estimate the causal relationships linking environmental variables measuring selection pressures and the structure of the genome, and how these relationships are affected by, for example, the ecological context or geographical constraints.

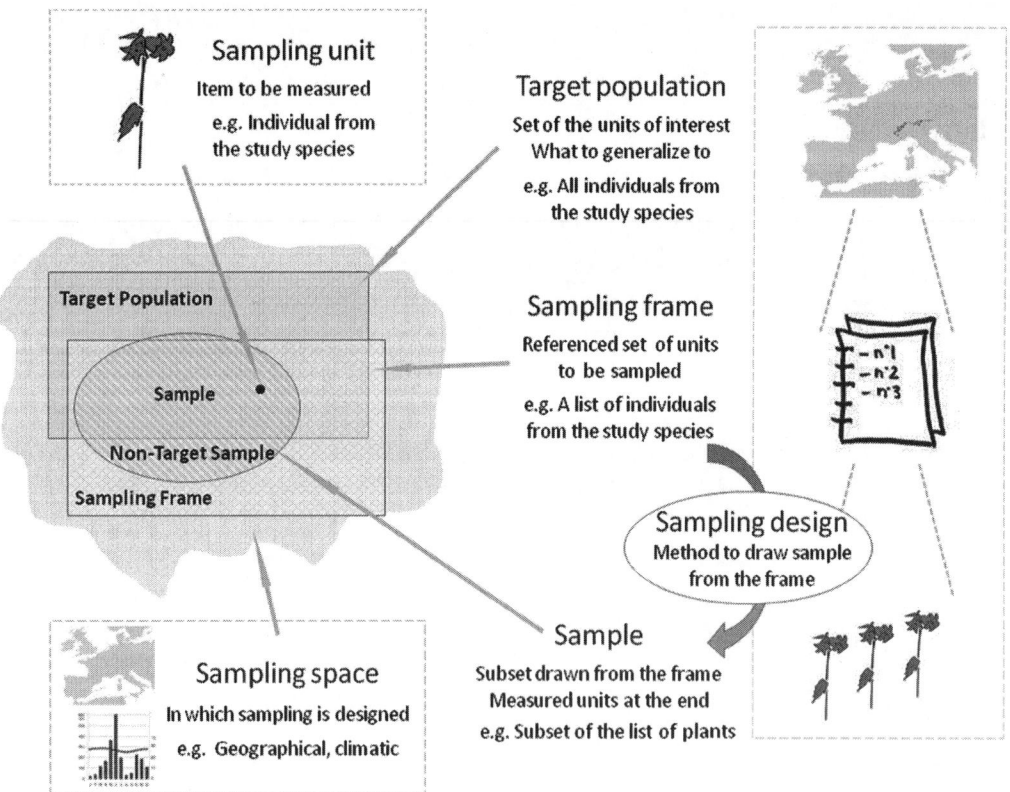

Fig. 1. Sampling in population genomics: concepts and vocabulary—adapted from ref. 14 for the specific objective of landscape genomics.

2. Conceptual Basis for Elaborating a Sampling Strategy in Landscape Genomics

2.1. State of the Art in Sampling in Population Genetics

Population genetics has long recognized the importance of sampling strategies (Fig. 1). Until recently, geneticists were, however, mostly interested in the effects of sampling effort—i.e., how many individuals/population to sample—and did not fully elaborate sampling designs—i.e., which individuals to sample, in particular, with respect to environmental drivers. For instance, Muirhead et al. (4) quantified the effect of the numbers of individuals sampled in populations, the number of populations surveyed, and the degree of genetic differentiation between populations for correct assignment of an individual to its source population. Their review highlighted that many studies failed to provide a rationale for their choice of sampling design and to discuss the limitations of these choices. Actually, Muirhead et al. (4) focused more on sample sizes than on sampling strategies.

Geneticists were, however, aware of the effect of using inappropriate sampling designs, especially haphazard/opportunistic sampling (5).

However, neighbour mating leading to the pattern of isolation by distance may create difficulty to identify the correct sampling strategy. For example, Schwartz et al. (6) used simulations to understand the effect of sampling schemes on population clustering. They simulated one population resulting from (a) individuals following a neighbour mating scheme leading to isolation by distance and (b) individuals randomly distributed in the landscape. From these simulated data, they applied six different sampling schemes: (1) Trapper sampling, 360 individuals drawn randomly across the entire simulated landscape, (2) Research sampling, individuals drawn in 10 clusters of 36, i.e., all individuals are collected from small portion of the landscape, (3) Corner sampling, similar to the previous one, but sampling in two distinct areas of the landscape, (4) Line transect sampling, 10 equally spaced lines of 36 individuals, (5) Multigeneration trapper sampling, 120 individuals sampled at each generation t_{18} to t_{20}, (6) Mixed sampling, samples are obtained from hunters, harvesters, and trappers and through research effort. When isolation by distance was present in the data, clustering methods always inferred a higher number of populations than expected whatever the chosen sampling scheme.

Finally, while we focus here mainly on the spatial aspects of sampling (i.e., where to sample individuals), the temporal aspects should also be considered (e.g., (7)). We do not address, however, how to optimize the temporal components of sampling, as little is actually known (8).

In practice, studies looking for loci linked to selection pressures have not used specific sampling strategy except that of Poncet et al. (9). They sampled *Arabis alpina* populations in the French

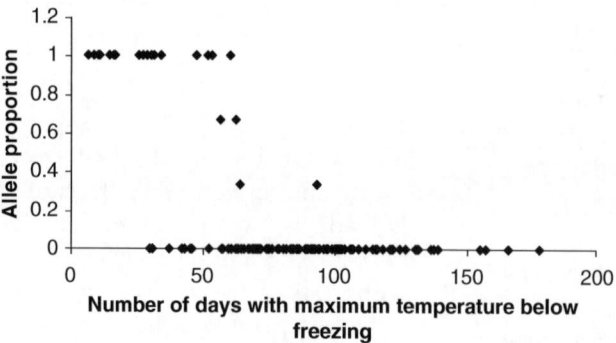

Fig. 2. Example of response curve (logistic regression) for allele proportion and temperature for the plant species *Arabis alpina* in the European Alps. Data used were described in Manel et al. (34).

Alps to maximize the variation of climate variables such as temperatures and precipitation, conditional on having a high probability of finding the species. This was done by building a species distribution model fitted using existing occurrence and climatic databases in the French Alps. Ninety-nine sampled locations were then chosen where the species probability of presence was high. Sampled individuals were then genotyped for 825 amplified fragment length polymorphism loci (AFLPs) and each locus correlated to the environmental variables. An example of the logistic response curve observed between a locus and temperature is shown in Fig. 2.

This methodology allowed to highlight that temperature and precipitation were the two major drivers of the pattern of adaptive genetic variation for this alpine plant.

2.2. A Major Issue: Individual Versus Population Sampling

As selective pressures mostly act on individuals, the fundamental sampling unit is the individual (Fig. 1). However, both natural constraints (such as fish in lakes, birds on islands, or butterflies on meadows) and processes occurring within groups of individuals (e.g., competitive interactions among individuals, local exchanges among demes) lead to consider study systems at higher hierarchical levels such as clusters of individuals or populations. It is indeed of interest to assess both within and between populations patterns and processes, so as to develop better predictive models of genetic structure. Studies of adaptive genetic variation can benefit by genotyping many populations in a broad range of conditions (10). However, it is not always possible to sample both broad range of conditions and a large number of individuals in each of these conditions. Landscape genomics will however probably gain from using an explicit individual approach, but still focusing on sampling a broader range of environmental conditions and taking into account spatial structure. However, this requires more tests about sampling; Only one individual in each location may give insight into adaptive patterns and hence variability between populations. The large number of markers associated with an advanced sampling design of individuals by location, together with advanced statistical modelling, should result in the highest accuracy of genotype–environment associations, especially by removing the effect of confounding variables resulting in spurious correlations (11).

2.3. State of the Art in Sampling in Ecology

Sampling is a key issue for answering most ecological and evolutionary questions. The importance of developing a rigorous sampling design tailored to specific questions has already been discussed in the ecological and sampling literature and has provided useful tools and recommendations to sample and analyze ecological data (12). However, sampling issues are often difficult to overcome in ecological studies due to apparent inconsistencies between theory and practice, often leading to the implementation of ad hoc sampling designs that suffer from unknown biases. Until recently, analyses of sampling surveys and statistical modelling have been seen as two distinct statistical schools. On the one hand, survey samplers were sceptical of the use of models, considering them as crude approximations that would lead to biased inferences. On the other hand, modellers thought that survey samplers were interested only in the estimation of population descriptive statistics (such as totals and means) but less in estimating causal relationships, the latter requiring models. Nowadays survey samplers acknowledge that models are often needed for analyzing survey samples, whereas modellers recognize the importance of incorporating the survey design in the statistical analysis (13). This merging of the different approaches, however, has not yet reached most of ecological, evolutionary, or genetics applications, and advanced sampling design together with appropriate analytical methods remain very rare. Albert et al. (14) illustrated some of the main sampling issues in ecology using simulations (see ref. 8 for how simulations can be designed for genetic data). They demonstrated the inefficiency of classical sampling designs (e.g., random sampling) to estimate response curves (e.g., relationships between traits and climatic gradients). In particular, they showed that response-surface designs (15), which have rarely been used in ecology (but see, ref. 16), are much more efficient than more traditional methods. In addition, they highlighted the consequences of selecting appropriate sampling space and sampling design, as well as of using prior knowledge, simulation studies, or model-based stratification on accurately estimating complex relationships. Better choosing a sampling design (and not only the sample size) could help better understanding ecological patterns and processes generating these patterns.

3. Methods

3.1. The Assumption to Test

In landscape genomics, sampling should aim at finding which loci are under selection or not, i.e., whether the correlation between allele frequency/proportion and the selection pressure (e.g., environmental gradient) is different from 0, and if this difference is not confounded by other factors such as historical changes or population heterogeneity (17, 18). Defining an appropriate sampling design will thus determine the patterns that are subsequently found and their interpretation. An appropriate design should allow for (1) efficient and unbiased estimation of parameters of interest and (2) testing underlying assumptions (or appropriate modelling when it is unclear which sets of assumptions hold). "All models are wrong, but some are better than others" (13) and a good design should also allow for choosing the better models.

Note that we do not focus on what type of models may be used to correlate allele frequencies to environment. For that a wide range of models can be used (e.g., (11, 19, 20)). The efficiency of a sampling design will depend on how the data are analyzed (i.e., which model is used), and using simulations to explore different sampling designs require that the modelling framework is also known before the data are collected.

3.2. Defining an Appropriate Sampling in Landscape Genomics

1. An often neglected aspect of sampling design is the definition of the target population, i.e., the statistical population of units that is the object of inference (Fig. 1). As landscape genomics use individuals as sampling units, a target population should include all individuals for which one want to generalize observed and estimated patterns, i.e., the causal relationship between environmental drivers and allele frequencies.

2. What is the appropriate space to use to develop efficient stratification ((14), Fig. 3)? The set of individuals that can be sampled/collected can indeed be stratified in different spaces (21): geographic space (e.g., latitude, longitude), topographic space (e.g., elevation, curvature), climatic space (e.g., temperature, precipitations), and biological space (e.g., habitat suitability, population growth rate, or density). As illustrated in Fig. 3, the geographical space, in which the researcher moves to collect individuals, is probably not the most appropriate sampling space in landscape genomics. Switching from one space to another can indeed have important consequences. For instance, the transect sampling in Fig. 3 leads to a complete coverage of the latitudinal gradient, but to an incomplete coverage of the altitudinal and temperature gradients and of the biological response. Sampling in the geographical space would then probably lead to problems of truncated gradients and would not allow uncovering the underlying patterns accurately. For

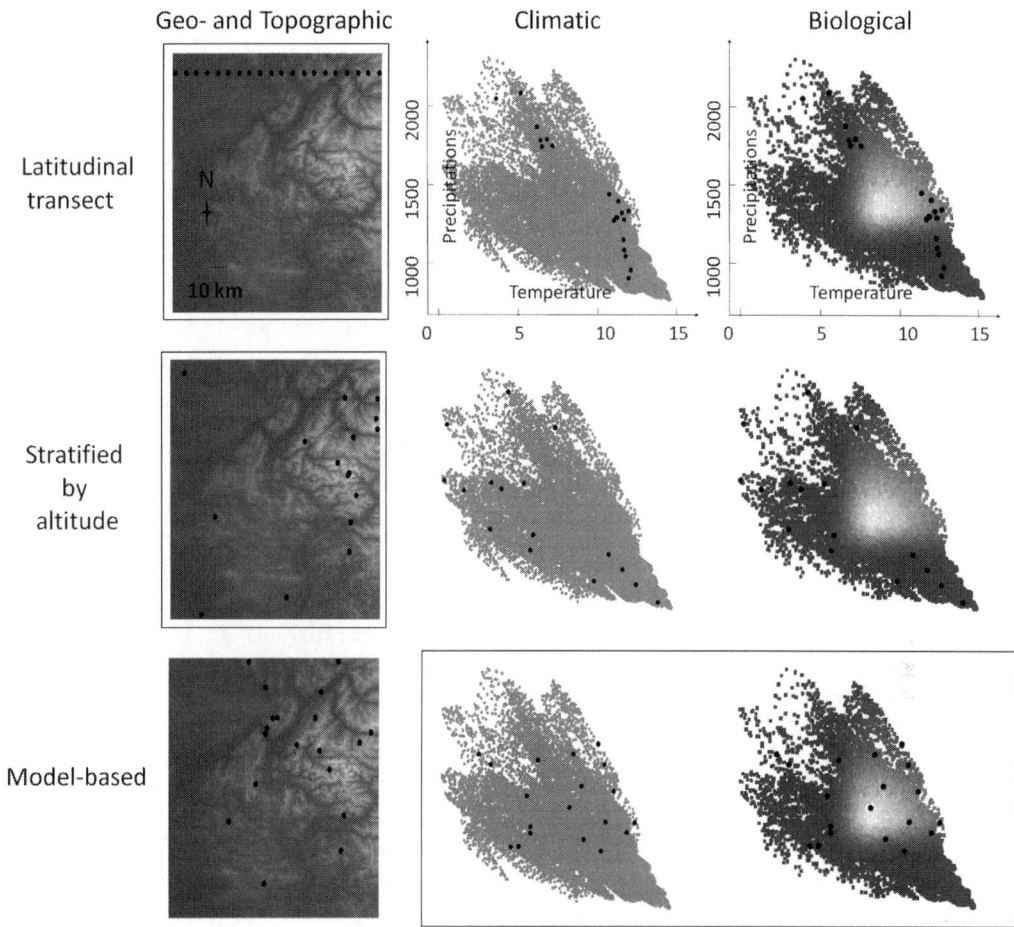

Fig. 3. Importance of the sampling space used to set up the sampling design and stratification. Three sampling designs are represented *from top to bottom*, a latitudinal transect sampling (drawn from the frame within the geographic space), stratification by altitude (drawn from the frame within the topographic space), and model-based stratification (drawn from the frame within the climatic space). The *black boxes* indicate in which space the sampling has been performed. Different spaces are represented (*from left to right*): the topographic space represented within the geographical map (*dark grey for low altitudes and light grey for high altitudes*), the climatic space represented by temperatures and precipitations, and the biological space represented by the frequency of allele or any other variable of interest mapped into the climatic space (*dark grey for low values and light grey for high values*). The black points (20 measurements) represent a sample drawn from the frame through the different sampling designs (one draw per line). Axes have not been repeated on all figures, they are the same for all lines. After Albert et al. (14).

landscape genomics, we thus believe that climatic or biological spaces should be the most appropriate since they are linked to environmental gradients that are of interest. Stratification in these spaces can also be combined with other variables affecting genetic structure such as population history.

3. Independence among sampling units or model residuals is often required at the analytical stage. As a consequence, it is important to avoid relatedness (genetic dependence) by choosing distant individuals (required to have biological knowledge

on the species) or to use appropriate models allowing the non-independence of the sampling units. For instance, Bradbury et al. (19) have developed a software that implements mixed linear models controlling for population structure. Non-independence can also be considered from a spatial point of view and is often described in terms of spatial autocorrelation. Appropriate methods allow considering spatial autocorrelation in the model (see ref. 22 for a review). In particular, Manel et al. (23) described how to consider spatial dependence in the data in the context of landscape genomics by using, for example, Moran's eigen vectors map (24, 25).

4. Previous knowledge on the study system, either using expert knowledge or the literature, can be used as a first basis: (1) to run a preliminary analysis for building a first response curve and deriving expectations on the patterns of interest; (2) to establish or improve a stratification in a biological space (see Subheading 2), for instance, by using a previously known distribution area (26, 27) or habitat suitability models (9, 28, 29) as a prior knowledge (e.g., to stratify along a habitat suitability gradient); (3) to run a simulation study to test various possible sampling designs and efforts (6, 14, 26).

These solutions have yet rarely been used in genetics. Among them, model-based (or climatic-based) stratifications appear as promising to derive sampling design strategies in landscape genomics. Such stratifications rely on the assumption that individuals are more adapted in the core of their species distribution (where the species is the most abundant) than in the margin. We then assume either that the relationship between the allele proportion and environmental variable (or habitat suitability) is logistic (Fig. 2) or Bell-shaped. This is the kind of hypotheses that are of interest and should be tested in landscape genomics, since they are not well established and may help to test if the relationship between environment and allele frequencies is the same in the core and in the margins of a species distribution (30–32).

3.3. Caveats Correlations

Artefactual correlations between allele frequencies and the environment can also result from the past demographic effects such as admixture after secondary contact or can also be confounded by population structure (18). As a consequence, it is important to know or to investigate the history and the structure of the studied population and to use appropriate methods (19, 20). Cushman and Landguth (11) illustrated how naive approaches can lead to spurious results, and how structural/causal modelling can lead to better inferences. In addition, if a large number of loci per individual is tested, appropriate approaches such as using false-discovery rate should be used when testing for significance (33).

References

1. Stinchcombe JR, Hoekstra HE (2008) Combining population genomics and quantitative genetics: finding the genes underlying ecologically important traits. Heredity 100:158–170

2. Joost S, Bonin A, Bruford MW et al (2007) A spatial analysis method (SAM) to detect candidate loci for selection: towards a landscape genomics approach to adaptation. Mol Ecol 16:3955–3969

3. Luikart G, England PR, Tallmon D et al (2003) The power and promise of population genomics: from genotyping to genome typing. Nat Rev Genet 4:981–994

4. Muirhead JR, Gray DK, Kelly DW et al (2008) Identifying the source of species invasions: sampling intensity vs. genetic diversity. Mol Ecol 17:1020–1035

5. Storfer A, Murphy MA, Evans JS et al (2007) Putting the 'landscape' in landscape genetics. Heredity 98:128–142

6. Schwartz MK, McKelvey KS (2009) Why sampling scheme matters: the effect of sampling scheme on landscape genetic results. Conserv Genet 10:441–452

7. Ehrich D, Yoccoz NG, Ims RA (2009) Multiannual density fluctuations and habitat size enhance genetic variability in two northern voles. Oikos 118:1441–1452

8. Epperson BK, McRae B, Scribner K et al (2010) Utility of computer simulations in landscape genetics. Mol Ecol 19:3549–3564

9. Poncet B, Herrmann D, Gugerli F et al (2010) Tracking genes of ecological relevance using a genome scan: application to *Arabis alpina*. Mol Ecol 19:2896–2907

10. Turner JRG (2010) Population resequencing reveals local adaptation of *Arabidopsis lyrata* to serpentine soils. Nat Genet 42:260–263. doi:10.38/ng.515

11. Cushman SA, Landguth E (2010) Spurious correlations and inference in landscape genetics. Mol Ecol 19:4179–4191

12. Williams BK, Nichols JD, Conroy MJ (2002) Analysis and management of animal populations. Academic, San Diego

13. Little RJ (2004) To model or not to model? Competing modes of inference for finite population sampling. J Am Stat Assoc 99:546–556

14. Albert C, Yoccoz N, Edwards T et al (2010) Sampling in ecology and evolution – bridging the gap between theory and practice. Ecography 33:1028–1037

15. Box G, Hunter W, Hunter J (2005) Statistics for experimenters. Design, innovation, and discovery, 2nd edn. Wiley, Hoboken

16. Inouye BD (2001) Response surface experimental designs for investigating interspecific competition. Ecology 82:2696–2706

17. Excoffier L, Foll M, Petit RJ (2009) Genetic consequences of range expansions. Ann Rev Ecol Evol S40:481–501

18. Excoffier L, Hofer T, Foll M (2009) Detecting loci under selection in a hierarchically structured population. Heredity 103:285–298

19. Bradbury PJ, Zhang Z, Kroon DE et al (2007) TASSEL: software for association mapping of complex traits in diverse samples. Bioinformatics 23:2633–2635

20. Yu JM, Pressoir G, Briggs WH et al (2006) A unified mixed-model method for association mapping that accounts for multiple levels of relatedness. Nat Genet 38:203–208

21. Hutchinson GE (1957) Concluding remarks. Cold Spring Harb Symp Quant Biol 22:145–159

22. Diniz-Filho JAF, Nabout JC, Telles MPD et al (2009) A review of techniques for spatial modeling in geographical, conservation and landscape genetics. Genet Mol Biol 32:203–211

23. Manel S, Joost S, Epperson B et al (2010) Perspectives on the use of landscape genetics to detect genetic adaptive variation in the field. Mol Ecol 19:3760–3772

24. Borcard D, Legendre P (2002) All-scale spatial analysis of ecological data by means of principal coordinates of neighbour matrices. Ecol Model 153:51–68

25. Dray S, Legendre P, Peres-Neto PR (2006) Spatial modelling: a comprehensive framework for principal coordinate analysis of neighbour matrices (PCNM). Ecol Model 196:483–493

26. Dengler J, Oldeland J (2010) Effects of sampling protocol on the shapes of species richness curves. J Biogeogr 37:1698–1705

27. Brito JC, Crespo EG, Paulo OS (1999) Modelling wildlife distributions: logistic multiple regression vs overlap analysis. Ecography 22:251–260

28. Singh NJ, Yoccoz NG, Bhatnagar YV, Fox JL (2009) Using habitat suitability models to sample rare species in high-altitude ecosystems: a case study with Tibetan argali. Biodivers Conserv 18:2893–2908

29. Guisan A, Broennimann O, Engler R et al (2006) Using niche-based models to improve the sampling of rare species. Conserv Biol 20:501–511

30. Bridle JR, Polechova J, Kawata M, Butlin RK (2010) Why is adaptation prevented at ecological

margins? New insights from individual-based simulations. Ecol Lett 13:485–494

31. Kawecki TJ (2008) Adaptation to marginal habitats. Ann Rev Ecol Evol S39:321–342

32. Petit RJ, Excoffier L (2009) Gene flow and species delimitation. Trends Ecol Evol 24: 386–393

33. Storey JD, Tibshirani R (2003) Statistical significance for genome wide studies. Proc Natl Acad Sci USA 100:9440–9445

34. Manel S, Poncet B, Legendre P et al (2010) Common factors drive adaptive genetic variation at different spatial scales in *Arabis alpina*. Mol Ecol 19:3824–3835

Chapter 2

OligoTag: A Program for Designing Sets of Tags for Next-Generation Sequencing of Multiplexed Samples

Eric Coissac

Abstract

Next-generation sequencing systems allow high-throughput production of DNA sequence data. But this technology is more adapted for analyzing a small number of samples needing a huge amount of sequences rather than a large number of samples needing a small number of sequences. One solution to this problem is sample multiplexing. To achieve this, one can add a small tag at the extremities of the sequenced DNA molecules. These tags will be identified using bioinformatics tools after the sequencing step to sort sequences among samples. The rules to apply for selecting a good set of tags adapted to each situation are described in this chapter. Depending on the number of samples to tag and on the required quality of assignation, different solutions are possible. The software oligoTag, a part of OBITools that computes these sets of tags, is presented with some example sets of tags.

Key words: Next-generation sequencing, Multiplexing, Sample, Tags

1. Introduction

High-throughput sequencers allow for easily and quickly generating a huge number of sequences. Currently, two systems are mainly used: the 454 GS FLX from Roche®; and the Solexa system from Illumina®;. In their current versions, the 454 GS FLX produces 1 million sequence reads per run and the Solexa machine produces 1 billion reads. The characteristics of these machines and of the sequences produced make these two technologies more complementary than concurrent. Many molecular ecological studies can take advantage of these new systems allowing the elaboration of large-scale experimental protocols. Many useful techniques for these studies rely on PCR (Polymerase Chain Reaction) amplicon sequencing. Before high-throughput sequencing technologies, these PCR-based techniques required a cloning step for building a

François Pompanon and Aurélie Bonin (eds.), *Data Production and Analysis in Population Genomics: Methods and Protocols*, Methods in Molecular Biology, vol. 888, DOI 10.1007/978-1-61779-870-2_2, © Springer Science+Business Media New York 2012

DNA library where each clone contained one DNA molecule synthesized during the PCR amplification. A subset of clones from the library could then be sequenced to estimate the global diversity included in the PCR amplicon. Such an approach limited the number of sequences that could be produced. Actually, in most of the studies, a few thousands of clones were typically analyzed (e.g., (1–5)). Now the new sequencing techniques allow direct sequencing of individual DNA molecules composing a PCR amplicon without any cloning step. This has many advantages, including protocol simplification, as cloning PCR amplicons is not so easy even with commercialized kits. The cloning was also a potential source of bias for library representativity. But the main advantage is certainly that these new machines allow parallel sequencing of a very large number of individual sequences. It is thus possible to reach a higher sequence depth leading to a better coverage of the sample (e.g., (6)). Depending on the machine used, one can physically divide one run into up to eight or sixteen areas, each of them receiving a sample. Per sequencing area we might obtain 75,000 and 50,000,000 sequence reads for a 454 GS FLS or a Solexa system, respectively. Such high coverages are excessive for specific experiments where a higher sample number and a smaller sequencing depth are required. The strategy used to reach this aim is then to mix several samples in one sequencing area. For allowing individual analysis of each sample, a short sample-specific oligonucleotide (i.e., a tag) is added at the extremity of each molecule of each sample before sequencing (e.g., (7)). This tag will allow sorting sequences corresponding to each sample after the sequencing step using appropriate bioinformatic tools.

2. Theoretical Background

2.1. The Simplest System

We can imagine several strategies for adding such tags at the end of the DNA molecules constituting a sample. They can rely on a ligation step adding some adaptors including the sequence tag or the tags can be directly added to the PCR primers during their synthesis. This last strategy requires to order sets of primer pairs differing only by their tags. Whatever the strategy selected, the most important decision to make is to define which sequences should be used as tag. In a wonderful world, one could address this question only by taking into account the number of samples that need to be identified. The number of different DNA words of length n can be calculated from the formula 4^n. For tagging N samples, you have just to select the smallest n as $4^n > N$. As example, to tag $N = 300$ samples, you will need to use tags of length 5 ($4^4 = 256$ and

$4^5 = 1,024$). Then, by enumerating the 300 first words of length 5 (*AAAAA, AAAAC, AAAAG*, etc.), you can obtain your tag list.

2.2. Dealing with Sequencing Errors

Unfortunately, we are not in a wonderful world, and sequencers produce sequences with errors. Thus, the tag attached to each molecule can be sometimes read correctly and sometimes read erroneously. We are faced with a piece of information transmitted through a noisy channel. The emitter, e.g., the experimentalist, designs or emits a correct tag. The transmitter device is composed of a chained set of complex operations: primer synthesis, PCR amplification, sequencing process. Finally, the receiver is once again the scientist. This metaphor allows for linking our problem to the transmission information theory, a well-known problem in computer science. If we build our tag set (i.e., a code in the information transmission theory) following the simplest model and if an error occurs during the message transmission, we will read a wrong code after reception. Without the possibility to detect this error, we will assign the attached sequence to the wrong sample (see Fig. 1). Such a code is called a "no-error tolerant code".

For detecting reading errors, we must use a subset of all the possible words. Once we choose one word as a tag, i.e., *acggt*, if we preclude all words differing by only one letter (e.g., *aGggt, acTgt, Gcggt*, etc.), we guarantee that reading a tag with one error gives an erroneous tag corresponding to no sample. Given the Hamming distance (d_H) equal to the number of differences between two words, we can define a code as a set of DNA words where for all possible word pairs (w_i, w_j), $d_H(w_i, w_j) \geq 2$. This new code is called a "one-error tolerant code" (see Fig. 2a).

Similarly we can build a "two-error tolerant code" by selecting a subset of words in such a way that for all possible word pairs (w_i, w_j), $d_H(w_i, w_j) \geq 3$. This new code will lead to a sample misassignation only if more than two bases of the tag are erroneously read. This new code has a second property. If we consider that we can produce no more than one error during the transmission process, such a code not only detects the error but also allows correcting it. This is possible because only one of the used tags is present at a Hamming distance of one from the tag read (see Fig. 2b). Thus, a two-error tolerant code can be considered as a "one-error autocorrective code."

$$acggt \qquad \Longrightarrow \qquad aGggt$$

a particular tag transmission chain an erroneously received tag

Fig. 1. Reading a tag of five nucleotides with one error can lead to sample missassignation. If all possible DNA tags composed of five nucleotides are assigned to a sample, we are not able to detect reading errors. We produce a no-error tolerant code. The capital letter corresponds to the reading error.

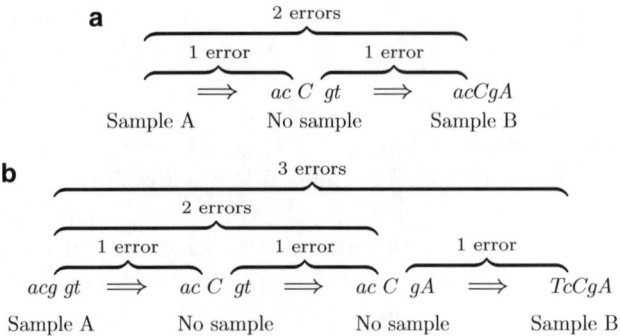

Fig. 2. Reading a tag of five nucleotides (**a**) If tag *acggt* is assigned to sample A and tag *accga* to sample B, when the tag *accgt* is read we can deduce that it is an error and discard the associated sequence. This corresponds to a one-error tolerant code. (**b**) If sample B is associated to tag *tccga* instead of *accga*. When the tag *accga* is read, we conclude to an error. This error can be explained as one reading error of sample B tag or as two reading errors of the sample A tag. This corresponds to a two-error tolerant code. If we assume that no more than one error is possible, only the first hypothesis is acceptable. We can keep the sequence and assign it to the sample B. Thus, we use the code as a "one-error autocorrective code".

2.3. Choosing a Set of Tags

The choice of an adequate tag system depends mainly on two parameters: the number of samples to tag and the number of expected errors that is a combination of the probability to misread a base of the tag and the total number of sequences produced by the sequencer. To objectively decide, we can develop a simple probabilistic model. If we consider a homogeneous misreading probability P_{mis} and tags of length l_{tag} then the probability $P_{1,e}$ to read a tag with e errors can be expressed as a binomial distribution (see Eq. 1).

$$P_{1,e} = \binom{l_{tag}}{e} P_{mis}^{e} \left(1 - P_{mis}\right)^{(l_{tag} - e)}.$$ (1)

The binomial part $\binom{l_{tag}}{e}$ estimates all the possible ways to position e errors in a tag of length l_{tag} and is computed using Eq. 2:

$$\binom{l_{tag}}{e} = \frac{l_{tag}!}{e! \left(l_{tag} - e\right)!}.$$ (2)

Using Eq. 1, it is easy to estimate how many tags would be read with $1, 2, 3,$ or 4 errors, when running a full GS-FLX 454 run or Solexa lane with one or 50 millions of reads, respectively (see Table 1). Even with a low error rate, the large number of reads leads to an expected large number of tags with up to three reading errors. This demonstrates the importance of taking into account errors when designing a set of tags. Table 1 shows that for a

Table 1
Estimated numbers of misread tags: count computation is done for a tag length $l_{tag} = 6$, the three used error rates correspond to values usually observed from 454 or Solexa runs. (e) is the number of errors in the tag

errors (e)	for a full 454 run 10^6 reads Error rate (P_{mis}) 0.20%	0.25%	0.30%	errors (e)	for a Solexa lane 50.10^6 reads Error rate (P_{mis}) 0.20%	0.25%	0.30%
1	11880	14813	17731	1	594023	740671	886580
2	59	92	27	2	2976	4640	6669
3	0.16	0.31	0.54	3	7	15	26
4	$2.4\ 10^{-4}$	$5.8\ 10^{-4}$	$1.2\ 10^{-3}$	4	$1.2\ 10^{-2}$	$2.9\ 10^{-2}$	$6\ 10^{-2}$

GS-FLX 454 run with $P_{mis} = 0.0025$, a tagging system of 6 nucleotides with a Hamming distance greater or equal to three will lead to misassign on average one sequence to a wrong sample every three runs while requiring discarding $14,905$ reads (approximatively, 1.5% of the sequences). Considering the same tagging system as a one-error autocorrective code, we correctly reassign $14,813$ of these reads to their respective samples but 92 reads with two errors are misassigned. Depending on the effect of such a level of misassignation, an autocorrective code can be considered as a good or a bad solution.

2.4. Tagging Both Ends of the PCR Amplicon

When sequencing a short enough PCR amplicon to be fully sequenced in a single read, it is possible to tag it at both ends (i.e., tagging the forward and the reverse primers). This double tagging reduces the sample misassignation probability. By checking the tags at each extremity, the only cause for misassignation is to observe the same errors on both tags. Let's consider that a reading error can change a nucleotide equiprobably into any of the three others, then we deduce Eqs. 3 and 4 from Eq. 1 to estimate the probability P_{1,e^2} of observing the same e errors on both extremities of a read. This probability function is used for estimating frequencies of sample misassignation with a double-tag system (see Table 2):

$$P_{1,e^2} = P_{1,e} \left(\frac{P_{mis}}{3} \right)^e \left(1 - \frac{P_{mis}}{3} \right)^{(l_{tag} - e)} \tag{3}$$

$$= \binom{l_{tag}}{e} e^{\frac{4}{3} P_{mis}} (l_{tag} - e)^{\left(2 - \frac{4}{3} P_{mis} \right)}. \tag{4}$$

When tags are added to both ends, even with only two differences between tags, almost no misassignation is possible (Table 2). By comparison, with a single-end tagging system, even with four differences between tags we cannot achieve the same level of

Table 2
Estimated numbers of sample misassignations with a double-tag system: computation is done for a double-tag system of length $l_{tag} = 6$ without autocorrection. The three used error rates correspond to values usually observed from 454 or Solexa runs.(e) is the number of errors in the tag

errors	for a full 454 run 10^6 reads Error rate (P_{mis})			errors	for a Solexa lane 50.10^6 reads Error rate (P_{mis})		
(e)	0.20%	0.25%	0.30%	(e)	0.20%	0.25%	0.30%
1	7	12	17	1	394	614	882
2	$2.6\ 10^{-5}$	$6.4\ 10^{-5}$	$1.3\ 10^{-4}$	2	$1.3\ 10^{-3}$	$3.2\ 10^{-3}$	$6.6\ 10^{-3}$
3	$4.7\ 10^{-11}$	$1.8\ 10^{-10}$	$5.3\ 10^{-10}$	3	$2.4\ 10^{-9}$	$9\ 10^{-9}$	$2.7\ 10^{-8}$
4	$4.7\ 10^{-17}$	$2.8\ 10^{-16}$	$1.2\ 10^{-15}$	4	$2.4\ 10^{-15}$	$1.4\ 10^{-14}$	$6\ 10^{-14}$

confidence (Table 1). But as the two copies of the tag can be misread independently, the frequency of discarded sequences is twice higher with a double-tag system than with a single-tag one (see Tables 1 and 2)

Like for a single-tag system, a double-tag system with a Hamming distance greater or equal to three can also be used as a one error autocorrective code. In such a system, a sequence with one or two erroneously read tags is reassigned to the good sample if when corrected, both tags match the same sample.

2.5. Lexical Constraints on Tag Design

A tag is a DNA word. To limit misreading, once the tag length is fixed, lexical rules can be set to restrict the usable words. For example, knowing the difficulties to read unambiguously homopolymers with the GS-FLX 454, we should exclude tags with more than h consecutive identical letters. According to the same principle, if a set of tags is designed to be linked to a PCR primer with a nucleotide X (*i.e.*, A, C, G, or T) at its 5' end we should avoid this nucleotide at the 3' end of the tag. Also, if a tag is linked to the primers during their synthesis, we can use only tags with a precise $G+C$ content to reduce the effect of the heterogeneity of the primer melting temperatures (Tm).

2.6. Building a Set of Tags

While some approximations were used to build efficiently a set of tags (8), the exact way for defining a set of tags relies on graph theory. In mathematics, a graph $\mathcal{G}(V, R)$ is defined by a set of nodes or vertices V and a relation R describing a set of edges E linking some node pairs. In our particular case, V is the set of all words matching our lexical constraints (length, $G+C$ content, maximum homopolymer length, etc.).

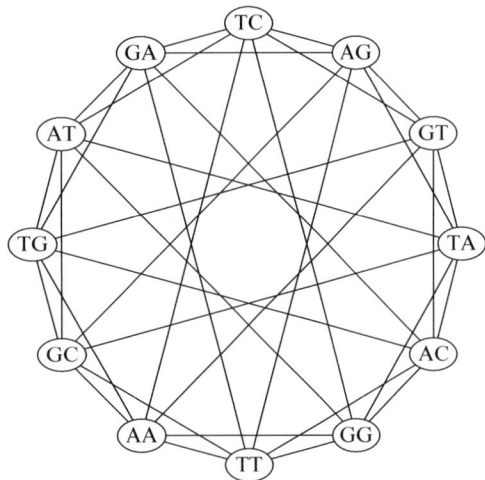

Fig. 3. Hamming graph for DNA words of length 2: This graph is built with all words of size two that do not begin by a C so $V=\{$ AA, AC, AG, AT, GA, GC, GG, GT, TA, TC, TG, TT $\}$. The relation R is defined for a Hamming distance between two words $d_H \geq 2$.

$$V = \{w|w \text{ is a DNA words of length } l \text{ matching lexical constraints}\}. \quad (5)$$

R the set of edges is composed of DNA word pairs with a Hamming distance (d_H) greater or equal to e (i.e., the minimum number of differences between two tags).

$$R = \{(w_i, w_j) \in V \times V \mid d_H(w_i, w_j) \geq e\}. \quad (6)$$

In such a graph (see Fig. 3), a usable set of tags T is defined as a subset of V forming a complete subgraph (see Eq. 7). In the graph theory, T defines a clique. Identifying the best set of tags is equivalent to looking for the largest clique in the graph $\mathcal{G}$. Unfortunately, this requires a computational time that increases exponentially with the size of the graph. So only approximate solutions are computable.

$$T \subset V \text{ in such a way that } \forall \{w_i, w_j\} \subset T \Rightarrow (w_i, w_j) \in R. \quad (7)$$

3. The *oligoTag* Program

OligoTag is part of OBITools (http://www.grenoble.prabi.fr/trac/OBITools), a set of UNIX command line programs dedicated to the analysis of the output from high-throughput sequencers.

3.1. Installing OBITools

To install OBITools, you need access to a Unix computer with Python language installed and a C compiler. It can be a PC with Linux or a Macintosh computer or any other Unix system. All the following commands must be entered in a unix terminal window. Usually, on a linux machine, you can start such a windows from the application/utilities menu. On Macintosh, a similar application is available in the Applications/Utilities folder.

3.1.1. Checking Prerequisite

To check if a C compiler is installed on your system, follow the instructions presented in Listing 1. If a gcc C compiler is installed on your computer, just by running the gcc command without argument you will have an error message "no input file" indicating that you have not specified a c file to compile.

unix–shell > gcc

i686–apple–darwin10–gcc –4.2.1: no input files

Listing 2 shows the result obtained if no C compiler is installed on your system: if a command is not installed on a system an error message "Command not found" is generated, then you must install a C compiler. If you are a Linux user, you must install the corresponding package from your package manager. If you are a Macintosh user, you need to install the "Developer tools" package available on the system DVD or on the Apple web site.

unix–shell > gcc

gcc: Command not found

You need Python 2.6 or 2.7, which should be available on all modern Unix system. To check your python version, follow the instruction presented in Listing 3. From a unix shell, you can run the python interpreter in interactive mode by typing the command python. This displays the python version, in our case python version 2.7. To quit python, just press keys Ctrl-D.

unix–shell > python

Python 2.7 (r27:82500, Jul 6 2010, 10:43:34)

[GCC 4.2.1 (Apple Inc. build 5659)] on darwin

Type " help", "copyright", "credits" or "license "for more information.

>>>

MacOSX users are invited to install a version of python downloadable from the python web site (http://www.python.org) even if a version of python is included by default in the system. Be careful, python 2.x and 3.x versions are almost incompatible, so don't use python 3.x with OBITools.

Finally, the SetupTool python package must be installed. First, you have to check the presence of the easy_install command as you did for gcc. If this command is absent, you can download it from the python package index web site (http://www.pypi.org) and follow the corresponding installation instructions.

Installing OBITools Package	OBITools can now be easily installed using the easy_install command (Listing 4). It can be necessary to begin this command line by the word sudo to access the administrator privilege.

unix–shell > easyinstall obitools

. . .

unix–shell >

3.2. OligoTag Options

Several options are available for specifying characteristics of the generated tag set. As in many unix programs, most of them exist in two forms. The short form corresponds to one letter preceded by a dash (e.g., -s). The long one is a full word preceded by a double dash (e.g., --oligo-size). Both forms of the same option are listed together. When an option requires a parameter like --oligo-size or -s, in the short form the parameter must follow directly the option (-s 5), whereas in the long form the option name and the parameter value must be separated by an equal sign (--oligo-size=5).

Length of the Tags

The length of the oligonucleotide is strongly related to the maximum size of the potentially identifiable tag set (see Subheading 2). The length must be an integer value greater or equal to 1. Values larger than 8 lead to huge memory usage and very long computation time (see option Subheading 3.2.9).

-s <###>, --oligo-size=<###>

<###> is an integer value corresponding to the generated length of the tags.

Size of the Set of Tags

This is the minimum number of tags required in the generated set of tags. These values must be set in relation with the option Subheading 3.2.1. Looking for a too large set with a too small size of tags leads either to no solution or to a very long computation time. To limit this effect, see also option Subheading 3.2.9.

-f <###>, --family-size=<###>

<###> is an integer value corresponding to the size of tag set to generate.

Minimum Hamming Distance Between Two Tags

This is the minimum Hamming distance d_H between two tags of the solution set. This distance is associated with the chance of misassigning a sequence to a sample (see the parts Subheadings 2.3 and 2.4). Increasing the distance reduces the probability of assignation errors but reduces the size of the tag set (see options Subheadings 3.2.1 and 3.2.2).

-d <###>, --distance=<###>

<###> is an integer value corresponding to the minimum distance between two tags.

Maximum Number of G or C Nucleotides per Tag

This option lexically constraints a tag to be acceptable in a set by limiting the sum of G and C nucleotides. This can be used to limit the nonequivalent impact of GC rich and AT rich tags on the primer melting temperature. This constraint reduces the maximum size of the set of potentially identifiable tags.

-g <###>, --gc-max=<###>

<###> is an integer value corresponding the maximum number of G or C nucleotides acceptable in a tag

Acceptable Tag Pattern

Using the IUPAC code (Table 3), you can specify exactly the pattern of the tags to generate. The pattern must have the same length than the oligonucleotide size (see options Subheading 3.2.1) and must be constituted of a series of one of the IUPAC codes. If you set the oligonucleotide size to 6 using the option -s 6 and specify a pattern GNNBNR using the option -a gnnbnr, you will only accept

Table 3
Nucleic IUPAC code used to represent nucleotides

Code	Nucleotide
A	Adenine
C	Cytosine
G	Guanine
T	Thymine
U	Uracil
R	Purine (A or G)
Y	Pyrimidine (C, T, or U)
M	C or A
K	T, U, or G
W	T, U, or A
S	C or G
B	C, T, U, or G (not A)
D	A, T, U, or G (not C)
H	A, T, U, or C (not G)
V	A, C, or G (not T, not U)
N	Any base (A, C, G, T, or U)

tags starting on their 5′ end with G, with no A at their fourth position and a purine (A or G) at their 3′ extremity. A too restrictive pattern can drastically reduce the maximum size of the potentially identifiable tag set.

-a <IUPAC pattern>, --accepted=<IUPAC pattern>

<IUPAC pattern> a string describing the IUPAC pattern of acceptable tags.

Non acceptable Tag Pattern   Reciprocally to the Subheading 3.2.5 option described above, you can specify a pattern indicating the tag that must not be include in a set using the IUPAC code (Table 3). Using this option can drastically reduce the maximum size of the potentially identifiable tag set.

-r <IUPAC pattern>, --rejected=<IUPAC pattern>

<IUPAC pattern> a string describing the IUPAC pattern of unacceptable tags.

Maximum Homopolymer Length   Homopolymers may cause many PCR and sequencing errors, especially when using the GS-FLX technology. To limit sample missassignation, it is reasonable to limit the length of homopolymers in tags to two. Only tags with no homopolymer longer than the specified limit will be retained in the tag set.

-p <###>, --homopolymer=<###>

<###> is an integer value corresponding the maximum length of an homopolymer.

Minimum Homopolymer Length   This reciprocal option of the previous one is normally less useful. Only tags with at least one homopolymer longer or equal to the specified limit can be retained in a tag set.

-P <###>, --homopolymer-min=<###>

<###> is an integer value corresponding the minimum length of an homopolymer.

Computation Time Out   Computation of a tag set is divided into two steps. During the first one, the Hamming distance graph is built according to all the options specified. Then a maximum clique algorithm looks for the cliques larger than the family size limit. If you ask for a too large tag set, this second part can be infinitely long. So it is useful to specify a time out limit (in seconds) for this search. In practice, it is really rare to find an interesting solution in more than ten minutes, so 600 s is a good compromise.

-T <###>, --timeout=<###>

<###> is an integer value expressed in seconds indicating the maximum time that the program can spend to look for a set of the required size. If this time is over, then the largest set of tags found is returned instead.

3.3. Running oligoTag

OligoTag is a unix command line program and must be used from a unix terminal windows. From a unix shell, by typing the following command line (Listing 5), we look for a solution corresponding to the graph presented in Fig. 3. The option $-s\,2$ indicates the size of the word (i.e., tag length). The option $-f\,1$ indicates the minimum size of the desired set of tags. The option $-d\,2$ indicates the minimum Hamming distance d_H. Finally, the $-r\,CN$ rejects all words matching a given pattern (here CN with N meaning $A, C, G, \{$ or $\}T$).

unix–shell > oligoTag –s 2 –f 1 –d 2 –r CN

Build good words graph . . .

Initial graph size : 12 edge count : 36

aa

gc

tg

unix–shell >

The proposed solution is reported in Fig. 4. As the exact solution of the problem defined in part Subheading 2.6 cannot be computed in a reasonable time, oligoTag cannot guaranty to find

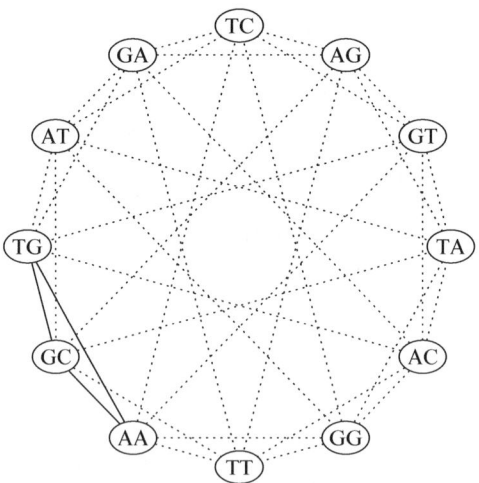

Fig. 4. Solution proposed by oligoTag for DNA word of length 2: Solution proposed by oligoTag is reported on this graph with the filled edges. This clique is maximum, you cannot add another vertex without rejection of the formula 2.7. In this particular case, this solution also corresponds to one of the largest cliques in this graph.

the largest tag set. OligoTag will just try to identify a clique that cannot be extended with a cardinality greater than a threshold defined via the option $-f$.

OligoTag is a standard unix command, so by using the > redirection character you specify to save the output of the oligoTag program to the mytag.txt text file. You are free to choose the output file name. The cat command allows you to read the content of the newly created file (Listing 6).

unix–shell > oligoTag –s 2 –f 1 –d 2 –r CN > mytag . txt

Build good words graph . . .

Initial graph size : 12 edge count : 36

unix–shell > cat mytag . txt

aa

gc

tg

unix–shell >

4. Examples of Precomputed Tag Lists

All these tag lists were computed with oligoTag specifying a minimum Hamming distance of 3 and no homopolymer longer than 2 (see Table 4).

5. Analyzing Tagged Sequences

The OBITools are also useful for analyzing tagged raw sequences. They allow identifying and trimming the amplification primers and tags. They allow assigning sequences to a sample according to its tag and distributing these annotated sequences according to their associated sample and experiment in several files. These tasks can be achieved using three OBITools programs: ngsfilter, obiannotate, and obisplit. Several other OBITools exist and are not presented here despite their potential utility. Like oligoTag, all other OBITools programs are unix command line tools. This allows chaining them using simple unix scripts and then automatizing the treatment by defining more or less complex pipelines.

5.1. The ngsfilter Program

Based on the description of the PCR primers used and of the tag associated to each multiplexed sample, ngsfilter looks for the forward and reverse primers, matches the flanked tag, and annotates the sequence with its experiment and sample names. At the same

Table 4
Example of tag lists computed with oligoTag: Each part of the table corresponds to a different tag (4 to 7) length and family size options. The oligoTag unix command and the associated options used to generate each of these sets are indicated

| Tag size : 4 | | | | Tag count : 11 | | | | |

unix command : oligoTag -d 3 -p 2 -s 4 -f 10

aaca	acac	ctcc	gctt	tatc	aggt	tccg	tgaa	gtga
attg	caag							

| Tag size : 5 | | | | Tag count : 33 | | | | |

unix command : oligoTag -d 3 -p 2 -s 5 -f 24

aacaa	aagcc	gactt	gtaat	cgagg	aatgg	cgcat	acaac	tctaa
gtcca	ggtta	attct	accgt	gtggc	caatc	gccag	acgta	tgatt
taacg	gcgct	agaca	catca	tacgc	ctgtt	tgtcc	agctc	aggag
ttaga	cttag	tagat	gatac	cctgc	atatg			

| Tag size : 6 | | | | Tag count : 108 | | | | |

unix command : oligoTag -d 3 -p 2 -s 6 -f 96

aacaac	aaccga	ccggaa	agtgtt	ccgctg	aacgcg	ggctac	ttctcg	ttctcg
tcactc	gaacta	ccgtcc	aagaca	cgtgcg	ggtaag	ataatt	cgtcac	cgtcac
ttgagt	aagcag	ttgcaa	cacgta	taacat	tgcgtg	ggtcga	cactct	cactct
cttggt	tccagc	acttca	gcgaga	tggaac	gtacac	aagtgt	tcttgg	tcttgg
aaggtc	ggcgca	tcgacg	cctgtc	agaaga	aatagg	ggttct	taatga	taatga
gtaaca	aatcct	agaccg	tggcgg	ctataa	aatgaa	cgaatc	agagac	agagac
ttcgga	cgacgt	ctcatg	tgtata	acaacc	tcagag	gtagtg	agcact	agcact
gcggtt	acacaa	gctccg	tacttc	gttgcc	gtatgt	gtcaat	agcctc	agcctc
tcgtta	tgtggc	ctctgc	atggat	acaggt	tccgct	gtccgg	cattag	cattag
gaagct	gatatt	agctgg	cgcgat	acattg	ccaagg	accata	aggatg	aggatg
gtctta	tatacc	acctat	aggtaa	attcta	gtgatc	gacggc	gtgcct	gtgcct
tatctg	cggcca	cctaat	acgcgc	gtgtag	ttcctt	cagagc	tgatcc	tgatcc

| Tag size : 7 | | | | Tag count : 316 | | | | |

unix command : oligoTag -d 3 -p 2 -s 7 -f 300

aacaaca	aacacac	tccgact	tgcctgc	cgtcgca	ggtcagt	ccgctag	acgcggc	acgcggc
gaagctg	aacaggt	gccggcc	ggtcgag	taagcct	aacattg	aaccaag	agtgaag	agtgaag
ccggatc	ggtctcc	aaccgcc	ccggcct	ttcggcg	agtgcca	taagtgg	acggcga	acggcga
tcctaag	aacctga	ggtgatc	acggtac	aacgagc	tggcacc	taatagt	ccgtaat	ccgtaat

(continued)

Table 4
(continued)

Tag size : 7   **Tag count : 316**

unix command : oligoTag -d 3 -p 2 -s 7 -f 300

gaatgat	aacgccg	caagtat	acgtacc	ggtccta	aacggaa	tcctgca	ggtgtga	ggtgtga
taattaa	ggttaca	tccttgt	aagagta	ccgttca	aactatt	tccgcgc	agttggc	agttggc
cctaacg	ggttcgg	gacataa	ataacgc	actaagt	gcgatcc	tcgagcg	tacagtc	tacagtc
ggttgtt	gtaacag	ctcacct	ataagtg	cctagga	aagaagg	gcctctg	tgtatag	tgtatag
aagacct	gtaagct	tggccgt	tcgcagg	gcgcctt	taccgat	tcgccac	gtaatga	gtaatga
gaccttc	cctcata	gtacaat	cctccgc	cacgctt	gacgcga	tcgctct	aagcatc	aagcatc
gtacctc	aagccaa	cgtgact	actctaa	taaggta	gacgtct	gtactcg	gcttgaa	gcttgaa
cctgcaa	aagctcg	gtagacc	tgtgctg	aaggaat	tactacc	tcggttg	cacttac	cacttac
gtagcgt	cctggtg	gtaggaa	tatactt	actggct	tactcta	ctagtta	cctgtgt	cctgtgt
atatact	gcgtgtc	gagaatt	gtatagg	gaatata	acttcag	taagaac	gtatcca	gtatcca
aagtcgc	caatgtc	gctactc	aagtgag	cgctcgt	cgaacat	ctattgt	agaaccg	agaaccg
gtattac	ctcaata	tagatac	gtcaacg	agaagga	aatacga	agaatac	cgaattg	cgaattg
tctacca	aatagcg	gtcagac	cagctgt	ggtgcat	aatatat	ttaccga	agacagg	agacagg
aatcact	ctccgaa	agacgat	atcctat	tagctta	aatcctg	taggaga	agactca	agactca
tctcgcc	gtccgtg	acaagcc	cagtatg	taggctc	agagatt	ctcgcgg	cgaggcc	cgaggcc
gtcgatt	gcttagc	tagtccg	gcttcct	gtcgtag	tcttatt	aatgtcc	gagttga	gagttga
agtaacc	atctctc	tagtgtt	aattaac	ttatctg	cagaggc	gtccaga	cgatgaa	cgatgaa
ttatgcc	agatcta	tataagc	gtctcat	ctctggc	ggaagtc	catattc	ctcttcg	ctcttcg
tatagaa	cgcaagg	ttacaa	catcagg	acaacaa	tacgatg	tgtggac	ctgacac	ctgacac
tgaatct	agcactt	catcgac	gtgaagc	agcagag	ttcatcc	gtgacta	acaatgg	acaatgg
ttccaac	acacaac	atgcagt	tgacgcg	agccata	gtgatat	acaccgt	cgagaga	cgagaga
acacgta	gtgccgg	ctgcgtt	gatggca	gactgcg	acagaca	atggcag	ctggacg	ctggacg
cggttgc	cattcat	cgcggat	agcggtc	acagctc	atggtct	acaggag	cgcgtca	cgcgtca
ttcgtga	agcgtgg	tcgaata	agctacg	tgatcac	cgctatc	ctgtaga	gtggttc	gtggttc
acatatg	agcttaa	atgttgg	cgtctat	cggaaca	ggcagca	acattat	tacacgg	tacacgg
tcaagtt	gtgtggt	accaatc	ttgagga	cttagat	ctacatg	aatgata	aggatgt	aggatgt
accatct	cttatca	gcactgc	accgcat	cggcctc	gttattg	ttgcgag	cttccag	cttccag
aggcgtg	attcttc	accgtta	gttcggc	acctaga	gtgcgca	ccatcga	tcagtcc	tcagtcc
aggtcat	ttggcca	cttgagc	attgctt	ccatgcg	aggtgca	acctgac	gttgccg	gttgccg
ttgtatc	ctgctcc	tgtaggt	caaccta	tgctgtg	acgactg	caacgct	gccacgt	gccacgt
taacaca	acgagat	ttgaact	taaccag	agtatta	taacggc	agtccac	gccgagg	gccgagg
caagcgc								

```
@HWUSI-EAS1510_0005:1:1:12765:8937#0/1_SUB_CONS
ccagctcagtggggcaagcctcagccgctatccgtgtctttgtaatctcatgggagaa
+HWUSI-EAS1510_0005:1:1:12765:8937#0/1_SUB_CONS
H:555999<9DAD33=\BGcK2EYDS==4^HET?TFOAJQLA4?3T:B:@B?B^V'?:
@HWUSI-EAS1510_0005:1:1:12866:6509#0/1_SUB_CONS
ccctctgctcagggcaatcctcagcaccaatcctttttttagtattcgaatgatgaac
+HWUSI-EAS1510_0005:1:1:12866:6509#0/1_SUB_CONS
I>;;>DAADD_L@aF:XBGSI?B^@=@27'UCN=F1T[;B_@B@FDB@B52@4@G441
```

Fig. 5. The first two sequences of a Solexa run in fastq format: each sequence is described by four lines. The first line starting with the @ character is the equivalent of a fasta title line. The next line gives the sequence. The third line starting with a + character repeats the title line, and the fourth line corresponds to the encoded quality line.

time, the primers and tag copies are trimmed out and only the amplified part of the sequence is kept in the output.

The input file can be a raw sequence file in fastq or fasta format. It can also be a pair of fasta and quality files as provided by GS-FLX sequencers. Depending on the input format, the output can be formatted in fasta or fastq. Fastq is an extension of fasta format including *per base quality values* (see Fig. 5) (9).

The Main ngsfilter Options

For simplicity reasons, only the most useful options of ngsfilter will be listed below.

Specifying the Input Format

Like all OBITools, ngsfilter usually automatically recognizes the input read file format. However, you can specify the encoding quality schema of the fastq files:

- --sanger: input file is in sanger fastq nucleic format (standard and default fastq format)

- --solexa: input file is in fastq nucleic format produced by a Solexa sequencer (Illumina 1.3+)

- --illumina: input file is in fastq nucleic format produced by an old Solexa sequencer (Solexa/Illumina 1.0 format)

For the fasta format, the --fna and the --qual=<filename> options allows specifying that the fasta file was produced by a GS-FLX sequencer and what the associated quality file (file with the .qual extension) is.

Specifying the Output Format

If the input file contains quality data, then the default output format is fastq, otherwise fasta format will be used. Fastq outputs are always encoded following the Sanger rules. You can force the output to fasta or fastq format using the following options, respectively : --fastq-output or --fasta-output. If the output if forced to fastq without providing quality data in the input, a default quality of 40 will be associated with each base.

```
Sol3-gh_R    EH1087     acacgctct    GGGCAATCCTGAGCCAA    CCATTGAGTCTCTGCACCTATC    F
Sol3-gh_R    EH0380B    acacgtcgt    GGGCAATCCTGAGCCAA    CCATTGAGTCTCTGCACCTATC    F
Sol3-gh_R    EH0379B    acactctgc    GGGCAATCCTGAGCCAA    CCATTGAGTCTCTGCACCTATC    F
Sol3-ITSA    EH1087     cagctgatg    GATATCCGTTGCCGAGAGTC    GCACGGCATGTGCCAAGG    F
Sol3-ITSA    EH0380B    cagctgtga    GATATCCGTTGCCGAGAGTC    GCACGGCATGTGCCAAGG    F
```

Fig. 6. Sample description file: The full set of multiplexed samples must be described in a tabular text file containing six columns. The first two describe, respectively, the experiment, and the sample, the third one the tag associated with this sample. The fourth and fifth columns indicate the forward and reverse PCR primers. The sixth column indicates if partial sequences (i.e., incomplete amplicons) are expected with T (True) or F (False).

Specifiying the Sample Description File

The file describing all the samples multiplexed in one sequencer lane must follow the format presented in Fig. 6. Its name is specified to ngsfilter using either the short -t <filename> option or its long version --tag-list=<filename>.

Running ngsfilter

In this example, ngsfilter is used to analyze a fastq file named myrun.fastq produced by a Solexa sequencer. By using the > redirection character, the output is saved to the good.fastq text file. During the execution of ngsfilter, a progress bar with an estimation of the remaining computation time is displayed

unix–shell > ngsfilter –t samples . tag —solexa myrun . fastq > good . fastq

myrun . fastq 100.0 % | #########################–] remain : 00:00:00

unix–shell >

5.2. The obiannotate Program

ngsfilter like other OBITools annotates the sequences in fasta or fastq format using a tag/value system. These tag/value pairs are included in the title line of each sequence (see Fig. 7). obiannotate is used to change the annotation associated to each sequence.

Running obiannotate

After ngsfilter, you can clean the file for keeping only the experiment and sample annotations. The option -k allows specifying which annotation must be kept (see Listing 8). Then sequences will look like those presented in Fig. 8. Cleaning the annotated sequences to keep only the experiment and sample annotations can be done following Listing 8.

unix–shell > obiannotate –k exper iment –k sample good. fastq > clean . fastq

good. fastq 100.0% | #########################–] remain : 00:00:00

unix–shell >

5.3. The obisplit Program

Obisplit allows distributing sequences from the file annotated and cleaned by ngsfilter and obiannotate into different files according to

```
@HWUSI-EAS1510_0005:1:2:13183:12244#0/1_CONS_SUB reverse_score=88.0; tag_length=9;
tail_quality=28.2; reverse_match=ccattgagtctctgcacctatc; direct_tag=acacgctct;
sample=EH1087; reverse_tag=acacgctct; reverse_primer=ccattgagtctctgcacctatc;
direct_score=68.0; cut=[28,42,1]; direct_match=gggcaatcctgagccaa;
direct_primer=gggcaatcctgagccaa; experiment=Sol3-gh_R; mid_quality=32.3703703704;
head_quality=40.4; avg_quality=32.8918918919;
tggctcagctgtgg
+
LLG4BGC@>BB5:C
@HWUSI-EAS1510_0005:1:2:14001:3874#0/1_CONS_SUB_CMP reverse_score=88.0; tag_length=9;
complemented=True; tail_quality=24.8; reverse_match=ccattgagtctctgcacctatc;
direct_tag=acactctgc; sample=EH0379B; reverse_tag=acactctgc;
reverse_primer=ccattgagtctctgcacctatc; direct_score=68.0; cut=[33,50,1];
direct_match=gggcaatcctgagccaa; direct_primer=gggcaatcctgagccaa; experiment=Sol3-gh_R;
mid_quality=40.3103448276; head_quality=43.7; avg_quality=38.7564102564;
atcttattctaaaatga
+
HOE71A?>QN5_\QUO]
```

Fig. 7. Output ngsfilter file formatted in fastq format: The first two sequences analyzed are provided. In contrary to the classical fastq format, the long title line is not repeated twice. Several fields were added to the title line. complemented=True in the second sequence indicates that the reverse complement of the sequence has been used to find the primers. head_quality is the average quality of the first 10 bases of the sequence. mid_quality is the average quality of the central part of the sequence. tail_quality is the average quality of the last 10 bases of the sequence. avg_quality is the average quality of the whole sequence. direct_primer is the true sequence of the identified forward primer. direct_score is the alignment score of sequence with the forward primer. direct_match is the real sequence of the forward primer identified on the sequence. reverse_primer is the true sequence of the identified reverse primer. reverse_score is the alignment score with the reverse primer. reverse_match is the real sequence of the reverse primer identified on the sequence. cut indicates where the original sequence was truncated. tag_length is the length of the tag found. direct_tag is the sequence of the tag found on the 5′ side of the forward primer. reverse_tag is the sequence of the tag found on the 5′ side of the reverse primer. sample is the sample id associated with the sequence. experiment is the experiment id associated with the sequence.

```
@HWUSI-EAS1510_0005:1:2:13183:12244#0/1_CONS_SUB sample=EH1087; experiment=Sol3-gh_R;
tggctcagctgtgg
+
LLG4BGC@>BB5:C
@HWUSI-EAS1510_0005:1:2:14001:3874#0/1_CONS_SUB_CMP sample=EH0379B; experiment=Sol3-gh_R;
atcttattctaaaatga
+
HOE71A?>QN5_\QUO]
```

Fig. 8. Two sequences cleaned by obiannotate.

the value of one of their annotations. This can be seen as a demultiplexing of the sequences. A first level of demultiplexing must be done on the base of the experiment: in Listing 9, the clean.fastq file is splitted in as many files as necessary, each file storing the subset of sequences with the same experiment value. All file names will begin by the prefix myrun_ and finish with the experiment value.

unix–shell > obisplit –t exper iment –p myrun clean . fastq

good . fastq 100.0% | ##########################–] remain : 00:00:00

unix–shell >

A second split of each of the created file can then be done according to the sample tag.

References

1. O'Brien HE, Parrent JL, Jackson JA et al (2005) Fungal community analysis by large-scale sequencing of environmental samples. Appl Environ Microbiol 7:5544–5550

2. Briée C, Moreira D, López-García P (2007) Archaeal and bacterial community composition of sediment and plankton from a suboxic freshwater pond. Res Microbiol 158:213–227

3. Fierer N, Morse JL, Berthrong ST et al (2007) Environmental controls on the landscape-scale biogeography of stream bacterial communities. Ecology 88:2162–2173

4. Nicol GW, Leininger S, Schleper C, Prosser JI (2008) The influence of soil ph on the diversity, abundance and transcriptional activity of ammonia oxidizing archaea and bacteria. Environ Microbiol 10:2966–2978

5. Zinger L, Coissac E, Choler P, Geremia RA (2009) Assessment of microbial communities by graph partitioning in a study of soil fungi in two alpine meadows. Appl Environ Microbiol 75:5863–5870

6. Teixeira LCRS, Peixoto RS, Cury JC et al (2010) Bacterial diversity in rhizosphere soil from antarctic vascular plants of admiralty bay, maritime antarctica. ISME J 4:989–1001

7. Cronn R, Liston A, Parks M et al (2008) Multiplex sequencing of plant chloroplast genomes using solexa sequencing-by-synthesis technology. Nucleic Acids Res 36:e122

8. Hamady M, Walker JJ, Harris JK et al (2008) Error-correcting barcoded primers for pyrosequencing hundreds of samples in multiplex. Nat Methods 5:235–237

9. Cock PJA, Fields CJ, Goto N et al (2009) The sanger fastq file format for sequences with quality scores, and the solexa/illumina fastq variants. Nucleic Acids Res 38:1767–1771

Chapter 3

SNP Discovery in Non-model Organisms Using 454 Next Generation Sequencing

Christopher W. Wheat

Abstract

Roche 454 sequencing of the transcriptome has become a standard approach for efficiently obtaining single nucleotide polymorphisms (SNPs) in non-model species. In this chapter, the primary issues facing the development of SNPs from the transcriptome in non-model species are presented: tissue and sampling choices, mRNA preparation, considerations of normalization, pooling and barcoding, how much to sequence, how to assemble the data and assess the assembly, calling transcriptome SNPs, developing these into genomic SNPs, and publishing the work. Discussion also covers the comparison of this approach to RAD tag sequencing and the potential of using other sequencing platforms for SNP development.

Key words: Single nucleotide polymorphism, Next generation sequencing, Roche 454, Non-model species, Transcriptome sequencing, Normalization, Transcriptome assembly, SNP calling, Bioinformatics

1. Introduction

Single nucleotide polymorphisms (SNPs) are single base pair differences between or among individuals at the same genomic position. Given their high density within genomes and simple mutational dynamic, SNPs have become popular molecular markers for a wide range of analyses, from QTL mapping and association studies to pedigree reconstruction and demographic analyses (1). Ongoing advances in next generation sequencing (NGS) now provide a quick and cost-effective means of developing SNP markers in novel populations and/or species (2).

This chapter is focused upon the development of SNPs in species having little to no previously existing molecular resources. The chosen method presented is the use of Roche 454 sequencing technology

François Pompanon and Aurélie Bonin (eds.), *Data Production and Analysis in Population Genomics: Methods and Protocols*, Methods in Molecular Biology, vol. 888, DOI 10.1007/978-1-61779-870-2_3, © Springer Science+Business Media New York 2012

to sequence the normalized transcriptome of a pooled set of individuals. The reason for this decision is simple. Nearly a dozen studies have been published over the past 2 years documenting the utility of this approach across diverse taxa to simultaneously obtain access to the genes, and their variation, in novel species ((2); see Note 1). Importantly, many of these studies come from research groups with little prior genomic experience, demonstrating the feasibility of tackling the bioinformatics challenges inherent in NGS data. Alternative routes to obtaining SNPs are certainly feasible, as other templates could be used (e.g. non-normalized transcriptome, genomic DNA, or restriction-digested genomic DNA) and other NGS technologies employed to good effect (2). In fact, before beginning any NGS sequencing for SNPs, a research group should directly consult with their chosen core sequencing facility, as well as colleagues who have thought through these issues, to assess what approach they should use as the NGS landscape is rapidly changing (in terms of sequencing performance and costs). All of this being said, new benchtop 454 sequencing machines have just hit the market (454 GS Junior), making this approach to obtaining SNPs more relevant than ever. This chapter is therefore an example of a proven method, which raises many of the issues research groups will face regardless of their technical approach.

Research groups planning to develop SNPs for their study system should have a very clear goal for their project, and work backwards from that position for the integration of technology. Importantly, issues of scale, a timeline of returnable results, and costs need to be considered. Projects that need <500 SNPs should be designed differently from those seeking 1,000s of SNPs, in order to save time and money. Two important issues to consider are target representation and sequencing coverage (or depth). Target representation refers to the amount of unique DNA material to be sequenced in the pool of DNA. Sequencing coverage refers to the amount of sequencing providing data, or coverage, over the same region (i.e. reads per bp). Consider a butterfly with a genome size of ~400 Mbp. With 1×10^6 reads, 400 bp each, which is a good result from full plate of 454 sequencing, this would roughly provide an average of 1.0× coverage of that genome (that is, on average, one read per bp of the genome). Regardless of the number of individuals pooled to make that genomic DNA pool, there would be little to no SNP information since on average the given sequencing coverage is expected to only provide on sequence per bp. This low coverage simply cannot sample the variation in pool of chromosomes, although since we are dealing with average, some regions will have many reads and thus some SNP information. However, sequencing the transcriptome of a pool of butterflies is very different. A rough estimate of transcriptome size is roughly 30×10^6 bp (~15,000 genes × ~2,000 bp). A full 454 sequencing plate would be expected to cover this target size with ~10× coverage, that is, reading about ten independent copies of each gene region. Thus, by reducing the representation of the genome for a pooled set of individuals allows for data to be gathered on the genetic differences among individuals for the same genetic regions, and this is the type of data that allows for the identification of SNPs in novel species. However, the reality of sequencing the transcriptome is much more complex than this simple example.

Transcriptome sequencing provides a highly informative approach to obtaining genomic sequence information, as it sequences the mRNA of genes expressed in the tissue studied. Coding regions of genes are by far the most studied gene regions of genomes and are readily compared among taxa. However, many research programs are not interested in coding DNA variation or comparative analyses per se, but are solely concerned about identifying SNP variation across their focal species' genome for rapid QTL mapping, association studies, or phylogeographic analysis. I recommend such research projects read and understand the RAD tag sequencing approach, which uses a genomic reduction technique for obtaining the same random regions of the genome across pools of individuals, coupled with sequencing these regions using the short-read high-throughput method of Illumina sequencing (Etter and Johnson, this volume; (2–5)). While there is much power in this RAD tag approach, readers should be aware of the limitations of this approach for association studies, as well as comparative analyses and inferences (see Note 2, Hohenhole et al., this volume). There are also significant challenges in data handling associated with the RAD tag approach, especially in the absence of a reference genome ((3, 5); but see ref. 4).

The transcriptome is a highly complex pool of mRNA from many different genes. For different genes, expression levels can span about four orders of magnitude in copy number in a given mRNA pool from a single tissue. In addition, the mRNA transcripts from a given gene can be highly diverse, as many genes are alternatively spliced (i.e. not every mRNA transcript has all the exons of a given gene, as some may be skipped, or differently combined; see Note 3). Assembling all of this data back together again is very challenging, especially when using the short read data of NGS platforms, which is why the long read lengths of the 454 sequencing technology is the preferred method for this approach. In summary then, depending upon the project in question, there will likely be a need for reducing the representation of the mRNA expression variation inherent in the transcriptome by normalizing the number of copies of the different mRNA sequences so that they differ <2-fold in their copy numbers. Normalization greatly increases the ability of sequencing to randomly access the mRNA from different genes, but getting many reads from each gene requires good coverage. Good coverage and normalization costs money. However, sequencing a non-normalized pool will result in many sequences from only several hundred genes, as it will primarily sequence the genes that are highly expressed, and will likely identify many SNPs that may be sufficient for a given research project's objectives. Whether to sequence a normalized or non-normalized mRNA pool is thus an important decision, which depends directly on the focus of a given project and the near term future needs.

What are the project goals? Opposite ends of the research spectrum for those reading this chapter are likely to be researchers seeking access to <500 SNPs versus those seeking access to many 1,000s of SNPs. Within these two polar end groups, some projects may require SNPs from random genes, others may wish these from very specific genes. SNPs may be desired from the family, population, or species level. All of these issues must be thoroughly considered and clearly answered before efficient project planning can begin in earnest. These issues will be addressed below in the appropriate section focused upon sampling design.

Further important points for discussion regarding SNP development are (1) the conversion of transcriptome SNP information into exonic models that will work with genomic DNA samples and (2) understanding the limitations of ascertainment biases in SNP identification. Once high-quality SNPs are identified in transcriptome sequencing, developing these into markers that can be used on genomic DNA can be challenging in novel species where the exon intron structure for genes is not known. There are several ways to overcome this barrier to implementation of SNPs in novel species, and these are discussed below. Ascertainment bias (6) concerns the non-random sampling of SNPs from a population and the use of these SNPs for gathering population level information. SNPs identified from transcriptome sequencing of a handful of individuals, who pass quality filtering and other thresholds, are most likely to be SNPs having a relatively high frequency in the population sampled. As a result, rare SNPs will be underrepresented when using these identified SNPs to process population samples. Therefore, the resulting data will be biased towards intermediate frequencies and thus analytical assumptions of an idealized site frequency spectrum of genetic variation will be violated (6). This is a significant problem in most SNP studies and correction methods have been developed for model genomic species (6). Ascertainment biases may be more difficult to overcome in novel species and thus studies planning to use information from the site frequency spectrum must seriously assess their experimental design and assumptions.

The final section of this chapter is focused upon getting a transcriptome sequencing project published. Since this will represent a sizable investment of money, time, and effort, a reasonable publication should be expected. However, unless clear contributions can be made to the literature and generate findings of interest to a broader readership, publication of the resulting effort will be relegated to a lower-tiered journal. Therefore, nesting a biological question, or study of some technical aspects of transcriptome assembly, into the experimental design should be considered.

2. Methods

2.1. Sample Choice and Preparation

With the complexities of the transcriptome in mind, experimental design should target sampling and sample preparation to maximize the discovery of SNPs either in many or in a few transcripts. These two approaches require different levels of sequencing depth of the mRNA pool, and thus cost different amounts of money.

We have already talked about how gene expression level, that is, the number of mRNA copies, can vary among genes by at least four orders of magnitude. This can also vary across time within one tissue, across tissues, among developmental stages, and individuals. However, not all genes are expressed all the time. A commonly used rule of thumb expects roughly 70% of genes within an organism to be expressed within a given tissue, and roughly 30% of these mRNA transcripts are likely to come only from <50 genes, which are likely to be involved primarily in housekeeping function (e.g. central metabolism, gene expression; (7)). In addition, some genes are expressed only in response to specific environmental conditions, such as infection.

454 sequencing is effectively random, and thus the proportion of any gene's mRNA in the sequencing pool should be reflected in the number of 454 reads of that gene. Therefore, 100,000 sequencing runs (i.e. a typical run on a 454 GS Junior) on a non-normalized pool from a single tissue type is likely to provide ~30,000 reads from the ~50 most highly expressed genes in the chosen tissue. Given the rate at which expression level drops, sufficient sequencing coverage (i.e. >20 reads/bp) for identifying high-quality and common SNPs may extend for several 100 genes or more. This is therefore a cheap and quick way to scan for and likely identify SNPs across roughly 100 different genes.

Are SNPs from transcritpome sequencing truly random? No, they are not. Only the genes expressed will be sequenced, and higher expressed genes will generally have more alleles sequenced, and thus their SNPs will be more readily identified. Higher expressed genes are usually those that have a very conserved rate of molecular evolution and can, depending upon the species studied, have higher levels of codon bias. Genes involved in housekeeping function, having high levels of codon bias, are not likely to have lots of SNP variation within them. While these issues may complicate quick and easy SNP detection, most studies find moderate SNP levels across genes from pooled population samples.

The process of normalizing the transcriptome aims to reduce the variation among mRNA transcripts among genes to within a 10× range (8). This equalization of expression among genes means that each new sequencing read will have a much higher probability of being from a new gene rather than another read from a highly expressed gene. While this dramatically increases the ability of

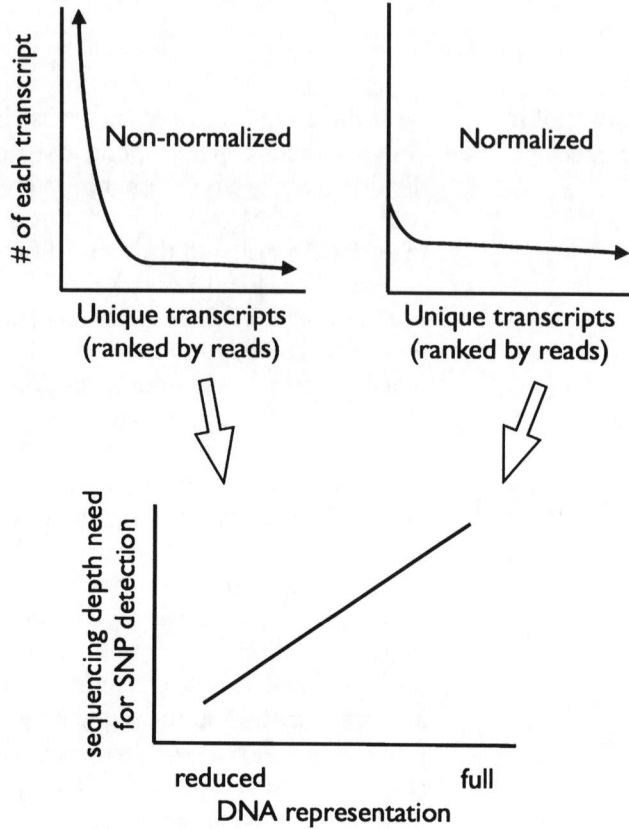

Fig. 1. The relationship between normalized and non-normalized cDNA pools and the sequencing depth per base pair needed to detect SNPs. Non-normalized pools have many copies of only a few unique transcripts, resulting in those transcripts being sequenced much more often than random. As a result, a reduced sequencing depth is needed to repeatedly sample the highly expressed genes, and this repeated sequencing provides the data needed to identify SNPs.

random sequencing to explore genes in the transcriptome pool of cDNA, it consequently requires a high level of sequencing coverage in order to get many reads from the same genes, which is required for good SNP identification (Fig. 1). With enough sequencing coverage, sequencing of the normalized transcriptome will produce 1,000s of SNPs for a very large fraction of the expressed genes. This approach can be extended to increasing the sequencing of unique transcripts by pooling many different tissue types, developmental stages, i.e. tissue types likely to differ in their expression profile (e.g. (9)). Thus, in order to maximize SNP identification across the transcriptome, one should collect mRNA from many different tissue types and time points, normalize, and sequence with high coverage. A desire to focus upon a specific subset of genes will require an understanding of the molecular biology of when and where they are expressed.

2.1.1. Normalization

There are several methods for normalization: do-it-yourself, commercial services, and commercial kits. Do-it-yourself approaches require a fair amount of molecular biology expertise (e.g. (10)). As a result, many labs doing 454 transcriptome sequencing utilize commercial companies. Be aware that companies vary in their turn-around time and quality. Be sure to consult with people who have had previous experience with the company rather than relying upon that entity's self-promotion alone. One company continually observed in the literature is Evrogen (http://www.evrogen.com/; e.g. (9, 11)). There are also good commercial kits available for normalization, which are relatively easy to follow and are a nice cost compromise between do-it-yourself approaches and commercial companies (e.g. Trimmer and Trimmer-Direct cDNA normalization kits, by Evrogen; (12)).

2.1.2. Accessing 100 SNPs with Low Time and Money Investment

As mentioned above, directly sequencing the non-normalized cDNA pool is a very good way to rapidly obtain sequence information for finding SNPs across ~100 genes and likely many more. Increasing the sequencing depth of unique transcripts can be accomplished by focused upon a single, low-complexity tissue type or organ, as this should further reduce the transcriptome representation. Recently, researchers have been exploring how well sequencing such pools using Illumina technology performs, and these advances should be closely followed.

2.1.3. Accessing 1,000s of SNPs for as Many Genes as Possible

In order to maximize SNP identification across a large and diverse fraction of the transcriptome, one should collect mRNA from many different tissue types and time points, normalize, and sequence with at least half and ideally a full plate run in order to obtain sufficient sequencing data for at large fraction of the transcriptome.

2.1.4. Modified Approach to Accessing 1,000s of SNPs for as Many Genes as Possible, Using Low Sequencing Coverage

One potentially powerful approach to obtaining excellent SNP data for many genes while using low sequencing coverage is to only sequence the transcriptome from either the 3′ or 5′ end of the mRNA transcript. On average, the 5′ UTR (untranslated region) is between 100 and 200-bp long across metazoa (13). Thus, a 400-bp read on average will cover the highly variable 5′ UTR and about 200 bp into the start of a given gene. 200 bp of coding DNA provides more than sufficient length for good annotation of most genes. Using the SMART™ and SMARTer™ PCR cDNA Synthesis (Clontech), full-length cDNA transcripts are made, and a specific primer is attached to the 5′ end. This procedure could be modified to selectively add a 454 sequencing primer solely to the 5′ end, followed by sequencing. Directional sequencing is now available for the 454 platform (e.g. ScriptSeq™ mRNA-Seq Library Preparation Kit by Tebu-bio, Inc.), but has only been used for quantitative RNA-Seq. However in the context of SNP discovery, such an approach would effectively reduce the representation of the mRNA pool by >>50%, since sequences are only going to be read from the terminal 5' end, and not from the 3' end or any other sheared gene product.

2.1.5. Focusing Upon Specific Genes

When and where specific genes are expressed can be a difficult question sometimes, as this can depend highly upon the turnover rate of the gene product in question. Thus, finding specific genes of interest can be a challenge, and special care must be taken to ensure that the tissue sampled is likely to have the mRNA of interest. Searches of the molecular biology literature of the closest related genomic reference species may provide sufficient insights into relevant expression patterns.

One alternative, or potentially complementary, approach is to use gene capture methods to specifically isolate genes of interest and concentrate them in a sample compared to the rest of the unwanted transcriptome. This enriched sample could then be directly sequenced, greatly increasing the information recovery per sequencing effort. Gene capture methods utilize the hybridization dynamics of double-stranded DNA by using a single strand of the gene that will be captured as "bait" to fish for the complement in the pool of the sample DNA. Such methods have their "bait" either attached to a microarray slide, a magnetic bead, or even free in solution (e.g. (14–16)). However, knowledge of which bait to use, and the DNA sequence of the bait itself, requires previous sequencing efforts. Also, these enrichment methods generally result in an enriched pool containing 60–80% targeted material, meaning that upwards of 40% of a given sequencing effort can be "off target".

2.2. Pooling and Barcoding

By combining mRNA from different individuals into a single pool for sequencing, the high quantity of sequencing reads generated by 454 sequencing will likely read many transcripts from the same gene regions from different individuals. This all depends upon the expression level of the gene, whether normalization has taken place or not, and the consistencies of expression level for the genes in question among the pooled individuals. These complexities make it difficult to calculate the expected level of coverage per gene per pooled individual.

Many individuals from a single population, or from multiple populations, or perhaps of two different phenotypes of interest can be pooled. Their mRNA may be from the same or differing tissues or developmental stages, or sexes. A positive relationship exists among several factors that will affect the range of SNPs identified: the number of individuals pooled, their genetic relationships (e.g. closely vs. distantly related individuals), the number of tissue types chosen, and the level of sequencing coverage.

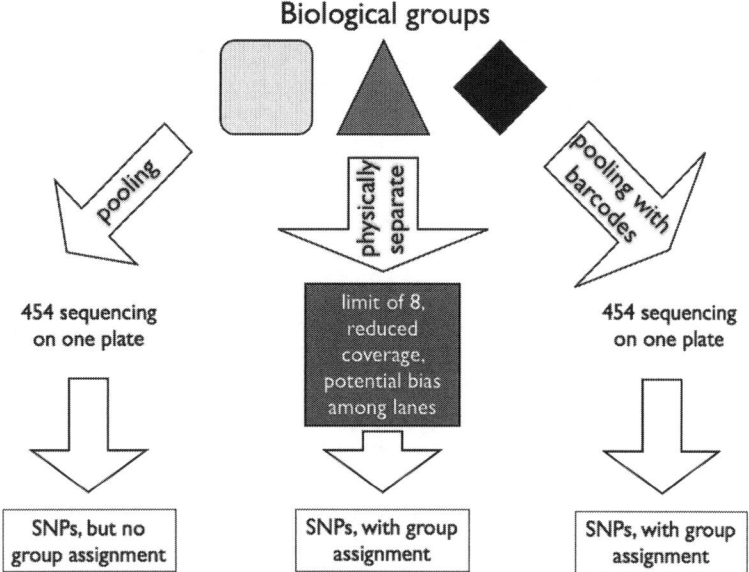

Fig. 2. Three ways to combine different biological groups for 454 sequencing. Biological groups, defined by phenotype or population, are represented here by three different shapes (*square*, *triangle*, and *diamond*). These groups of mRNA can be directly pooled and sequenced, uncovering SNP information that cannot be reassigned to the groups without additional work. The groups can also be sequenced after being placed into physically separate wells of a single 454 plate. Such plates allow for up to eight independent groups. Finally, the mRNA can be barcoded, wherein small stretches of DNA, usually 4-bp long, are attached to each fragmented DNA strand before sequencing. Each group has a unique DNA sequence (i.e. barcode) that will be included in the sequence data and later used to sort the sequence data back into groups.

2.2.1. How to Pool Individuals

There is no exact math for estimating the number of individuals to include, given the complexity of the transcriptome. Try and use as many individuals as you can from the groups of interest to the project, as this will ensure a good sampling of the allelic diversity within the group. Given the sensitivity of the method to low-quality mRNA, quality control assessment of each individual's mRNA prior to pooling is strongly suggested (for example, using the Bioanalyzer by Agilent is highly recommended).

2.2.2. Barcoding Samples

While sequencing a pooled sample of individuals is an excellent method of randomly sampling the variation within the pool, this can also erase information of importance such as which individuals were from a given population or of a certain phenotype (Fig. 2). There are several ways around this problem, and incorporating these changes into experimental design can produce data that can quickly move a research group forward (e.g. (11, 17)). First, there are physical barriers separating the different biological samples, and currently for 454 sequencing these allow for 2, 4, 8, and 16 independent groups to be analyzed. Second, there are molecular markers called barcodes, which can be added to the mRNA samples and directly sequenced on each mRNA strand (18). These are identified later during post-sequencing analysis. Commercial kits are available (e.g. Nextera™ by Epicenter Biotechnologies, Inc.) and software packages that can identify and group samples by barcodes are emerging. Choosing among these methods will likely depend upon the familiarity of your local core facility with the different methods, experimental design, bioinformatic considerations, and cost, since tagging costs money and reads per tag inherently have lower coverage.

2.3. 454 Sequencing

Until recently, nearly all 454 Life Science sequencing was performed by a core facility on a university campus or at a commercial company. In 2010, 454 Life Science released a bench-top version of the 454 sequencing machine, bringing 1/10th of the throughput to individual labs (i.e. 0.1×10^6 reads instead of 1×10^6). This exemplified the only constant in NGS technology, which is that change happens and, as advances in machines and chemistry continue, the result is a diversity of sequencing options and template requirements. Direct, detailed discussions with one or more technicians at the chosen facility is strongly encouraged to ensure your sample is suitably prepared for the platform options you are considering. If normalizing, make sure that the final normalized product is compatible with these requirements. Much like the conflict between software and operating systems on computers, there can potentially be incompatibility issues, whether real or imagined (see Note 4).

2.3.1. How Much to Sequence? Tradeoffs for Consideration

There are fundamental tradeoffs in sequencing considerations: coverage versus cost, coverage versus depth, biological grouping versus coverage, etc. More data results in better coverage of the transcriptome and facilitates the identification of high-quality SNPs from more genes, but increased coverage costs money although price/bp is dropping every year. Biological grouping of your data can in theory provide greater insights and potentially lead to faster identification of the SNPs of interest. However, each biological group is essentially splitting your coverage. For example, splitting a single full plate of 454 sequencing into four groups will ideally lead to four groups of sequences having 250×10^3 reads each. While all of this data can be combined for transcriptome assembly, the identification of SNPs in each of the groups requires deep coverage. Analysis will then center upon the question of whether the SNP differences identified among the groups accurately reflect the SNP frequency in those groups. Given the likely low coverage of each gene within each of the four groups, there will likely be a very high false positive rate of identified SNP differences. Such an approach will, however, generate a list of candidates SNPs that can be screened in a larger and more representative biological sample. Thus, biological grouping can certainly produce high-quality SNPs, but their biological relevance (i.e. frequency in the biological groups) needs to be verified. In other words, it is unlikely that such pooling will produce both high-quality SNPs and verify their frequencies in the biological groupings.

2.4. Assembling the Data

Transcriptome assembly using next generation data, that is, aligning all of the fragmented sequencing reads back into their pre-fragmented full-length transcripts, is computationally challenging and a developing field. Over the past several years a number of assembly software packages focused on transcriptome assembly have arisen (e.g. Abyss, Mira3, Oasis, Newbler, Ngen, CLC). The first four programs use the Linux operating system and command line operation, while the latter two provide a GUI interface, with the last two working across the major OS platforms being expensive commercial products (with 30-day free trial periods). Only recently has their relative performance in transcriptome assembly begun to be assessed (19), and such assessment requires additional attention. These programs also differ in performance and RAM requirements, with software using the older overlap-layout-consensus (OLC) methods (Mira, Ngen, CAP3) generally having a high RAM demand compared to software using the newer de Bruijn graph methods (Abyss, Oasis, and CLC; see Note 5).

2.4.1. A Robust Assembly Strategy

Recent comparative analysis of assembly performance has two important conclusions. First, there is significant difference among the software available, with the two commercial software programs (CLC and Ngen) performing best although several of the freeware programs are not far behind (19). Second, rather than focusing upon a single one of these programs, combining the results from different assemblies, by assembling their output together, produces the most robust results (19). Importantly, a combination of two freeware programs (MIRA and Newbler 3.5) performs nearly as well as combinations using the commercial software packages, or the commercial packages themselves (Table 11 in ref. (19)). One package has gone a bit farther, using four rounds of assembly with MIRA, followed by a clustering of these using CAP3, then a blasting of the resulting contigs among themselves to find possible contigs that should be joined (iAssembler; http://bioinfo.bti.cornell.edu/tool/iAssembler/). Users with bioinformatics background will certainly benefit from the freeware programs available. For those new to this field, buying and using the commercial software is a smart investment, as 1–2 months of salary for a well-trained bioinformatician is the equivalent of the software cost for a GUI program that a capable Ph.D. student can master in a similar amount of time.

2.4.2. Handling the Data: Locally or on a Server

The initial bottleneck many novice labs will encounter is not having sufficient computer infrastructure to handle the data and analyses. Important issues to consider are whether you will be analyzing the data on a local computer or a server, how much RAM these computers have, and what programs you wish to run. Generally speaking, at least 8 gigs of RAM are necessary to run analyses using de Bruijn graph methods, but this low amount of RAM can cause programs using OLC methods with moderate amounts of data to crash. However, there are a growing number of computer servers available to the academic community, providing an attractive alternative to performing assemblies locally (e.g. (20)). Servers differ in their performance, ease of use, software available, and length of time that can be allocated to a given project. They generally do not allow for commercial software to be run and may not have the latest version of your desired software (though many will happily upgrade if you ask them). Some servers are free (e.g. Bioportal; http://www.bioportal.uio.no/) while most charge for their services by CPU time used (e.g. (20)). Finally, regardless of where assemblies are performed, results will still need to be assessed locally.

The best way to assess these issues is to perform an assembly dry run. Download datasets that are likely to be similar to the ones you will generate (search the Short Read Archive (SRA) with a species name: http://www.ncbi.nlm.nih.gov/sra; e.g. "papilio"), and try assembling them (after cleaning and trimming the data). Commercial software packages are generally provided free for 30-day trial licenses. There is simply no better way to assess your infrastructure than trying to replicate previous studies.

2.4.3. Factors Affecting Transcriptome Assembly

Beyond the issues of RNA quality, sequencing coverage and methodology, researchers may be concerned about how genetic variation within their RNA pools may affect assembly. Generally speaking, genetic variation among individuals from a single population should not be extensive enough to affect assembly. While most assembly programs allow the user to define cutoff levels for sequence comparison and cluster inclusion, setting these values too low will cause problems within gene families, as allele from multiple independent genes may end up being grouped rather than having groups consist solely of allelic variation at a single locus (i.e. a paralogous clustering of reads rather than an orthologous clustering; see Dlugosch and Bonin, this volume). But these issues are also a matter of scale, as instances of gene clustering will certainly occur for genes of recent duplication origins, but such instances are not likely to be very common. While studies are beginning to start comparing data from different populations, and populations from very different regions could be substantially diverged, but in these cases population specific data is usually identifiable (e.g. barcoded). SNP variation and copy number variants are of low concern, however, compared to the single largest effect on transcriptome assembly, alternative splicing, which is likely to affect SNP calls unless steps are taken to minimize such impacts. Alternative splicing is an inherent and unavoidable biological reality affecting a very large component of the metazoan transcriptome.

2.5. Assessing Your Resulting Assembly

Obtaining the good assembly with a given dataset is central to obtaining many SNPs that can be used in downstream applications. For example, identifying SNPs in small contig fragments that have many errors will increase the false positive rate of SNP identification (calling SNPs when none actually exist). Even if SNPs of high quality are identified, if regions near them are of low-quality information, this can make it difficult to develop molecular assays for these SNPs (as such assays usually require 20–50 bp of good sequence information on both sides of a given SNP location). In sum, long and accurate contigs are necessary for identifying the region flaking SNPs, which will be needed for designing molecular assays.

There are many different ways to assess how well a given dataset has been assembled. At one end, a user could compare all of the assembled contigs back to the predicted gene set generated by a whole genome sequencing project. For research groups without such genomic resources, assessment must use indirect approaches. In general, cDNA sequencing projects in novel species identify roughly 4–6,000 genes that find somewhat good matches to the large protein databases, and these are predominantly housekeeping genes since they have a slow rate of molecular evolution. Of these, around 2,000 have good annotation information regarding biological function. Most transcriptome studies focus upon the following metrics to assess assembly: total number of reads used in assembly, the number of contigs >100-bp long (or some other number), the N50 of the assembly (i.e. the length of the middle contig when ranking all contigs based on length), the maximum contig length, the summed contig length (e.g. (19)).

While these metrics will certainly help users determine the relative assembly performance due to changes in parameters or methods, other metrics should also be employed. The predicted gene set from the closest related species having whole genome sequence is a valuable resource (Fig. 3). First, such a dataset should be blasted against itself to identify and remove alternative splice variants and recent gene duplicates (for every such comparison with >95% DNA identity over 100 bp, keep the longest gene). Such filtering can remove nearly a one-third of some predicted gene sets. Second, blasting your assembly against the resulting filtered predicted gene set then provides an independent, robust, and minimum estimate of the number of genes represented in the assembly, since many genes will be covered by several of the assembled contigs. Generally a blast cutoff level of an evalue $<1 \times 10^{-5}$ is used. This approach is suggested as a comparison to simply blasting an assembly against one of the large genomic databases (Fig. 3). Since many databases contain entries of the same gene from different species, when blasting your assembly against these databases, it is possible that contigs in your assembly that are from different parts of the same gene but have not assembled together (which is common) could all have

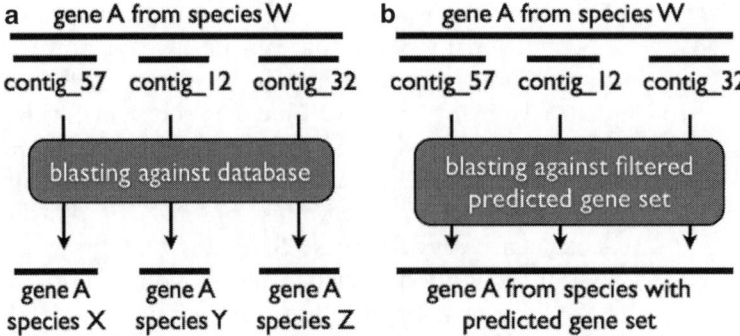

Fig. 3. Diagram showing the origins of potential biases in unigene estimates of assembled transcriptomes, and how to overcome this bias. In both scenarios, A and B, three contigs have been assembled from the 454 sequencing data that are all from gene A from species W. Either insufficient data or assembling performance resulted in the failure to join all three of these contigs. In (**a**) blasting these three contigs against one of the large protein databases (e.g. UniProt -Universal Protein Resource) can result in each contig hitting the correct gene, but from different species. Given the different IDs of the resulting hits, these will be counted as different genes in many analyses. Contrast this with (**b**), wherein each of the three contigs is compared to a predicted gene set from one species. If this gene set is not filtered to remove alternative spice variants, it is likely that the different contigs might find their best hit with different alternative splice isoforms, thereby inflating the number of unigenes hit. By filtering the predicted gene set to remove the alternative splice isoforms, the three contigs will find their best match to a single gene and thereby show a correct lower estimate of the number of unigenes contained in the assembly.

their top blast hit to same gene, but from different species (since different regions of a gene will have different rates of evolutionary change). This will result in a spurious increase in your estimated number of unique genes in your assembly. Third, extract the length of the aligned region from the blast results of your assembly against the predicted gene set, and sum this for those contigs hitting the same predicted gene (although be sure to only count the length covered from the predicted gene set). This sum can then be divided by the predicted gene length, and this ratio can then be plotted against the predicted gene length, providing a graph of the % of each gene covered against the expected length of that gene (e.g. (9, 17, 21, 22)). Using metrics based on this graph, sum of the length of these contigs, and the number of contigs needed for a given amount of coverage of the predicted genes, are potentially much more informative than the metrics presented in the previous paragraph. Again, the alternative splicing occurring within the assembly should be removed, as this will likely be creating many small contigs that are not assembling as a whole, since more assembly programs cannot effectively handle alternative splicing events.

One can also assess the coverage of the transcriptome by comparing the number of genes hit in different well-defined functional groups of genes likely to be present in the mRNA pool sequenced (17, 22).

2.6. Calling SNPs from Transcriptome Data

Visualizing the alignment of individual reads against their consensus contig demonstrates different approaches to SNP identification (Fig. 4), which can be set at different levels of stringency in order

```
                        "loose"          "strict"         7 bp window flanking SNP
                           ↓                ↓               ↙
Contig_34:    CTTGTKGCAT ACAATCGCGGCCTGT GTGTTT-TATGGGATTTTG
Read_27463    .....T.... ..............  ..............
Read_38910    ...T.....  .............. | .............
Read_97293    ..         ..............|......A.........
Read_00342    .G....     |........T.....  .............. ...
Read_01293    ...G....   |........T.....| ...........
Read_01293                ..............|.............
```

Fig. 4. SNP identification using different criteria, when looking at the assembled reads in a contig. The contig represents the consensus sequence of all the individual sequencing reads for this genetic region, with high-quality SNPs coded as degenerate. For the "loose" criterion, at least two individual reads of good quality are needed for each alternative allele, for a minimum of four total reads. For the "strict" criterion, a minimum of six reads total is needed, with a minimum allele frequency of 25%. A 7-bp window flanking the "strict" SNP is outlined, showing how there are no additional variants within that window, suggesting that the identified SNP is derived from the alignment of orthologous sequencing reads without effect from alternative splicing.

to control the false positive rate (i.e. SNPs identified that are not real). For several researchers, the minimal requirement for calling an SNP is having at least two independent sequencing reads per alternative allele (not including the contig, which is essentially the consensus sequence), with a maximum of two alternative states (e.g. (9, 17, 21)). This can be defined as a "loose" criterion (Fig. 4; (21)). Stringency can be further increased by additionally requiring SNPs to be in regions not containing any other SNPs, indels, or homopolymer runs (i.e. long stretches of a given nucleotide, such as the poly A tail of mRNA) within a given number of base pairs from the identified SNP (and this should be bidirectional; e.g. (9)). This reduces the probability of SNPs being identified in regions of low sequence quality, alternatively spliced exons, or incorrectly grouped alleles from paralogous genes. While regions of low quality should have already been filtered out or masked prior to and during assembly, 454 sequencing has inherent errors associated with homopolymer runs and thus, these should be avoided during SNP identification. Alternatively, spliced exons can be very small and may end up looking somewhat similar to SNPs. However, in general splicing is likely to result in a series of clustered SNPs (e.g. Fig. 4 in ref. (9)). This clustering of SNPs is also likely to occur when alleles from paralogous genes are clustered together, as such alleles should differ from each other by numerous SNPs, but this is dependent upon the age of the duplication even for the paralogs. Researchers should also keep in mind that these criteria will necessarily bias the identification of SNPs away from genes that might be inherently highly polymorphic. Further increases in stringency can be obtained by requiring a minimum number of reads for the SNP site and a minimum % of reads for the minor allele to a specified level (e.g. a minimum of six reads at the SNP site, 25% of which corresponding to the minor allele; (21)). This last criterion is considered "strict", as it scales with the number of reads per SNP (Fig. 4), which is important as some genes will have 100s of reads and thus detecting SNPs by chance increases proportionally (i.e. detecting two similar errors at the same site increases with the number of reads; (21)).

Nearly every publication uses a different combination of transcriptome assembly and SNP calling software, and thus there is no consensus on methodology (but here is a list of many such studies: (2)). Yet, nearly all have some level of filtering mentioned in the paragraph above and generally approach SNP calling by analyzing the ace file output from alignment programs (e.g. GigaBayes; http://bioinformatics.bc.edu/marthlab/GigaBayes). Several SNP calling methods do use the quality scores of the individual reads themselves, but these generally map reads back to genomic backbones that are assumed to be highly accurate. Similarly, two individual reads of high quality differing at a given site could indicate an SNP, but this should be considered as more relaxed than the "loose" standard mentioned above. The coding position of SNPs can also be inferred from alignment information from different blast results (e.g. (9, 21)).

2.7. From SNPs In Silico to Data in Hand

After identifying SNPs in silico, users generally validate some fraction. Then, the in silico predictions are used to design a molecular protocol for scoring large numbers of SNPs from many samples (e.g. (23)). Several recent publications have quickly and easily validated their SNPs by designing PCR primers flanking SNPs and sequencing through these regions from a population sample of individuals (e.g. (24)). Pooling DNA from individuals could also be used and these products directly sequenced, as polymorphic sites should be visible in electropherograms if SNPs are common, thereby decreasing the number of sequencing reactions needed for validation. Validation of SNPs usually sees around an 80–90% success rate, and this missing 20% could be due to differences in population sampling or technical effects, or true false positives (e.g. (23)).

A set of contigs can then be created that include only regions of high-quality sequence reads and what are considered high-quality SNPs. Generally there is a requirement of roughly 20–100 bp on either side of a SNP for the design of probes for SNP analysis, depending upon method. This requirement highlights the importance of good, accurate contig assembly.

All of the work mentioned up until this point has been based upon cDNA, that is, transcribed genes that do not include introns. Introns are pervasive in eukaryotic genes, with the genomic length between the start and stop codons of a gene often being increased by several orders of magnitude due to long, intervening introns.

Currently very little is known about the rate at which exon boundaries decay over time across different clades of organisms, although intron lengths appear to be well conserved out to around 40 million years of divergence, at least among *Drosophila* species (25). The extent to which exon boundaries are conserved is an important issue, since most species for which SNPs are going to be developed using the approach outlined here, lack a genome. If exon boundaries are well conserved over many millions of years, the exon boundaries in a novel species can be inferred by using the boundaries in the orthologous gene of a distantly related species whose genome has been sequenced. The accuracy of this inference of exon boundaries across evolutionary time depends upon the gene, the evolutionary distance, the phylogenetic clade in question, and is likely related to the amino acid divergence rate of the genes (i.e. the molecular evolutionary constraint). For example, orthologous genes from humans and mice that are 75–80% identical at the amino acid level share 95% of their intron locations (Fig. 4 in ref. (25)). These species likely diverged 75 million years ago (26). For genes having an amino acid identity of 65% identical or less, the location of introns that are the same in both species starts to decrease below 90%. Thus, developing SNPs from transcriptome data in eutherian mammals could make very good use of exon boundaries from the existing sequenced genomes, especially if contigs are chosen for genes that are >75% identical at the amino acid level between the compared species.

Researchers should also be aware that the average exon size in animals is around 100 bp in length (25). Some SNP assay methods require at least 50 bp on either side of a given SNP for probe/primer development, necessitating that researchers thoroughly consider the issue of exon boundaries in order to accurately develop SNPs across 1,000s of genes. Recent studies have clearly made this development (e.g. (23)), but other labs may consider an additional round of whole genome sequencing to generate sequencing information that extends into the intron regions flanking exons of interest.

2.8. Publishing the Work

As the number of transcriptome sequencing projects increases, new thresholds are emerging for these publications in the journals where such work has commonly found a home (e.g. BMC Genomics). Such publications today require making a significant contribution to the literature in addition to generating a transcriptome sequence database, by addressing either a specific biological question or issue, and/or technical issues regarding assembly. This roughly parallels the field of whole genome sequencing, as novel genomic sequences need more than just a description of the genes identified and their sizes in order to be accepted for publication in the top-tier journals. Many transcriptome publications are therefore sequencing samples containing some type of biological grouping in order to identifying SNP variation potentially under directional selection (e.g. (11, 17, 27)), population informative or useful in QTL mapping (23), or comparative analysis of diversity (21). Transcriptome papers have also focused upon assembly performance (19). Ideally, like many of these aforementioned examples, combination of these foci will become the norm in de novo transcriptome sequencing papers.

3. Notes

1. A fairly comprehensive list has just been published, showing at least 12 studies that have used NGS to identify SNPs, the vast majority of which have used 454 sequencing (2).

2. A quick mention about the potential limitations of the RAD tag approach for association studies, as well as comparative analyses and inferences, when working with species not having a well-assembled genome. RAD tags provide the powerful ability to deeply sequence pools of DNA, at specific regions in the genome (3). This method works by digesting genomic DNA with a specific restriction endonuclease that cuts DNA upon recognition of a specific, short DNA sequence, resulting in many thousands of DNA fragments that will be similar when performed on the genomic DNA from different individuals. By attaching sequencing primers and a barcode to these cut regions, a pool of many individuals can be combined and sequenced using NGS. The result is the generation of reads from the same genomic region, obtained from known individuals, for thousands of regions throughout the genome. This sequencing thereby provides SNP data for these regions that can be associated with specific individuals. The sequenced regions, however, are randomly distributed and thus for even moderately sized genomes, the vast majority of these markers will not be in coding regions of genes. Thus, RAD tags provide insights into patterns of genomic variation, but it is very difficult to connect those patterns to any specific genes or gene order without specifically being able to map these reads to a genome, or making a linkage map of these reads themselves. In contrast, obtaining SNP data for specific genes allows one to potentially associate SNPs with specific causal changes within or near genes having a causal effect. Additionally, SNPs in one species can be compared to the orthologous gene in a related species, and if that species has a map for those genes and you can assume similar gene order, these SNPs can be placed in a chromosomal order. Recent empirical study in sticklebacks provides a nice example of these issues in the context of SNP association studies. After deep sequencing using the RAD tag approach on a species with

a genome, most of the RAD tag markers were located in genomic regions roughly 1 million bp in size, which themselves contained around 100 genes (5). Consider these results without a genome. Identifying associations between SNPs and phenotypes of interest would be extremely difficult with even moderate levels of recombination, but even once such associations are found, extending such observations to potentially causal genetic changes will be extremely difficult.

3. Alternative splicing can create dramatically different mRNA variants from the same genomic region. For ecologists, a good review was written by J. Marden (28), and for additional information regarding potential interactions with SNPs, see ref. 29.

4. For example, companies and core facilities generally guarantee their sequencing results once they have done their own quality control on your sample. However, if you are giving them material that is different from their requirements or standard protocol, they may consider that you sample enough of a departure from their standard protocol that should something go wrong, they assume no responsibility (regardless of whether the fault lies with them or your sample).

5. Assembly programs can be separated into two main classes, those that use the OLC versus de Bruijn graph method. The OLC approach was originally designed to handle thousands of long sequence reads generated by Sanger sequencing. This method works by computing all of the pairwise "overlaps" between the reads in the dataset, with these overlaps stored as connections in a network graph, which is then used to compute a layout of the reads for the formation of a contig and thereby the consensus sequence. Given all of the pairwise comparisons that need to be performed, this method performs most efficiently with a limited number of sequencing reads that provide extensive overlap. In contrast, the de Bruijn graph method breaks down sequencing reads into smaller fragments called k-mers, where k is the length of these fragments (usually between 10 and 25 bp). The de Bruijn graph method identifies overlaps of length $k-1$ among these k-mers and not between the actual reads. This dramatically reduces the redundancy in short-read datasets, especially when handling expression data or genomic data with a high repetitive content. However, choice of the parameter k has a significant impact on assembly performance and generally the optimal value must be determined empirically, as it is influenced by read length and error rate, yet independent of sequence depth. The result is a dramatic difference in the computational requirements when handling of NGS datasets, although the results from de Bruijn graph methods are not necessarily better. Hybrid approaches are

perhaps the best (19), and individual reads likely will need to be mapped back to the resulting consensus sequences for SNP calling.

Acknowledgements

Funding was provided by the Finnish Academy Grant No. 131155. I thank J. Marden for the opportunity to pursue SNPs using 454 sequencing technology.

References

1. Schlötterer C (2004) The evolution of molecular markers – just a matter of fashion? Nat Rev Genet 5:63–69

2. Ekblom R, Galindo J (2011) Applications of next generation sequencing in molecular ecology of non-model organisms. Heredity 107:1–15

3. Baird NA, Etter PD, Atwood TS et al (2008) Rapid SNP discovery and genetic mapping using sequenced RAD markers. PLoS One 3:e3376

4. Emerson KJ, Merz CR, Catchen JM et al (2010) Resolving postglacial phylogeography using high-throughput sequencing. Proc Natl Acad Sci USA 107:16196–16200

5. Hohenlohe PA, Bassham S, Etter PD et al (2010) Population genomics of parallel adaptation in threespine stickleback using sequenced RAD tags. PLoS Genet 6:e1000862

6. Clark A, Hubisz M, Bustamante CD, Williamson SH, Nielsen R (2005) Ascertainment bias in studies of human genome-wide polymorphism. Genome Res 15:1496–1502

7. Weber APM, Weber KL, Carr K, Wilkerson C, Ohlrogge JB (2007) Sampling the arabidopsis transcriptome with massively parallel pyrosequencing. Plant Physiol 144:32–42

8. Bonaldo MF, Lennon G, Soares MB (1996) Normalization and subtraction: two approaches to facilitate gene discovery. Genome Res 6:791–806

9. Vera JC, Wheat C, Fescemyer HW et al (2008) Rapid transcriptome characterization for a non-model organism using 454 pyrosequencing. Mol Ecol 17:1636–1647

10. Vogel H, Wheat CW (2011) Accessing the transcriptome: how to normalize mRNA pools. In: Orgogozo V, Rockman MV (eds) Molecular methods for evolutionary genetics. Humana, New York

11. Galindo J, Grahame JW, Butlin RK (2010) An EST-based genome scan using 454 sequencing in the marine snail *Littorina saxatilis*. J Evol Biol 23:2004–2016

12. Buggs R, Chamala S, Wu W et al (2010) Characterization of duplicate gene evolution in the recent natural allopolyploid *Tragopogon miscellus* by next-generation sequencing and Sequenom iPLEX MassARRAY genotyping. Mol Ecol 19:132–146

13. Pesole G, Mignone F, Gissi C et al (2001) Structural and functional features of eukaryotic mRNA untranslated regions. Gene 276:73–81

14. Gnirke A, Melnikov A, Maguire J et al (2009) Solution hybrid selection with ultra-long oligonucleotides for massively parallel targeted sequencing. Nat Biotechnol 27:182–189

15. Mamanova L, Coffey AJ, Scott CE et al (2010) Target-enrichment strategies for next-generation sequencing. Nat Methods 7:111–118

16. Ng SB, Turner EH, Robertson PD et al (2009) Targeted capture and massively parallel sequencing of 12 human exomes. Nature 461:272–276

17. Babik W, Stuglik M, Qi W et al (2010) Heart transcriptome of the bank vole (*Myodes glareolus*): towards understanding the evolutionary variation in metabolic rate. BMC Genomics 11:390

18. Parameswaran P, Jalili R, Tao L et al (2007) A pyrosequencing-tailored nucleotide barcode design unveils opportunities for large-scale sample multiplexing. Nucleic Acids Res 35:e130

19. Kumar S, Blaxter ML (2010) Comparing *de novo* assemblers for 454 transcriptome data. BMC Genomics 11:571

20. Langmead B, Schatz MC, Lin J, Pop M, Salzberg SL (2009) Searching for SNPs with cloud computing. Genome Biol 10:R134

21. O'Neil ST, Dzurisin JD, Carmichael RD et al (2010) Population-level transcriptome sequencing of nonmodel organisms *Erynnis propertius* and *Papilio zelicaon*. BMC Genomics 11:310

22. Wheat C (2010) Rapidly developing functional genomics in ecological model systems via 454 transcriptome sequencing. Genetica 138:433–451

23. Hubert S, Higgins B, Borza T, Bowman S (2010) Development of a SNP resource and a genetic linkage map for Atlantic cod (*Gadus morhua*). BMC Genomics 11:191

24. Hyten D, Song Q, Fickus E et al (2010) High-throughput SNP discovery and assay development in common bean. BMC Genomics 11:475

25. Yandell M, Mungall CJ, Smith C et al (2006) Large-scale trends in the evolution of gene structures within 11 animal genomes. PLoS Comput Biol 2:e15

26. Hugall AF, Foster R, Lee MS (2007) Calibration choice, rate smoothing, and the pattern of tetrapod diversification according to the long nuclear gene RAG-1. Syst Biol 56:543–563

27. Elmer K, Fan S, Gunter H et al (2010) Rapid evolution and selection inferred from the transcriptomes of sympatric crater lake cichlid fishes. Mol Ecol 19:197–211

28. Marden J (2008) Quantitative and evolutionary biology of alternative splicing: how changing the mix of alternative transcripts affects phenotypic plasticity and reaction norms. Heredity 100:111–120

29. Cartegni L, Chew SL, Krainer AR (2002) Listening to silence and understanding nonsense: exonic mutations that affect splicing. Nat Rev Genet 3:285–298

Chapter 4

In Silico Fingerprinting (ISIF): A User-Friendly In Silico AFLP Program

Margot Paris and Laurence Després

Abstract

The Amplified fragment Length Polymorphism (AFLP) is one of the cost-effective and useful fingerprinting techniques to study non-model species. One crucial AFLP step in the AFLP procedure is the choice of restriction enzymes and selective bases providing good-quality AFLP profiles. Here, we present a user-friendly program (ISIF) that allows carrying out in silico AFLPs on species for which whole genome sequences are available. Carrying out in silico analyses as preliminary tests can help to optimize the experimental work by allowing a rapid screening of candidate restriction enzymes and the combinations of selective bases to be used. Furthermore, using in silico AFLPs is of great interest to limit homoplasy and amplification of repetitive elements to target genomic regions of interest or to optimize complex and costly high-throughput genomic experiments.

Key words: AFLP fingerprint, In silico genotyping, Whole genome data, Homoplasy, Restriction enzymes

1. Introduction

The Amplified fragment Length Polymorphism (AFLP (1)) is one of the cost-effective and useful fingerprinting techniques to study non-model species. AFLP is based on the selective polymerase chain reaction (PCR) amplification of subsets of genomic restriction fragments. Genomic DNA is digested in thousands of fragments using restriction enzymes, and a subset of fragments is amplified by PCR using primers with one to four selective bases, thereby reducing the number of fragments on the profile. Fragments are separated by their length using electrophoresis, and discrete peaks can be visualized on a typical AFLP profile. Each discrete peak position is scored and characterized as a dominant biallelic locus (coded 0/1) in a 50–500-bp range.

François Pompanon and Aurélie Bonin (eds.), *Data Production and Analysis in Population Genomics: Methods and Protocols,*
Methods in Molecular Biology, vol. 888, DOI 10.1007/978-1-61779-870-2_4, © Springer Science+Business Media New York 2012

Recently, many authors have focused on improving the reliability and the accuracy of the AFLP technique, from the molecular steps to the data analysis. First, the AFLP protocol has to be carefully chosen depending on the study species; the initial AFLP protocol described for plants by Vos et al. (1) has been already successfully modified for the study of more challenging organisms like vertebrates (2) or insects (3). To control the quality of the AFLP procedure (contaminations, reliability of the method, or genotyping errors (4, 5)), negative controls and sample replicates are now included in most experiments (6–14). Then, several marker selection algorithms have been developed to optimize the challenging step of AFLP marker scoring by discarding biases due to subjective and unreliable personal procedures (15–17). Finally, statistical analyses appropriate for dominant markers have to be applied and many methods are now available to assess genetic diversity and population structure from AFLP data sets and to detect AFLP markers linked to selection (see ref. 18 for a review). More recently, a Bayesian method taking into account the distribution of band intensities in populations has been developed to allow the analyses of AFLPs as codominant markers (19). This method improves considerably the estimation of population structure and inbreeding coefficients from AFLP data sets and allows reaching a precision for these estimates very close to that obtained with SNPs (19).

Another crucial step in the AFLP procedure consists in the choice of restriction enzymes and/or selective bases that will generate AFLP profiles with an adequate number of peaks (typically between 20 and 100) with homogeneous length distribution and homogeneous fluorescence. Indeed, one of the major flaws of AFLPs is the presence of homoplasious peaks in the profiles that are due to co-migrating fragments of the same length (20–23). Here, we present the user-friendly program ISIF (22) that allows carrying out in silico AFLPs on species for which whole genome sequences are available. ISIF program is freely available at http://www-leca.ujf-grenoble.fr/logiciels.htm. It works in a Windows® environment and requires The Microsoft .NET Framework version 2.0 (freely available at http://www.microsoft.com/downloads). The program performs in silico AFLPs from any sequences by simulating the AFLP procedure step by step. First, it identifies all the restriction sites along the sequence and produces the pool of all possible restriction fragments. From those, it selects the final set of fragments that exhibit the selective bases used for the amplification. Finally, it determines the length of all the peaks of the AFLP profile, with the adaptor and primer lengths added when specified by the user. ISIF can provide the sequences of the virtual fragments for any known sequence, and for any restriction enzyme and selective bases combinations. Furthermore, it provides for each AFLP fragment the position along the genome. It, therefore, allows quickly detecting homoplasious fragments. ISIF program is also very useful for a rapid screening of candidate restriction enzymes and of the combinations of selective bases to be used in order to optimize the experimental work. Indeed, testing many primer combinations before the genotyping can help:

- Selecting enzymes and selective bases providing AFLP profiles with the appropriate number of peaks

- Choosing primer combinations that provide AFLP profiles with homogeneous length distribution

- Choosing primer combinations with low homoplasy rate

- Detecting and discarding primer combinations amplifying repetitive elements in the genome, such as transposable elements

- Combining primer pairs in order to maximize the distribution of the AFLP fragments throughout the genome

- Targeting genomic regions of interest by using primer pairs generating AFLP fragments in these regions

- Optimizing complex and costly high-throughput genomic experiments, such as Diversity Arrays Technology (DArT (24, 25)), pyrosequencing of AFLPs (26, 27), or Restriction-site Associated DNA (RAD (28, 29))

2. Program Usage

2.1. Reference Sequences Import

The program performs in silico AFLPs from all sequences written in capital or small letters saved as plain text without line numbers and spaces, such as text files. Import reference sequence files using the "+" button in the middle of the user interface of ISIF program (Fig. 1). The names of the imported files are indicated in the white square on the left side of the user interface. For genomes divided in several chromosomes or contigs, one separate file per chromosome/contig should be imported. Use the "–" button to remove the selected files.

2.2. Restriction Sites' Specification

ISIF can perform in silico AFLPs with any classical restriction enzymes (i.e., enzymes that cleave only once, and inside the recognition site). Restriction sites of restriction enzymes have to be specified in the "Left Cut" and in the "Right Cut" columns, on the right side of the user interface. Each line corresponds to one restriction enzyme/site. "Left Cut" column corresponds to the part of the sequence in 5′ of the enzyme cleavage location, and the "Right Cut" column corresponds to the 3′ part of the sequence after the cleavage location. For example, for the EcoRI enzyme restriction site 5′G↓AATTC3′ (↓ indicates the cleavage location), "G" corresponds to the "Left Cut" and "AATTC" corresponds to the "Right Cut" (see Note 1).

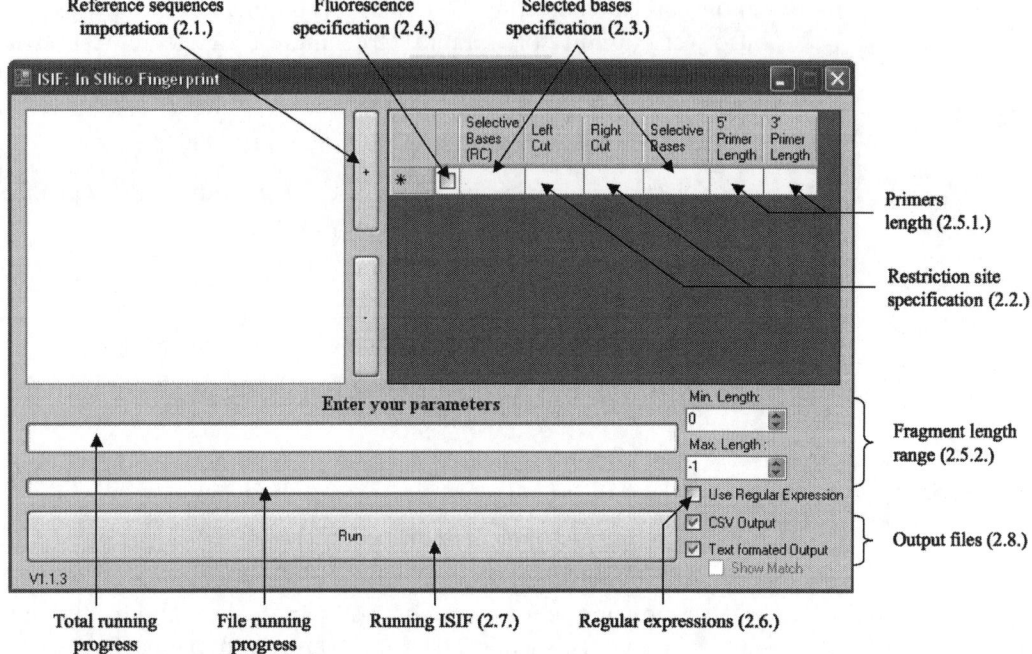

Fig. 1. ISIF user interface and parameters. For each parameter, the corresponding chapter number is indicated in *bracket*.

2.3. Selective Bases' Specification

Selective bases used in combination with a specific enzyme have to be specified in the same line of this enzyme restriction site, in the "Selective Bases" column (see Note 1). In silico AFLPs are performed on one 5′–3′ DNA strand of the reference genome (Fig. 2a, b). However, AFLP restriction sites are palindromic and both sides of the cutting sites are ligated with AFLP adaptors and potentially amplified. Therefore, to properly simulate AFLP procedure, the reverse complement sequences of the selected bases have to be specified in the "Selective Bases (RC)" column. They correspond to the selective bases sequences at the 3′ extremity of the AFLP restriction fragments of the reference genome (Fig. 2).

2.4. Fluorescent Enzymes' Specification

Two restriction enzymes are used in classical AFLP protocols and only one is favored during the amplification and the detection steps. This is achieved by using during the amplification step of the AFLP procedure a fluorescent primer in excess, which is associated with the enzyme restriction site that is favored. The favored enzyme has to be indicated in the enzyme line by checking the appropriate box (see Fig. 1). Only fragments cleaved at least in one extremity by this favored enzyme are presented in the ISIF output files.

2.5. Fragment Length

ISIF calculates and provides the fragment length of restriction fragments, from the cleavage site in 5′ to the cleavage site in 3′ (Fig. 2b). During the AFLP procedure, adaptors specific of each enzyme restriction site are ligated to the restriction fragments. After this ligation step, the fragments are amplified using primers,

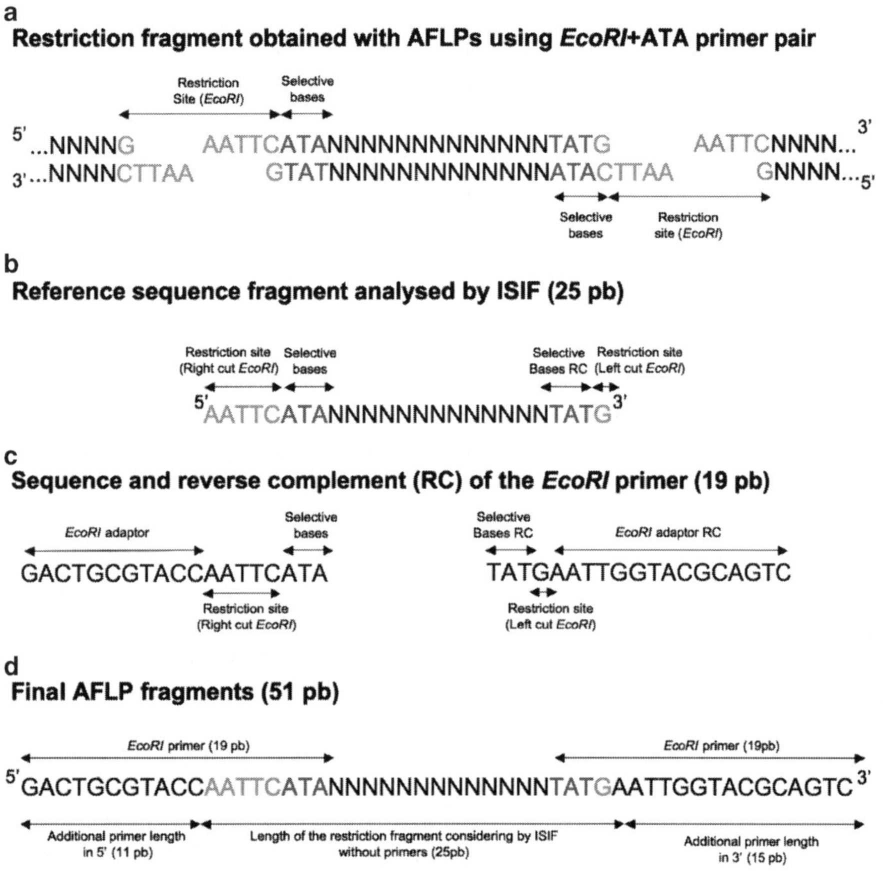

a
Restriction fragment obtained with AFLPs using *EcoRI*+ATA primer pair

Restriction
Site (*EcoRI*)   Selective
bases

5'...NNNNG   AATTCATANNNNNNNNNNNNNNNNTATG   AATTCNNNN...3'
3'...NNNNCTTAA   GTATNNNNNNNNNNNNNNNNATACTTAA   GNNNN...5'

Selective   Restriction
bases   site (*EcoRI*)

b
Reference sequence fragment analysed by ISIF (25 pb)

Restriction site   Selective   Selective   Restriction site
(Right cut *EcoRI*) bases   Bases RC   (Left cut *EcoRI*)

5'AATTCATANNNNNNNNNNNNNNNNTATG3'

c
Sequence and reverse complement (RC) of the *EcoRI* primer (19 pb)

EcoRI adaptor   Selective   Selective   *EcoRI* adaptor RC
bases   Bases RC

GACTGCGTACCAATTCATA   TATGAATTGGTACGCAGTC

Restriction site   Restriction site
(Right cut *EcoRI*)   (Left cut *EcoRI*)

d
Final AFLP fragments (51 pb)

EcoRI primer (19 pb)   *EcoRI* primer (19pb)

5'GACTGCGTACCAATTCATANNNNNNNNNNNNNNNNTATGAATTGGTACGCAGTC3'

Additional primer length   Length of the restriction fragment considering by ISIF   Additional primer length
in 5' (11 pb)   without primers (25pb)   in 3' (15 pb)

Fig. 2. Example of ISIF procedure and fragment length calculation for an AFLP restriction fragment amplified by the primer *EcoRI*+ATA. "RC" abbreviation corresponds to "Reverse Complement".

the sequence of which corresponds to the adaptor, plus the restriction sites and some supplementary selective bases (see Fig. 2c). ISIF can calculate and provide the length of final AFLP fragments by adding the primer length to restriction fragments (Fig. 2d, see Note 2).

2.5.1. Adding the Primer Length to Obtain the Final Length of the AFLP Fragments

The additional lengths of the primers have to be indicated in the "5′ primer length" and "3′ primer length" columns (Fig. 1). For each restriction enzyme, the additional lengths due to the primers have to be calculated as follows (Fig. 2d):

(a) 5′ primer length = total primer length – enzyme right cut length – selective bases length

(b) 3′ primer length = total primer length – enzyme left cut length – selective bases length

2.5.2. Selecting Fragment Length Range

By default, all fragments are presented in ISIF output files. Furthermore, as AFLP method focuses generally on fragments ranging between 50 pb and 500 pb, minimum and maximum lengths can be added in ISIF parameters using the "Min. Length:" and the "Max. Length:" options (see Fig. 1). If no primer length is indicated (see Subheading 2.5.1), ISIF output files provide only restriction fragments in the selected length range. If primer lengths are indicated, ISIF output files provide only final AFLP fragments in the selected length range.

2.6. Use of Regular Expressions

ISIF does not take into account the IUPAC nucleotide code for unknown degenerated bases, such as N, R, or H. However, it is possible to specify these degenerated bases using regular expressions (see Note 3). ISIF allows the use of regular expressions by checking the box "Use regular expression." First, this option can be useful to perform in silico AFLPs on reference sequences containing genetic polymorphism (see Note 4). Indeed, heterozygosity is important when using dominant markers such as AFLPs because both homozygote and heterozygote status lead to AFLP peaks. Second, by using regular expressions, in silico AFLPs can also be performed with restriction enzymes containing degenerated bases (see Note 5). Such restriction enzymes can be used in classical AFLPs or in other restriction-based genotyping methods, such as DArT (24, 25).

2.7. Running ISIF

When all parameters are specified, press the "Run" button to start ISIF. The running progress is indicated for the total analyses, as well as for each of the reference file analyses (Fig. 1). For indication, performing in silico APFLs on a computer with an Intel® Pentium® D CPU 2.80 GHz and 2.00 Go of RAM, the running time is about 2 min for the *Arabidopsis thaliana* genome (genome size of 120 Mb) and 10 min for the *Aedes aegypti* genome (genome size of 1,310 Mb).

2.8. Program Output

ISIF provides two different output files for each of the reference sequences: a "CSV reference-file-name" and a "Text reference-file-name" file. Uncheck the box "csv Output" or "text formatted Output" in the user interface (Fig. 1) when an output file is not wanted. The Text-formatted file provides the following parameters for each restriction fragment (Fig. 4):

SEQUENCE No.

Starting Cut:	5′ restriction site (and corresponding selective bases)
Ending Cut:	3′ restriction site (and corresponding selective bases)
Start Index:	Starting position in the reference genome (in pb)
End Index:	Ending position in the reference genome (in pb)
Length:	Total fragment length (restriction fragment length)
Fragment:	Restriction fragment sequence

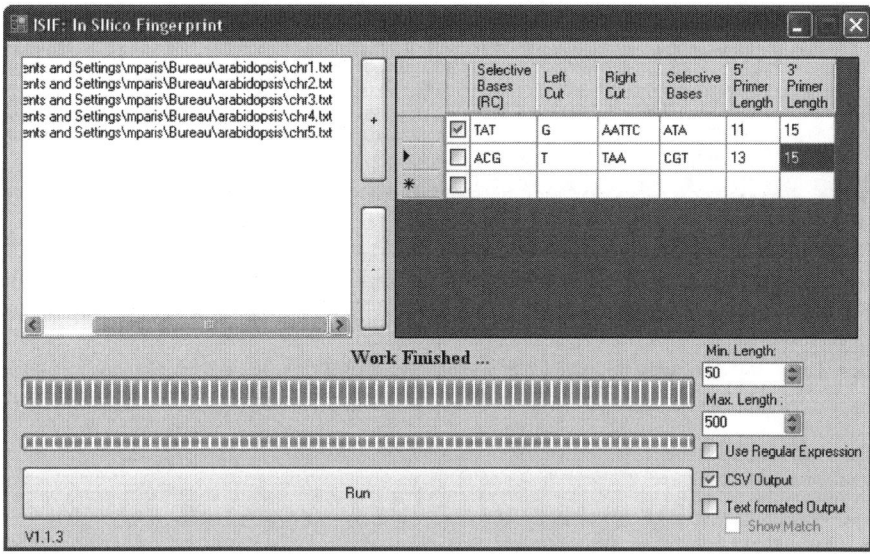

Fig. 3. Example of ISIF parameters to perform in silico AFLPs on the 5 *Arabidopsis thaliana* chromosomes using the primer pair *EcoRI*+ATA/*MseI*+CGT.

The CSV output file provides one line per restriction fragment using ";" as separator, and is compatible with spreadsheet editors or R program (30) for analyses (Fig. 5). The following parameters are presented in this order: starting cut, ending cut, start index, end index, total fragment length, restriction fragment length, and restriction fragment sequence.

3. Example

Figure 3 presents an example of program parameters for performing in silico AFLPs with the restriction enzyme and selective bases pair *EcoRI*+ATA/*MseI*+CGT. The first and the second lines correspond to the *EcoRI* and *MseI* enzyme parameters, respectively. The "fluorescent enzyme box" is checked only for *EcoRI*, and only fragments cleaved at least in one extremity by this enzyme are presented in the ISIF output files. The 19-pb primers GACTG-CGTACCAATTCATA and GATGAGTCCTGAGTAACGT were used in this example to amplify *EcoRI* and *MseI* fragments, respectively. Therefore, the additional length of primers were 11 pb in 5′ and 15 pb in 3′ for the *EcoRI* enzyme (Fig. 2d) and 13 pb in 5′ and 15 pb in 3′ for the *MseI* enzyme. In this example, final AFLP fragments range from 50 to 500 pb. Figures 4 and 5 present, respectively, the "Text" and the "CSV" output files obtained for the chromosome 1 of *Arabidopsis thaliana*. Using this primer pair, two AFLP fragments of 118 and 76 pb were obtained for this chromosome (Figs. 4 and 5).

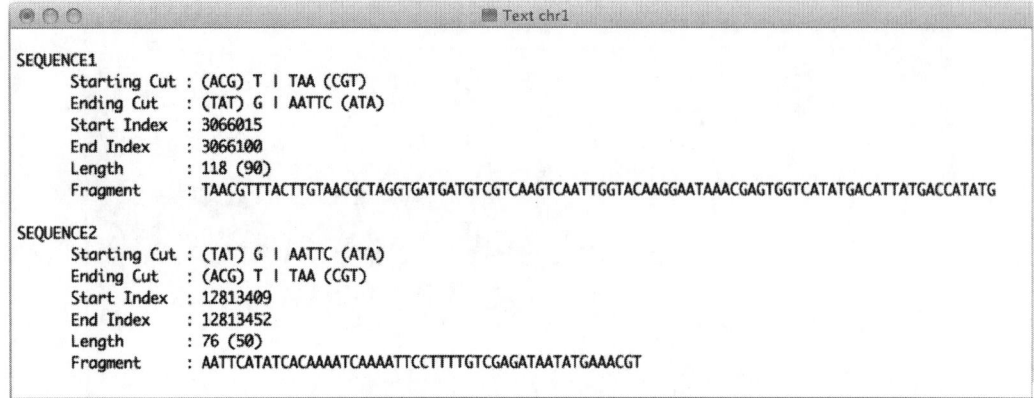

Fig. 4. "Text" output file provided by ISIF for in silico AFLPs on the chromosome 1 of *Arabidopsis thaliana* using the primer pair *EcoRI*+ATA/*MseI*+CGT.

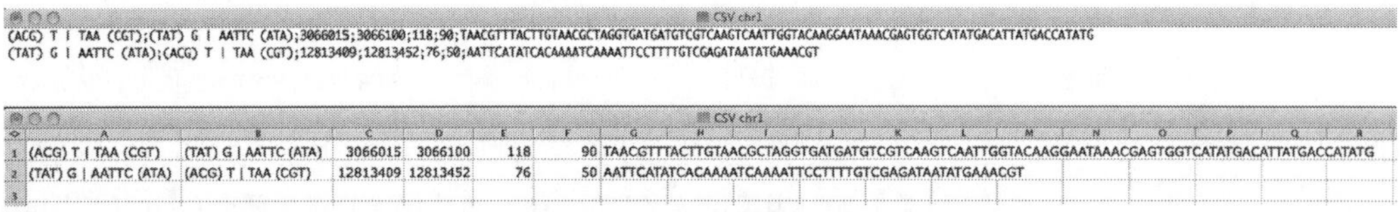

Fig. 5. "CSV" output file provided by ISIF for in silico AFLPs on the chromosome 1 of *Arabidopsis thaliana* using the primer pair *EcoRI*+ATA/*MseI*+CGT (screen shots from both text and spreadsheet editors).

4. Notes

1. ISIF distinguishes between capital and small letters. Therefore, restriction site sequences and selective bases have to be written in capital or small letters in order to match the reference genome format.

2. If no additional primer lengths are specified in ISIF parameters, the total fragment length corresponds to the restriction fragment length in output files.

3. Using regular expression, the special character "." denotes any single character and corresponds to the nucleotide code N in sequence data sets (IUPAC nucleotide code). The bracket expression "[ ]" matches a single character that is contained within the brackets. For example, the bracket expression "[AG]" denotes "A" or "G" and corresponds to the IUPAC nucleotide code R; and the bracket expression "[ACT]" denotes "A," "C," or "T" and corresponds to the IUPAC nucleotide code H.

4. To perform in silico AFLPs on reference sequences containing degenerate bases (IUPAC nucleotide code), these possible

bases have to be included in the restriction site and the selective bases' specifications using regular expressions. For example, the regular expression "[GSKBDVN]" represents all the possibilities to get a G in the reference sequence. On diploid species, avoiding the use of degenerated bases coding for more than two bases (B, D, H, V, and N) can help to limit the biases due to errors or uncertainties in sequences and to focus on polymorphisms. In this case, the specifications for the restriction enzyme EcoRI are "[GRSK]"for the "Left Cut" and "[ARWM][ARWM][TYWK][TYWK][CYSM]" for the "Right Cut" instead of "G" and "AATTC" (see Fig. 3).

5. In silico AFLPs can be performed with restriction enzymes containing degenerated bases using regular expressions. For example, the restriction enzyme Bsp1286I with the restriction site 5′GDGCH↓C3′ was used for the genotyping of *Aedes aegypti* mosquito strains (24, 25). Considering reference sequences containing no degenerated bases, "G[AGT]GC[ACT]" corresponds to the "Left Cut" and "C" corresponds to the "Right Cut" of the restriction site specifications for this enzyme.

Acknowledgments

This work was supported by a grant from the French Rhône-Alpes region (grant 501545401) and by the French National Research Agency (project ANR-08-CES-006-01 DIBBECO).

References

1. Vos P, Hogers R, Bleeker M et al (1995) AFLP – a new technique for DNA-fingerprinting. Nucleic Acids Res 23:4407–4414

2. Bonin A, Pompanon F, Taberlet P (2005) Use of amplified fragment length polymorphism (AFLP) markers in surveys of vertebrate diversity. Mol Evol 395:145–161

3. Paris M, Boyer S, Bonin A et al (2010) Genome scan in the mosquito *Aedes rusticus*: population structure and detection of positive selection after insecticide treatment. Mol Ecol 19: 325–337

4. Pompanon F, Bonin A, Bellemain E, Taberlet P (2005) Genotyping errors: causes, consequences and solutions. Nat Rev Genet 6:847–859

5. Bonin A, Bellemain E, Eidesen PB et al (2004) How to track and assess genotyping errors in population genetics studies. Mol Ecol 13: 3261–3273

6. Conord C, Lemperiere G, Taberlet P, Despres L (2006) Genetic structure of the forest pest *Hylobius abietis* on conifer plantations at different spatial scales in Europe. Heredity 97: 46–55

7. Fink S, Fischer MC, Excoffier L, Heckel G (2010) Genomic scans support repetitive continental colonization events during the rapid radiation of voles (Rodentia: Microtus): the utility of AFLPs versus mitochondrial and nuclear sequence markers. Syst Biol 59: 548–572

8. Karrenberg S, Favre A (2008) Genetic and ecological differentiation in the hybridizing campions *Silene dioica* and *S. latifolia*. Evolution 62:763–773

9. Meyer CL, Vitalis R, Saumitou-Laprade P, Castric V (2009) Genomic pattern of adaptive divergence in *Arabidopsis halleri*, a model

species for tolerance to heavy metal. Mol Ecol 18:2050–2062

10. Mraz P, Gaudeul M, Rioux D et al (2007) Genetic structure of *Hypochaeris uniflora* (Asteraceae) suggests vicariance in the Carpathians and rapid post-glacial colonization of the Alps from an eastern Alpine refugium. J Biogeogr 34:2100–2114

11. Nosil P, Egan SP, Funk DJ (2008) Heterogeneous genomic differentiation between walking-stick ecotypes: "isolation by adaptation" and multiple roles for divergent selection. Evolution 62: 316–336

12. Poncet BN, Herrmann D, Gugerli F et al (2010) Tracking genes of ecological relevance using a genome scan in two independent regional population samples of *Arabis alpina*. Mol Ecol 19:2896–2907

13. Puscas M, Choler P, Tribsch A et al (2008) Post-glacial history of the dominant alpine sedge *Carex curvula* in the European Alpine System inferred from nuclear and chloroplast markers. Mol Ecol 17:2417–2429

14. Tilquin M, Paris M, Reynaud S et al (2008) Long lasting persistence of *Bacillus thuringiensis subsp israelensis* (*Bti*) in mosquito natural habitats. PLoS One 3:10

15. Arrigo N, Holderegger R, Alvarez N (2012) Automated scoring of AFLPs using RawGeno v 20, a free R CRAN library. In: Bonin A, Pompanon F (eds) Data Production and Analysis in Population Genomics, Methods in Mol Biol Series, Humana Press

16. Arrigo N, Tuszynski JW, Ehrich D et al (2009) Evaluating the impact of scoring parameters on the structure of intra-specific genetic variation using RawGeno, an R package for automating AFLP scoring. BMC Bioinformatics 10:33

17. Herrmann D, Poncet BN, Manel S et al (2010) Selection criteria for scoring amplified fragment length polymorphisms (AFLPs) positively affect the reliability of population genetic parameter estimates. Genome 53:302–310

18. Bonin A, Ehrich D, Manel S (2007) Statistical analysis of amplified fragment length polymorphism data: a toolbox for molecular ecologists and evolutionists. Mol Ecol 16:3737–3758

19. Foll M, Fischer MC, Heckel G, Excoffier L (2010) Estimating population structure from AFLP amplification intensity. Mol Ecol 19:4638–4647

20. Caballero A, Quesada H (2010) Homoplasy and distribution of AFLP fragments: an analysis in silico of the genome of different species. Mol Biol Evol 27:1139–1151

21. Caballero A, Quesada H, Rolan-Alvarez E (2008) Impact of amplified fragment length polymorphism size homoplasy on the estimation of population genetic diversity and the detection of selective loci. Genetics 179:539–554

22. Paris M, Bonnes B, Ficetola GF et al (2010) Amplified fragment length homoplasy: in silico analysis for model and non-model species. BMC Genomics 11:287

23. Vekemans X, Beauwens T, Lemaire M, Roldan-Ruiz I (2002) Data from amplified fragment length polymorphism (AFLP) markers show indication of size homoplasy and of a relationship between degree of homoplasy and fragment size. Mol Ecol 11: 139–151

24. Bonin A, Paris M, Despres L et al (2008) A MITE-based genotyping method to reveal hundreds of DNA polymorphisms in an animal genome after a few generations of artificial selection. BMC Genomics 9:459

25. Bonin A, Paris M, Tetreau G et al (2009) Candidate genes revealed by a genome scan for mosquito resistance to a bacterial insecticide: sequence and gene expression variations. BMC Genomics 10:551

26. Paris M, Meyer C, Blassiau C et al (2012) Two methods to easily obtain nucleotide sequences from AFLP loci of interest. In: Bonin A, Pompanon F (eds) Data Production and Analysis in Population Genomics. Methods in Mol Biol Series, Humana Press

27. Van Orsouw NJ, Hogers RJ, Janssen A et al (2007) Complexity reduction of polymorphic sequences (CRoPS (TM)): a novel approach for large-scale polymorphism discovery in complex genomes. PLoS One 2:11

28. Baird NA, Etter PD, Atwood TS et al (2008) Rapid SNP discovery and genetic mapping using sequenced RAD markers. PLoS One 3:10

29. Hohenlohe PA, Bassham S, Etter PD et al (2010) Population genomics of parallel adaptation in threespine stickleback using sequenced RAD tags. PLoS Genet 6:2

30. R Development Core Team (2005) R: a language and environment for statistical computing. R Foundation for Statistical Computing, Vienna, Austria. http://www.R-project.org, ISBN 3-900051-900007-900050

Part II

Producing Data

Chapter 5

Diversity Arrays Technology: A Generic Genome Profiling Technology on Open Platforms

Andrzej Kilian, Peter Wenzl, Eric Huttner, Jason Carling, Ling Xia, Hélène Blois, Vanessa Caig, Katarzyna Heller-Uszynska, Damian Jaccoud, Colleen Hopper, Malgorzata Aschenbrenner-Kilian, Margaret Evers, Kaiman Peng, Cyril Cayla, Puthick Hok, and Grzegorz Uszynski

Abstract

In the last 20 years, we have observed an exponential growth of the DNA sequence data and simular increase in the volume of DNA polymorphism data generated by numerous molecular marker technologies. Most of the investment, and therefore progress, concentrated on human genome and genomes of selected model species. Diversity Arrays Technology (DArT), developed over a decade ago, was among the first "democratizing" genotyping technologies, as its performance was primarily driven by the level of DNA sequence variation in the species rather than by the level of financial investment. DArT also proved more robust to genome size and ploidy-level differences among approximately 60 organisms for which DArT was developed to date compared to other high-throughput genotyping technologies. The success of DArT in a number of organisms, including a wide range of "orphan crops," can be attributed to the simplicity of underlying concepts: DArT combines genome complexity reduction methods enriching for genic regions with a highly parallel assay readout on a number of "open-access" microarray platforms. The quantitative nature of the assay enabled a number of applications in which allelic frequencies can be estimated from DArT arrays. A typical DArT assay tests for polymorphism tens of thousands of genomic loci with the final number of markers reported (hundreds to thousands) reflecting the level of DNA sequence variation in the tested loci. Detailed DArT methods, protocols, and a range of their application examples as well as DArT's evolution path are presented.

Key words: DArT, Molecular marker, Complexity reduction, Microarrays, Diversity

1. Introduction

The promise of Human Genome Program to offer not only understanding of the genetic bases of diseases but also development of personalized medicine was a driving force behind technological

François Pompanon and Aurélie Bonin (eds.), *Data Production and Analysis in Population Genomics: Methods and Protocols,*
Methods in Molecular Biology, vol. 888, DOI 10.1007/978-1-61779-870-2_5, © Springer Science+Business Media New York 2012

progress in the area of genome sequencing and other types of DNA analysis, e.g. molecular markers. These technological and conceptual developments invigorated interest in using similar approaches in crop improvement. However, as the cost of developing the technologies similar to those used in biomedicine was prohibitive for the under-resourced fields of agriculture research, most of the marker technologies in crops were dominated by low input, but also low-throughput (and therefore high per-data point cost) methods like Amplified Fragment Length Polymorphism (AFLP) (1) or Simple Sequence Repeats (SSR) (2). At the same time, it was becoming apparent that most of the economically important variation in crops (e.g. yield) is governed by complex genetic networks which will demand cheap profiling of the whole genome for the molecular breeding to compete effectively with phenotypic selection. Importantly, with most breeding populations in most crops showing levels of Linkage Disequilibrium (LD) often orders of magnitude higher compared to human populations, it was clear that the density of markers for crops does not have to match the density recently developed for human genome profiling. In this conceptual framework, Diversity Arrays Technology (DArT) was developed to provide a practical and cost-effective whole-genome profiling capability: typing of thousands of markers in a single assay with technology development at a fraction of time and cost compared to main SNP typing platforms.

The historical perspective of technology development and the important aspects of intellectual property were recently presented (3), so this chapter focuses only on the technology itself and its applications.

1.1. Basic Principles

The principle of DArT is presented in Fig. 1 and described in some detail in the following sections. Development of DArT is initiated by defining a "metagenome"—a group of DNA samples representative of the germplasm anticipated to be analyzed with the technology. This group of samples usually corresponds to the primary gene pool of a crop species, but can be restricted to the two parents of a cross (if rapid creation of a linkage map is the goal) or expanded to secondary and even tertiary gene pools. All DArT arrays are developed with major focus on capturing allelic diversity of the organism of interest in order to limit the potential for ascertainment bias (the bias introduced from using markers developed from a small sample of the genotypes that are being studied). Significant ascertainment bias is routinely observed in SNP panels, often developed from very small number of genotypes. Recent population genetics and phylogenetic study using DArT arrays developed for *Eucalyptus* (as reported in ref. 4) indicated clearly that DArT markers were free from any detectable ascertainment bias (5).

Such DNA collection representing the gene pool of interest is processed using a complexity reduction method, a process which

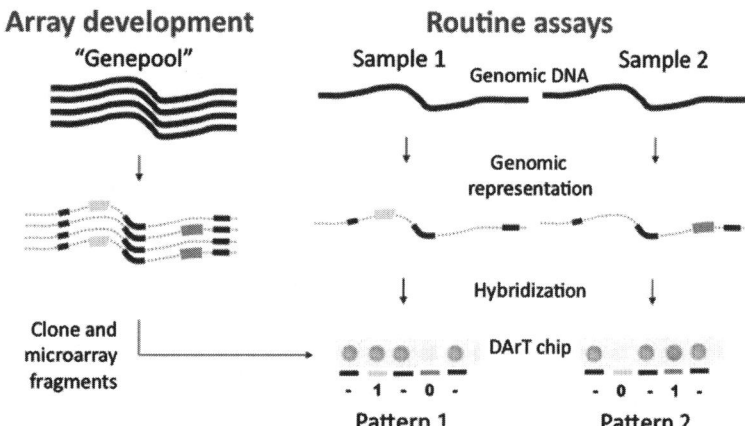

Array development
"Genepool"

Routine assays

Sample 1 Genomic DNA Sample 2

Genomic
representation

Hybridization

Clone and
microarray
fragments

DArT chip

- 1 - 0 -   - 0 - 1 -

Pattern 1   Pattern 2

Fig. 1. Principles of DArT array development and routine genotyping.

reproducibly selects a defined fraction of genomic fragments. The resulting collection of genomic fragments, called representation, is then used to construct a library of clones in *E. coli*. The process of cloning in bacteria performs a function of "individualization" of probes from the representation. Other methods than cloning can be utilized to perform the individualization step, but until recently *E. coli* cloning has been the most cost-effective. The inserts from individual clones are amplified and used as molecular probes on DArT arrays. It is important to point here that the process of "individualization" of markers through cloning contrasts DArT to technologies like AFLP, which do not allow characterization of markers beyond sizing. The main advantage is the ability to sequence markers in a high-throughput, cost-effective manner. The fact that the marker sequence is easily accessible offers new opportunities, like physical localization of the marker in the genome.

DArT reveals DNA polymorphism by querying representations of genomic DNA samples for the presence/absence of individual fragments (Fig. 1).

1.2. Complexity Reduction Methods

The critical aspect of complexity reduction methods deployed in DArT is a high degree of reproducibility in sampling genomic fragments (both qualitative and quantitative). Even though the number of potential methods of sampling polymorphic sites in the genome could be enormous, the methods used usually rely on restriction-enzyme (RE) digestion, adapter ligation, and amplification of adapter-ligated fragments. The main reason for choosing RE-based methods over other methods is the high level of precision (selectivity and reproducibility) of restriction enzymes. While the initial proof-of-concept paper (6) used an AFLP-like method to create representations for DArT, we soon abandoned the use of selective primers to generate DArT representations for two reasons. First, we found that genomic representations produced exclusively by RE digestion were more polymorphic and reliable than those produced by RE digestion in combination with selective primers. Second, the proprietary status of AFLP in some countries imposes some restriction on the availability and delivery of DArT methods using this type of complexity reduction.

We have tested a large number of RE combinations for their ability to detect DNA polymorphism. Most of our current methods are based on *Pst*I enzyme for precisely the same reason for which it was used in the past to generate Restriction Fragment Length Polymorphism (RFLP) probes: its propensity to generate low-copy probes. *Pst*I is sensitive to CXG methylation and does not cut the highly methylated, predominantly repetitive fraction of plant genomes (7). However, for most plant species, the number of fragments generated by *Pst*I digestion, adapter ligation, and amplification is too large for efficient DArT assays. We, therefore, use *Pst*I in combination with one or more frequently cutting enzymes while only ligating adapters to the ends created by *Pst*I. Wenzl et al. (8) tested nine such "co-digesting" enzymes and found statistically significant differences in their ability to reveal DNA polymorphism in DArT assays of barley. Interestingly, the two best enzymes (*Taq*I, *Bst*NI) for barley also performed the best for cassava (9) and hexaploid wheat (10). The ability of these two enzymes to digest fragments with methylated CXG (*Bst*NI) and CG (*Taq*I) motifs could account for their superior performance in DArT assays.

There are no simple rules governing the choice of enzymes for complexity reduction, and for each species a test of several combinations is strongly recommended. In such test, the initial evaluation is performed on agarose gel electrophoresis, and representations (amplification products) containing identifiable individual bands (i.e. likely to correspond to repetitive sequences) are rejected. Therefore, one can use a range of frequently cutting restriction enzymes to test for their ability to remove such bands from representations.

If there is some sequence information available for the species of interest, one can try complexity reduction methods other than the combinations of restriction enzymes described above. One of the methods efficiently applied on DArT platform is a Miniature Inverted-repeat Transposable Element (MITE)-based method (11), taking advantage of the fact that MITE elements insert themselves preferentially into low-copy sequences. It is absolutely critical for any complexity reduction method to significantly enrich for low-copy sequences in order to perform well in DArT. Otherwise, the choice of the method is limited mostly by researcher's creativity.

1.3. Genomic Representations

The choice of a complexity reduction method defines the genomic representations used in DArT assays. One of the most critical parameters is the fraction of the genome captured by the representation or, more precisely, the number and type of restriction fragments captured in the representation. A good representation contains mostly low-copy sequences as repetitive sequences interfere with DArT assays and do not contribute polymorphic probes. The selection of unique and low-copy sequences is an inherent feature of the hybridization-based assay format of DArT and distinguishes DArT from mobility-based assay technologies, such as AFLP.

Another important feature of genomic representations is sequence complexity, which is determined by the number of unique fragments and their average length. Most of the methods successfully tested to date produce probes in the range of 300–1,000 bp with a median size around 500 bp. Our initial estimates of the numbers of unique fragments in representations of plant genomes (at least for the initial platform comprising glass microarrays, fluorescently labelled targets, and a confocal laser scanner) were in a 5,000–10,000 range (12). However, after sequencing a large collection of DArT markers in a number of genomes, it becomes apparent that the true complexity of representations is indeed rather in a 20,000–50,000 range. For example, in oats, 19,000 DArT clones were assayed for polymorphism resulting in identification of over 2,000 markers representing 1,774 unique sequences (13). Based on the frequency of polymorphic markers on the array (10%) and the number of markers available in the $PstI/TaqI$ representation (>3,000 based on fitting the marker sequence redundancy data to a Poisson distribution), there are more than 30,000 unique fragments in this representation. Similar results were obtained in several other genomes and confirmed by applying Next-Generation Sequencing (NGS) to DArT representations (Nicolas Tinker and DArT PL, unpublished data).

1.4. DArT Assay

Most of DArT assays are performed by hybridizing *targets* (fluorescently labelled genomic representations) derived from specific samples to arrays containing a large collection of probes amplified from bacterial clones from a representation of the gene pool of interest (Fig. 2). In most applications, each probe is printed on the slide in replication. Each target is hybridized together with a *reference* DNA fragment to the microarray. We routinely use the multiple cloning site of the vector into which the fragments of the genomic representation were cloned as a reference. After hybridization, the slides are washed and scanned with an imaging system that measures the fluorescence intensity of the dyes used in the assay (we usually use FAM-labelled reference and Cy3- and Cy5-labelled targets). The images are analyzed with DArTsoft (see below) using Reference image to automatically identify the spots (features) and background. The hybridization intensity in the reference channel provides also quality control for each array element and is used as a denominator for the hybridization signal in the target channels. This approach reduces the platform noise caused by spot-to-spot

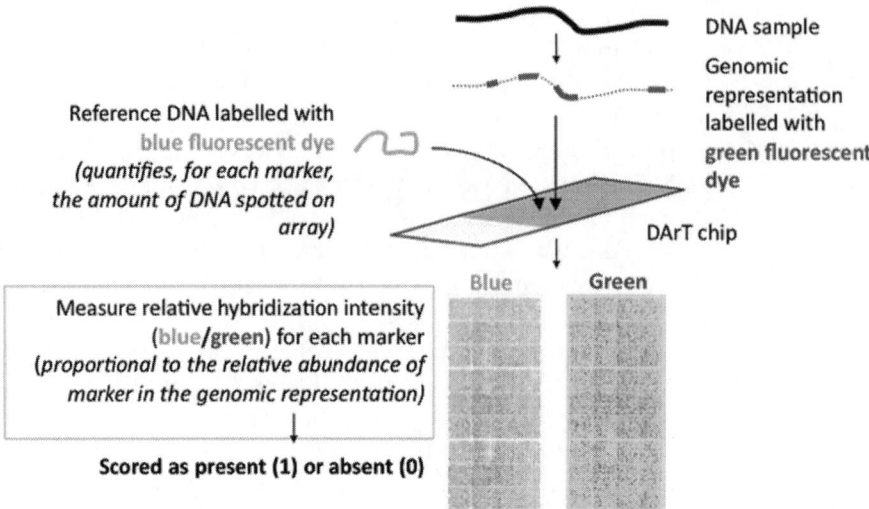

Fig. 2. The current format of DArT assay.

differences in the amount of DNA spotted on the array. Polymorphic clones are identified as those showing significant differences in hybridization signal intensity among the samples tested. Statistical methods deployed in DArTsoft are used to convert the hybridization signal intensities of polymorphic clones into scores. We routinely discover markers in experiments involving several hundreds of accessions.

DNA quality and amount are important issues when discussing DArT assays. The DNA needs to be suitable for digestion by restriction enzymes and ligation to adaptors, but also to be free from polymerase inhibitors so that the targets are prepared efficiently. Although more work is required to prepare it, investing in high-purity DNA is advantageous in the long term: it ships and stores better as well as provides higher quality data.

In order to ensure high quality of data produced on DArT platform, at least a subset of samples is processed with replication in each experiment. In fact, all marker discovery experiments (but not regular genotyping) are performed with full technical replication. This allows us to use technical reproducibility as one of the parameters in selecting high-quality markers.

To perform a DArT assay, we require only 1 μl of DNA at 50–100 ng per μl. Higher concentration can be actually detrimental to the assay, as it is sometimes associated with higher concentration of enzyme inhibitors. For smaller genomes, even less DNA is sufficient, as we showed clearly that high-quality data can be produced with less than 1,000 cell equivalents (6).

1.5. DArT Markers

DArT markers are bi-allelic markers that are either dominant (present vs. absent) or hemi-dominant (present in double dose, single dose, or absent). The type of marker classification depends both on the statistical methods applied (only more sophisticated statistical methods are capable of detecting allele dose difference reliably) and the specific marker (for some markers, the distribution of signal is strictly bimodal and such markers can only be classified in a dominant manner). The type of complexity reduction method applied in DArT determines the type of polymorphism detected by DArT markers. For example, the use of restriction enzymes that are sensitive to DNA methylation (for example *Pst*I) will identify markers reflecting both sequence variation (SNPs, InDels) and DNA methylation polymorphism. Some markers, therefore, may not be stable as a result of the dynamic nature of methylation states in some areas of the genome. However, the vast majority (97%) of DArT markers from a *Pst*I/*Bst*NI representation scored identically across a large number of DNA samples extracted from a single barley cultivar grown in a range of environments, and practically all of these markers behaved in a Mendelian manner in a mapping population (8). Recent data from *Arabidopsis thaliana* indicate that approximately 90% of DArT markers from a representation generated with methylation-sensitive *Pst*I detect SNP variation rather than methylation variation (14).

1.6. Comparison of DArT with Other Genotyping Technologies

DArT both shares some elements with other genome profiling technologies as well as has a number of distinct features. In its principle, DArT is most similar to RFLP, albeit performed in "reverse" (probes attached to solid support) and in very highly parallel manner. Indeed, tens of thousands of probes can be interrogated on the current platforms and even larger numbers of probes can be assayed in parallel on the novel, picoliter-volume platforms. There are also a number of technologies using DArT principles and methods (as described in ref. 6, US Patent no 6,713,258 B2. and related patent family/applications) under different names. For example, "Restriction Site Tagged (RST) microarrays" were reported for profiling of microbes (15), although the only discriminating factor from DArT is a slight modification of complexity reduction method. Recently, another variation on DArT with complexity reduction (and name) very similar to "RST microarrays" called Restriction site-Associated DNA (RAD) was reported (16). In addition, a DArT-like approach including a Suppression Subtractive Hybridization (SSH) step was used to identify orchid species (17).

1.6.1. Technology Development

As DArT markers are discovered and typed using the same assay, the technology eliminated the main cost in any other genome profiling technology establishment: assay development. While the same applies to AFLP technology, the genome coverage of a single AFLP assay is approximately two orders of magnitude lower than that of a DArT assay.

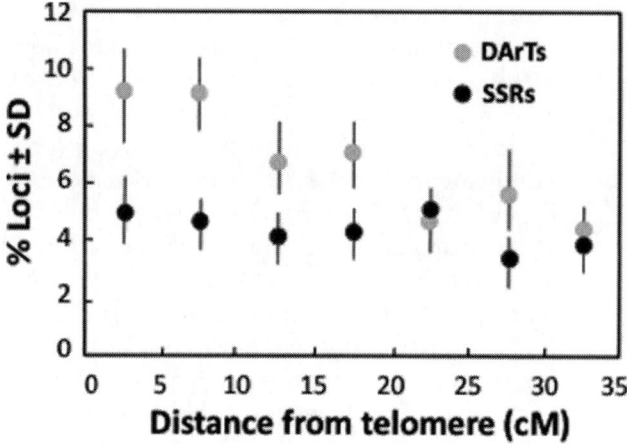

Fig. 3. Average distribution of DArT marker along chromosome arms in barley.

1.6.2. Polyploidy

In contrast to other SNP-typing techniques, DArT performs well in species of virtually any ploidy level, including sugarcane ($x=5–14$). The good performance is at least partly due to the fact that SNP detection in DArT is mediated by the high fidelity of restriction enzymes rather than primer annealing. The higher the level of ploidy (especially in more recent polyploids), the more competition there is between the primers used in SNP-typing assays and target sequence(s). As DArT is usually using only a single primer matching perfectly the adapter sequence, primer competition is eliminated from the assay.

1.6.3. Marker Distribution Across the Genome

The SNP and InDels surveyed by DArT are in theory randomly selected across the whole genome. However, the "methylation filtration" effect arising from using *Pst*I (a methylation-sensitive restriction enzyme) partly enriches genomic representations for hypo-methylated "gene space" regions in barley (Fig. 3). Similar picture of DArT marker distribution was observed in several other genomes for which high-density DArT maps have been already developed, like wheat (10), sorghum (18), and rye (19).

1.7. Example Applications of DArT in Genetics, Breeding, and Biodiversity Research

While detailed review of DArT applications is outside the scope of this paper, we mention below a few recent journal reports and developments utilizing DArT.

1.7.1. Animal DArT

One of the first papers presenting application of DArT in animals reported discovery of large number of markers for the two laboratory strains of mosquito under intense insecticide selection (11). A number of these markers were apparently linked to genes responsible for development of resistance to the selective agent, representing a clear example of DArT's ability to rapidly identify genomic regions responsible for complex phenotypes in non-model animal organisms.

1.7.2. Quantitative BSA-DArT

The quantitative nature of DArT assays was captured in the development of a Bulked Segregant Analysis (BSA) based on DArT platform (20). In the reconstruction experiment, it was shown that hybridization intensities on DArT arrays are proportional to the abundance of alleles in DNA pools. This feature allows not only rapid development of genetic markers linked to genes responsible for "simple" phenotypes (e.g. disease-resistance genes), but also measurement of seed purity and composition of seed mixtures (Diversity Arrays Technology Pty Ltd., unpublished).

1.7.3. Physical Mapping

While main applications of DArT are in genetic analysis and breeding, the technology performs well also in genomic studies, including construction of physical maps of complex organisms. Recently, a high-density genetic map of the largest chromosome 3B was linked with the physical map based on Bacterial Artificial Chromosome (BAC) clones (21) and the process was facilitated by hybridization of BAC pools to DArT arrays containing many thousands of markers. As the mapped DArT markers are progressively sequenced (completed for >5,000 wheat markers already), they provide unique anchors into both genetic and physical maps of various species.

1.8. DArT and NGS Technologies

The recent, spectacular progress in genome sequencing technologies opens a path to access genetic information from DNA sequence in novel ways. Marker discovery, for which the sequencers were used in the past, is not any longer the main cost of developing genome-profiling capacity on hybridization platforms. Therefore, most recently, a number of methods emerged to bypass assay development step and extract molecular marker information directly from the sequence data.

These new methods vary mainly in the way nucleic acid templates are sampled for sequencing. One of the main efforts in "genotyping by sequencing" (GBS) seems to be devoted to sequencing of genomic representations like complexity-reduced fractions of plant genomes. We have developed effective methods extracting marker data from sequencing of DArT representations (Diversity Arrays Technology Pty Ltd., unpublished). Our methods combine efficient preparation of sequencing templates, economical multiplexing achieved through various levels of barcoding of individual samples, and novel classification algorithms with the use of both leading commercial sequencing platforms and "open-access" instrumentation. An example of such "open-access" platform is the "Polonator" system developed by Dover Systems and Prof. Church's group from Harvard Medical School (http://www.polonator.org/).

While the technological progress is pushing towards novel (e.g. NGS) platforms for DNA profiling, there is still a clear market for cost-sensitive applications, which are not benefiting from increased numbers of markers. Even the most advanced NGS platforms provide limited "signal strength" (read depth) to obtain reliable allele frequency data in specific genomic locations. DArT arrays are, therefore, expected to be in use for quite some time in all applications requiring reproducible measurement for limited number of assays (e.g. seed purity determination, genetic ID, etc.).

2. Materials

2.1. Development of a New DArT Array and Construction of a DArT Library

- Genomic DNA (20–100 ng/µL).
- Sterile H_2O (Sigma).
- NaCl (1.2 M).
- $MgCl_2$ (60 mM).
- 10× RE buffer (100 mM Tris–OAc, 500 mM KOAc, 100 mM $Mg(OAc)_2$, 50 mM DTT, pH 7.8).
- 10× PCR buffer (100 mM Tris–Cl, pH 8.3, 500 mM KCl, 1.5 mM $MgCl_2$, 0.1% gelatine; Sigma).
- ATP (50 mM).
- dNTPs (10 mM, Sigma).
- *Pst*I adapter (5 µM; 5′-CACGATGGATCCAGTGCA-3′ annealed to 5′-CTGGATCCATCGTGCA-3′).
- *Pst*I primer (10 µM; 5'-GATGGATCCAGTGCAG-3′).
- BSA (10 mg/mL, NEB).
- *Pst*I (20 U/µL; NEB).
- Secondary enzyme (20 U/µL; NEB), e.g. *Taq*I.
- T4 DNA ligase (30 Weiss units/µL; NEB).
- REDTaq (1 U/µL; Sigma).
- Agarose.
- LB agar plates with 100 µg/mL ampicillin and 40 µg/mL X-Gal.
- Freezing medium (0.7× LB medium containing 4.4% glycerol, 8.21 g/L K_2HPO_4, 1.80 g/L KH_2PO_4, 0.50 g/L Na_3-citrate, 0.10 g/L $MgSO_4 \times 7\ H_2O$, 0.90 g/L $(NH_4)_2SO_4$, 100 mg/L ampicillin, 100 mg/L kanamycin).
- pCR2.1-TOPO vector solution (Invitrogen).
- OneShot electrocompetent *E. coli* cells (Invitrogen).
- Thermocycler.

- Incubator.
- Electoporator.
- Material for gel electrophoresis (tray, comb, electrophoresis power supply, etc.).

2.2. Preparation of DArT Arrays

- Deionized water.
- 77% ethanol.
- Agarose.
- Tris.
- HCl.
- $(NH_4)_2SO_4$.
- $MgCl_2$.
- DTT (100 μM).
- EDTA (100 μM).
- DArT-spotter solution (50% DMSO, 1.5 M sorbitol, 0.1 M triethanolamine–HCl, 0.5% [w/v] dextran, 0.02% [w/v] CHAPS).
- dNTPs (10 mM; Sigma).
- M13f primer (10 μM; 5′-GTTTTCCCAGTCACGACG TTG-3′).
- M13r primer (10 μM; 5′-TGAGCGGATAACAATTTCAC ACAG-3′).
- Taq polymerase (any provider, but titrate concentration in pilot experiments).
- SuperChip poly-L-lysine slides (Thermo Scientific).
- Plate centrifuge.
- TissueLyser (Qiagen).
- Microwave.
- Desiccator.
- Incubator.
- Thermocycler.
- Arrayer (e.g. MicroGrid arrayer from Genomics Solutions).
- Material for gel electrophoresis (tray, comb, electrophoresis power supply, etc.).

2.3. Genotyping Assay

- See above Subheadings 2.1 and 2.2.
- Isopropanol.
- EDTA (0.5 M, pH 8.0).
- DTT (100 mM).
- 1× SSC.

- SDS (0.1%).

- 10× NEB2 buffer (100 mM Tris–Cl, 100 mM MgCl$_2$, 500 mM NaCl, 10 mM DTT, pH 7.9).

- dATP, dCTP, dGTP, dTTP solutions (10 mM).

- Random decamers (500 μM).

- Cy3-dUTP (1 mM; Amersham) or Cy5-dUTP (1 mM; Amersham).

- 3:1 mixture of aminoallyl–dUTP (GeneCraft) and dTTP.

- 5-[and-6]-carboxyfluorescein, succinimidyl ester (Invitrogen).

- ExpressHyb solution (Clonetech).

- Denatured herring-sperm DNA (10 mg/mL; Promega).

- Klenow exo⁻ (500 U/μL; NEB).

- 6-cm-long glass coverslips.

- Scalpel with blades.

- 25-slide racks.

- Hybridization chambers.

- Desiccator.

- Water bath.

- Confocal laser (e.g. LS300 scanner from TECAN).

---

3. Methods

3.1. Development of a New DArT Array and Construction of a DArT Library

A DArT array for a new species is developed once from a collection of genotypes that capture the genetic diversity of the species' gene pool. Alternatively, arrays can be designed to maximize the number of polymorphic markers for a subset of genotypes, for example the parents of a breeding program or a mapping population.

A DArT library is a collection of cloned fragments derived from a representation of a species' gene pool. As described in the introduction, a common method to prepare representations involves digestion of DNA with restriction enzyme(s), adapter ligation, and amplification of small adapter-ligated fragments.

3.1.1. Digesting and Ligating Genomic DNA to Adapters

1. Prepare digestion/ligation mix containing (volumes per group of ten samples):

   (a) 52.4 μL H$_2$O (Sigma)

   (b) 7 μL 10× RE buffer (100 mM Tris–OAc, 500 mM KOAc, 100 mM Mg(OAc)$_2$, 50 mM DTT, pH 7.8)

   (c) 0.7 μL 10 mg/mL BSA (NEB)

   (d) 1.4 μL 50 mM ATP

(e) 0.7 μL 5 μM *Pst*I adapter (5′-CACGATGGATCCAG TGCA-3′ annealed to 5′-CTGGATCCATCGTGCA-3′)

(f) 0.7 μL *Pst*I (20 U/μL; NEB)

(g) 0.7 μL secondary enzyme (20 U/μL; NEB), e.g. *Taq*I

(h) 1.4 μL T4 DNA ligase (30 Weiss units/μL; NEB)

2. Mix 6.5-μL aliquots of digestion/ligation mix with 0.5 μL of genomic DNA (20–100 ng/μL; use a mixture of DNA samples of different genotypes or pool representations of individual genotypes after amplification).

3. Incubate at 37°C (2 h) and 60°C (2 h).

3.1.2. Amplifying Genomic Representations

1. Prepare amplification mix containing (volumes per group of ten samples):

(a) 400 μL H_2O (Sigma)

(b) 50 μL 10× PCR buffer (100 mM Tris–Cl pH 8.3, 500 mM KCl, 1.5 mM $MgCl_2$, 0.1% gelatine; Sigma)

(c) 10 μL 10 mM dNTPs (Sigma)

(d) 20 μL 10 μM *Pst*I primer (5′-GATGGATCCAGTGCAG-3′)

(e) 20 μL REDTaq (1 U/μL; Sigma)

2. Mix 50-μL aliquots of amplification with 0.5 μL of digestion/ligation reaction.

3. Amplify as follows:

(a) Step 1: 94°C for 1:00

(b) Step 2: 94°C for 0:20

(c) Step 3: 58°C for 0:40

(d) Step 4: 72°C for 1:00

(e) Step 5: Go to step 2(b) (29 more times)

(f) Step 6: 72°C for 7:00

(g) Step 7: Hold at 10°C

4. Analyse 5-μL aliquots of amplicons on a 1.2% agarose gel: the samples should show indistinguishable smears without any discernible bands.

3.1.3. Cloning Representation Fragments

1. If representations were prepared from individuals, pool them to create a "gene pool representation".

2. Assemble a ligation reaction as follows:

(a) 4.0 μL gene pool representation

(b) 1.0 μL of 1.2 M NaCl, 60 mM $MgCl_2$

(c) 1.0 μL pCR2.1-TOPO vector solution (Invitrogen)

3. Incubate at room temperature for 15–30 min.

4. Electroporate 2.0 µL of the ligation mix into a suspension of OneShot electrocompetent *E. coli* cells according to the protocol of the provider (Invitrogen).

5. Incubate electroporated cells in Eppendorf tubes at 37°C for a maximum of 15 min.

6. Plate cells on LB plates supplemented with 100 µg/mL ampicillin and 40 µg/mL X-Gal.

7. Incubate at 37°C overnight.

8. Pick white colonies into individual wells of 384-well microtitre plates filled with ampicillin/kanamycin-supplemented freezing medium (0.7× LB medium containing 4.4% glycerol, 8.21 g/L K_2HPO_4, 1.80 g/L KH_2PO_4, 0.50 g/L Na_3-citrate, 0.10 g/L $MgSO_4 \times 7\ H_2O$, 0.90 g/L $(NH_4)_2SO_4$, 100 mg/L ampicillin, and 100 mg/L kanamycin) (see Note 1).

9. Cover plates with lids and seal edges with Parafilm.

10. Incubate at 37°C overnight.

11. Store at –80°C for several years.

3.2. Preparation of DArT Arrays

The cloned inserts are amplified and spotted on microarray slides (DArT arrays).

3.2.1. Day 1

1. Prepare insert-amplification mix containing:
   (a) 50 mM Tris
   (b) 6 mM HCl
   (c) 16 mM $(NH_4)_2SO_4$
   (d) 1.5 mM $MgCl_2$
   (e) 200 µM dNTPs
   (f) 0.2 µM M13f primer (5′-GTTTTCCCAGTCACGAC GTTG-3′)
   (g) 0.2 µM M13r primer (5′-TGAGCGGATAACAATTTCA CACAG-3′)
   (h) Taq polymerase (any provider, but titrate concentration in pilot experiments)

2. Distribute 25-µL aliquots into 384-well PCR plates.

3. Transfer approximately 0.5 µL of bacterial culture per well into PCR plates.

4. Amplify inserts as follows:
   (a) Step 1: 95°C for 4:00
   (b) Step 2: 57°C for 0:35
   (c) Step 3: 72°C for 1:00
   (d) Step 4: 94°C for 0:35
   (e) Step 5: 52°C for 0:35

(f) Step 6: 72°C for 1:00

(g) Step 7: Go to step 4(d) (34 more times)

(h) Step 8: 72°C for 7:00

(i) Step 9: Hold at 10°C

5. Analyze amplified inserts on a 1.2% agarose gel (see Note 2).

6. Let amplification reactions dry at 37°C for 20 h.

3.2.2. Day 2

1. Add 35 μL of 77% ethanol (room temperature) per well, seal plates, and incubate for 1.5 h at room temperature.

2. Spin plates at >3,000 ×g for 40 min at 30°C.

3. Unseal plates, shake ethanol off, and spin plates upside down at 20 ×g for a few seconds (see Note 3).

4. Let pellets dry at room temperature.

5. Add 25 μL of DArT-spotter solution (50% DMSO, 1.5 M sorbitol, 0.1 M triethanolamine–HCl, 0.5% [w/v] dextran, and 0.02% [w/v] CHAPS) (see Note 4).

6. Spin plates briefly and incubate at room temperature for 24 h.

3.2.3. Day 3

1. Dissolve inserts by agitation in a TissueLyser (Qiagen).

2. Print inserts on SuperChip poly-L-lysine slides (Thermo Scientific) (we use a MicroGrid arrayer from Genomics Solutions).

3. Unload slides from microarray into 25-slide racks and store them in a dry place at room temperature for 24–72 h.

3.2.4. Days 4–6

1. Boil deionized water in a plastic beaker in a microwave oven.

2. Immerse slides in water once it has cooled down to 92°C.

3. Incubate slides for 2 min, gently moving them to loosen air bubbles trapped on the slides' surface.

4. Transfer slide racks into a solution of 100 μM DTT and 100 μM EDTA and gently agitate.

5. Dry slides by centrifugation at 500 ×g for 7 min.

6. Continue to dry slides in a desiccator under vacuum for 30 min.

7. Store slides in an airtight container with desiccant for up to 2 months.

3.3. Genotyping Assay

Unknown DNA samples are genotyped by generating genomic representations and querying the presence versus absence of individual fragments in these representations. The latter is achieved through hybridization to a DArT array containing a collection of individualized fragments from a representation of the species' gene pool.

Targets are the fluorescently labelled genomic representations that are hybridized to DArT arrays. They are prepared in a three-step process which includes DNA digestion and adapter ligation, amplification of small adapter-ligated fragments, and labelling of amplicons with a fluorescent dye.

1. Prepare and quality-control representations from genomic DNA samples of unknown individuals as described above.

2. Add 1× volume (45 μL) of isopropanol per sample and mix by pipetting.

3. Spin at >3,000×g for 40 min at 30°C.

4. Discard supernatants and add 100 μL of 77% ethanol (4°C) per sample.

5. Centrifuge as above.

6. Discard supernatants and let pellets dry at room temperature.

7. Prepare labelling mix containing (volumes per group of 100 samples):

    (a) 369 μL H_2O (Sigma)

    (b) 50 μL 10× NEB2 buffer (100 mM Tris–Cl, 100 mM $MgCl_2$, 500 mM NaCl, 10 mM DTT, pH 7.9)

    (c) 50 μL 500 μM random decamers

    (d) 10 μL 10 mM dATP

    (e) 10 μL 10 mM dCTP

    (f) 10 μL 10 mM dGTP

    (g) 1 μL 10 mM dTTP

8. Add 5 μL of labelling mix per sample.

9. Spin briefly and denature representations in a PCR machine at 95°C for 3 min; leave samples in PCR machine until temperature has reached 25°C.

10. In the meantime, prepare Cy-dye mix containing (volumes per group of 100 samples):

    (a) 420 μL H_2O (Sigma)

    (b) 50 μL 10× NEB2 buffer (see above)

    (c) 10 μL 1 mM Cy3-dUTP or 1 mM Cy5-dUTP (Amersham)

    (d) 20 μL Klenow exo⁻ (500 U/μL; NEB)

11. Remove samples from PCR machine and spin briefly.

12. Add 5 μL of Cy-dye mix per sample.

13. Spin briefly, wrap samples in aluminium foil, and incubate at 37°C for 3 h.

14. Store labelled representations (= targets) at −20°C until they are hybridized to the arrays.

1. Prepare labelled polylinker region of pCR2.1-TOPO as follows.

   (a) Amplify polylinker from self-ligated vector, using the insert-amplification protocol described above, in the presence of a 3:1 mixture of aminoallyl-dUTP (GeneCraft, Köln, Germany) and dTTP.

   (b) Chemically cross-link and purify the amplified fragment to 5-[and-6]-carboxyfluorescein and succinimidyl ester (Invitrogen) using the protocol of the provider.

2. Prepare hybridization mix and store at –80°C; mix:

   (a) 500 volumes of ExpressHyb solution (Clonetech)

   (b) 50 volumes of 10 mg/mL denatured herring-sperm DNA (Promega)

   (c) 2 volumes of 0.5 M EDTA (pH 8.0)

   (d) 2.5 volumes of 0.5–1 mg/mL FAM-labelled pCR2.1-TOPO polylinker

3. Prepare five slide-washing solutions:

   (a) Solution 1: 1× SSC, 0.1% SDS

   (b) Solution 2: 1× SSC

   (c) Solution 3: 0.2× SSC

   (d) Solution 4: 0.02× SSC

*3.3.3. Hybridization
of Targets to DArT Arrays*

Cy3- or Cy5-labelled targets are hybridized to DArT arrays together with an FAM-labelled polylinker fragment of the vector that was used to clone the representation fragments (pCR2.1-TOPO) (see Note 5). When amplifying the inserts spotted on the DArT array, the polylinker is co-amplified in two pieces at the ends of each insert and can be used to quantify the amount of DNA in each spot on the array.

1. Thaw a 7-mL aliquot of hybridization mix at 65°C for 1 h.

2. Distribute a set of 96 barcoded slides with same array design into 12 hybridization chambers, each holding 8 slides.

3. Transfer labelled targets (10 μL each) into 12 strips of 8 PCR tubes each.

4. Add 60 μL of hybridization mix to each target and mix thoroughly.

5. Denature on a PCR machine at 95°C for 3 min, and keep the samples in the PCR machine at 55°C with the lid closed.

6. Unload one strip at a time and pipette the eight samples on the surface of eight slides in a single hybridization chamber without introducing air bubbles (cover each sample with a 6-cm-long glass coverslip immediately after depositing it on the slide surface).

7. Seal hybridization chamber and immerse in a 62°C water bath; continue with the next PCR strip, etc.

8. Incubate chambers in the water bath overnight (approximately 16 h).

3.3.4. Washing and Scanning the Slides

1. Add DTT to solutions 1–4 (100 µM final concentration).

2. Immerse four 25-slide racks in solution 1.

3. Remove one chamber at a time from the water bath and open.

4. Lift coverslip with a scalpel blade, one slide at a time, immediately immerse slide in solution 1 to wash off excess liquid, and transfer to a rack in solution 1.

5. Once all 96 slides have been collected in solution 1, follow the following washing procedure.

6. Move racks up and down during 1 min, incubate without moving for another 4 min, and transfer rack to solution 2.

7. Move racks up and down during 1 min, incubate without moving for another 4 min, and transfer rack to solution 3.

8. Move racks up and down during 1 min, incubate without moving for another 1 min, and transfer rack to solution 4.

9. Move racks up and down during 30 s, and transfer rack to a solution of 100 µM DTT in water.

10. Move racks up and down during 30 s.

11. Remove racks from solution and blot them dry on KimWipe paper.

12. Dry slides by centrifugation at $500 \times g$ for 7 min.

13. Continue to dry slides in a desiccator under vacuum for 30 min.

14. Depending on the fluorescent labels used, scan slides on a confocal laser scanner at 488 nm (FAM), 543 nM (Cy3), and/or 633 nM (Cy5). We use a Tecan LS300 scanner with an autoloader for up to 200 slides for this purpose.

4. Data Analysis with the DArT Software

In order to efficiently support DArT technology, several software tools and systems have been developed and implemented. Because of the very specific needs and novelty of DArT, most of the system components have been developed "in-house". The main components of the system are DArTsoft, a genotypic data analysis program, and DArTdb, a laboratory management system.

We also integrated with those tools several peripheral software tools to deal more efficiently with processing and Quality Controlling data as well as implementing DArT markers in creating genetic maps, association analysis, and in breeding applications. In this chapter, we cover only the technology-specific elements of this informatic infrastructure, DArTsoft and DArTdb.

4.1. DArTsoft

DArTsoft is a purpose-built standalone application. It has been developed with a high volume data analysis in mind, so whole data extraction and analytical process is highly automated and driven by an effective Graphical User Interface. There are two major components in the package: a microarray image analysis component and a polymorphism and scoring analysis component.

4.1.1. Microarray Image Analysis

This task can be performed on local TIF files using the formats of printmaps—automatic outputs of various microarray printers. Spots on images are localized in a fully automatic manner mostly in large batches of many hundreds of images. The results are local data files containing relative hybridization intensities, row intensities, background values, and over 20 spot quality parameters. It also produces several control images, which can be used to visualize analysis results.

The algorithm for image segmentation uses a very efficient "seeded region growing" approach to locate spots. It allows finding spots very accurately even if there are some irregularities in spot sizes and shapes. The algorithm can also deal with image rotations. It has been integrated into DArTdb and interfaced there, so images stored in the database can be analyzed and the resulting data stored directly in database.

4.1.2. Polymorphism and Scoring Analysis

This task can be performed in both local and database modes. The local mode uses the results of local image analysis. The database mode requires connection to DArTdb and uses database tables, where row image results are stored. This assures fully automatic and correct clone and sample tracking. It is absolutely critical when handling big volumes of data.

The polymorphism analysis identifies polymorphic markers and scores them. It classifies relative hybridization intensities for all spots on the array. Scoring of markers is done in a dominant (bimodal distribution of intensities) or hemi-dominant (trimodal distribution) manner. The program reports scores confidence estimates, call rate, and Polymorphism Information Content (PIC) values. Our technology is quite unique in its quantitative approach to evaluating marker quality and the confidence associated with each individual genotype call (score). DArTsoft computes quality and confidence estimates, and these features will be critical to improve the performance of downstream data analysis software packages. DArTsoft has also an analysis results import function into Excel for easy data viewing and filtering.

4.2. DArTdb

DArT PL has developed DArTdb, a dedicated Laboratory Information Management System (LIMS) for DArT. It is a purpose-built storage, data management, and analysis system, which has been designed to strictly follow a technology workflow. Thus, DArTdb records all the data the technology relies on and provides a set of tools and utilities to manage them. Data relevant in the process include both technology development and a provision of the genotyping service, both on microarray and NGS platforms.

At the design stage of the system, the following constraints had to be taken into account.

- High throughput of the technology produces big amounts of data.

- Risks of sample and clone-tracking errors have to be eliminated through integration of the laboratory hardware into the system.

- Identifications of the sample and clone plates as well as clone assays have to be seamless.

- Access to the data has to be easy, quick, and secure.

Following the above guidelines, DArTdb is an access-controlled application with three different levels of user's permissions to the functionality and data: administrators, power users, and users. This access control, along with multilayer back-up/archive system, assures high standard of data security.

DArTdb is tightly integrated with our online ordering system as well as with DArTsoft. This assures direct access to any data in a secure and integrated way, eliminating potential problems during data transfers and conversions. Barcoding system of all plates and clone assays assures precise identification of all clones, samples, and hybridizations. DArTdb produces automatically all relevant statistics of activities captured by the system. Table 1 presents the recent status of DArTdb without commercial component of the system.

5. Notes

1. Filter-sterilize and store two 20× stock solutions: solution 1 with K_2HPO_4, KH_2PO_4, and $(NH_4)_2SO_4$; solution 2 with Na_3-citrate and $MgSO_4$. Autoclave LB medium and add solutions 1 and 2, glycerol, ampicillin, and kanamycin.

2. We perform quality control only on a sample of inserts (usually 48 taken across the 384 plate using multichannel pipettor) and accept the plates for array printing if success rate of insert production is above a selected threshold (usually >90% or 95% of well amplifies, single insert per well). For routine genotyping arrays, the QC procedures are much more stringent than those

Table 1
DArTdb statistics produced on March 24, 2011

Item	Value
Libraries	616
Library plates	7,645
Total wells	2,872,896
Destroyed library plates	2,043
All clones	948,555
NOT random clones	58,223
Array designs	447
Print runs	2,396
Slides printed	196,706
Experiments	2,328
Hybridizations	172,465
Scanned slides	171,167
Archived slides	139,618
Single images	471,205
Image size (gigaBytes)	3,853
Analyzed slides (pairs)	295,270
Data points analyzed	3,444,215,620
Extract plates	2,790
Extracts	237,808
Target plates	4,783
Targets	322,892
Products	111
Ordered extracts	127,328

for the marker discovery arrays in order to ensure that the probes identified as polymorphic on the discovery arrays are properly represented on genotyping arrays.

3. This is a critical step as insufficiently hard spin or excessively hard shaking of plates increases the risk of losing the pellet and therefore failing the subsequent steps of DArT assays.

4. This composition of spotting buffer was developed specifically for the poly-L-coated slides and may not perform equally well with other surface modifications. The DArTspotter performs well even in relatively long printruns as optimization was

directed at reduction of a gradient apparent during prolonged printing process.

5. Alternatively, the polylinker can be labelled with one of the two Cy dyes while labelling the genomic representation with the other.

Acknowledgements

We gratefully acknowledge contributions from all our visitors (http://www.diversityarrays.com/visitors.html) and funding organizations (http://www.diversityarrays.com/investors.html).

References

1. Vos P, Hogers R, Bleeker M et al (1995) AFLP: a new technique for DNA fingerprinting. Nucleic Acids Res 23:4407–4414

2. Weber J, May PE (1989) Abundant class of human DNA polymorphisms which can be typed using the polymerase chain reaction. Am J Hum Genet 44:388–396

3. Kilian A (2009) Case study 9: Diversity Arrays Technology Pty Ltd.: applying the open source philosophy in agriculture. In: Van Overwalle G (ed) Gene patents and collaborative licensing models: patent pools, clearinghouses, open source models and liability regimes. Cambridge University Press, Cambridge

4. Sansaloni CP, Petroli CD, Carling J et al (2010) A high-density Diversity Arrays Technology (DArT) microarray for genome-wide genotyping in *Eucalyptus*. Plant Methods 6:16

5. Steane DA, Nicolle D, Sansaloni CP et al (2011) Population genetic analysis and phylogeny reconstruction in *Eucalyptus* (Myrtaceae) using high-throughput, genome-wide genotyping. Mol Phylogenet Evol 59:206–224

6. Jaccoud D, Peng K, Feinstein D, Kilian A (2001) Diversity arrays: a solid state technology for sequence information independent genotyping. Nucleic Acids Res 29:e25

7. Rabinowicz PD, Schutz K, Dedhia N et al (1999) Differential methylation of genes and retrotransposons facilitates shotgun sequencing of the maize genome. Nat Genet 23:305–308

8. Wenzl P, Carling J, Kudrna D et al (2004) Diversity Arrays Technology (DArT) for whole-genome profiling of barley. Proc Natl Acad Sci USA 101:9915–9920

9. Xia L, Peng K, Yang S et al (2005) DArT for high-throughput genotyping of cassava (*Manihot esculenta*) and its wild relatives. Theor Appl Genet 110:1092–1098

10. Akbari M, Wenzl P, Caig V et al (2006) Diversity Arrays Technology (DArT) for high-throughput profiling of the hexaploid wheat genome. Theor Appl Genet 113:1409–1420

11. Bonin A, Paris M, Després L et al (2008) A MITE-based genotyping method to reveal hundreds of DNA polymorphisms in an animal genome after a few generations of artificial selection. BMC Genomics 9:459

12. Kilian A, Huttner E, Wenzl P et al (2005) The fast and the cheap: SNP and DArT-based whole genome profiling for crop improvement. In: Tuberosa R, Phillips RL, Gale M (eds) Proceedings of the international congress in the wake of the double helix: from the green revolution to the gene revolution, May 27–31 2003, Bologna, Italy

13. Tinker NA, Kilian A, Wight CP et al (2009) New DArT markers for oat provide enhanced map coverage and global germplasm characterization. BMC Genomics 10:39

14. Wittenberg AHJ, van der Lee T, Cayla C et al (2005) Validation of the high-throughput marker technology DArT using the model plant *Arabidopsis thaliana*. Mol Genet Genomics 274:30–39

15. Zabarovsky ER, Petrenko L, Protopopov A et al (2003) Restriction site tagged (RST) microarrays: a novel technique to study the species composition of complex microbial systems. Nucleic Acids Res 31:e95

16. Miller MR, Dunham JP, Amores A et al (2007) Rapid and cost-effective polymorphism identification and genotyping using restriction site associated DNA (RAD) markers. Genome Res 17:240–248

17. Li TX, Wang J, Bai Y et al (2004) A novel method for screening species-specific gDNA probes for species identification. Nucleic Acids Res 32:e45

18. Mace ES, Xia L, Jordan DR et al (2008) DArT markers: diversity analyses and mapping in *Sorghum bicolor*. BMC Genomics 9:26

19. Bolibok-Bragoszewska H, Heller-Uszyn´ska K, Wenzl P et al (2009) DArT markers for the rye genome–genetic diversity and mapping. BMC Genomics 10:578

20. Wenzl P, Raman H, Wang J et al (2007) A DArT platform for quantitative bulked segregant analysis. BMC Genomics 8:196

21. Paux E, Sourdille P, Salse J et al (2008) A physical map of the 1-Gigabase bread wheat chromosome 3B. Science 322:101–104

Chapter 6

Two Methods to Easily Obtain Nucleotide Sequences from AFLP Loci of Interest

Margot Paris, Claire-Lise Meyer, Christelle Blassiau, Eric Coissac, Pierre Taberlet, and Laurence Després

Abstract

Genome scans based on anonymous Amplified Fragment Length Polymorphism (AFLP) markers scattered throughout the genome are becoming an increasingly popular approach to study the genetic basis of adaptation and speciation in natural populations. A shortcoming of this approach is that despite its efficiency to detect signatures of selection, it can hardly help pinpoint the specific genomic region(s), gene(s), or mutation(s) targeted by selection. Here, we present two methods to be undertaken after performing an AFLP-based genome scan to easily obtain the sequences of AFLP loci detected as outliers by population genomics approaches. The first one is based on the gel excision of the target AFLP fragment, after simplification of the AFLP fingerprint and separation of the fragments by migration. The second one is a combination of classical AFLP protocol and 454 pyrosequencing.

Key words: AFLP fingerprint, Profile simplification, Gel excision, 454 pyrosequencing, Sequence assignment

1. Introduction

The Amplified Fragment Length Polymorphism (AFLP) (1) fingerprinting technique has many applications, such as the study of population genetic structure or diversity, phylogenetic inference, and gene or QTL mapping. The major advantage of this method is that it allows studying at a low cost a large number of loci in model as well as in non-model species for which genomic information is scarce, including plants, fungi, bacteria, insects, and other animals (2, 3). Furthermore, the AFLP technique produces very informative and reliable genetic data, especially when adequate

François Pompanon and Aurélie Bonin (eds.), *Data Production and Analysis in Population Genomics: Methods and Protocols*, Methods in Molecular Biology, vol. 888, DOI 10.1007/978-1-61779-870-2_6, © Springer Science+Business Media New York 2012

precautions are used during all the steps of the AFLP protocol in order to limit contamination (4), genotyping errors (5), or peak homoplasy (6).

Recently, AFLPs have become the markers of choice in "population genomics" studies aiming at detecting adaptive genes in non-model species. By using wide multi-locus screening of the genome to identify loci that show adaptive genetic variation (i.e., "outlier" loci), population genomics is a powerful approach to understand mechanisms of selection and species adaptation at the genomic level. AFLP-based genome scans have been used successfully to detect outlier loci linked to adaptation to altitude (7, 8) or to host plants (9–12), floral divergence (13), insecticide resistance (14), adaptation to soil type (15), domestication (16), or ecotype divergence (17–19). The detection of outliers represents only a first step in a comprehensive understanding of how selection shapes genomic divergence between populations. The further step is to identify precisely the genomic regions, genes, and, ultimately, the mutations involved in adaptation. However, the lack of sequence information given by anonymous AFLP loci remains a major limitation to undertake this second step, and so far, most studies using population genomics approaches conclude that a substantial proportion of the genomes analyzed show potential signatures of selection (about 5% of the analyzed loci, see ref. 20 for a review) but fail to go further in the characterization of adaptive genes. Here, we present two reliable methods that can be used to sequence AFLP outliers detected by genome scans. The first one is derived from a relatively simple method (21) based on the isolation of the AFLP fragments of interest by a direct excision from gel followed by sequencing of the excised fragments. An ingenious step of AFLP profile simplification allows the extraction and sequencing of the fragments of interest without ambiguity. This procedure was applied successfully for the sequencing of loci linked to adaptation to soil type in *Arabidopsis halleri* ((15), Meyer et al., unpublished data) and male sterility and self-incompatibility in *Cichorium intybus* (Blassiau et al., unpublished data; Gontié et al., unpublished data).

The protocol consists of three main steps (Fig. 1) and may be adapted according to the complexity of the AFLP fingerprint containing the fragment of interest. The first step aims to decrease the number of AFLP fragments generated by the initial *Eco*RI + 3/*Mse*I + 3 primer combination in order to more easily isolate the marker of interest from other AFLP fragments after migration on a gel. It consists in designing all the 12 possible *Mse*I + 6 primers from the initial *Mse*I + 3 nucleotides, and check which one allows amplifying the target AFLP fragment in simpler fingerprint (see ref. 21). This step can be skipped when the marker is already separated from other fragments in the initial AFLP profile by at least 10 bp. The two last steps are the excision of the marker from the less complex fingerprint obtained with the *Mse*I + 6 primer and the direct sequencing of the excised fragment, respectively.

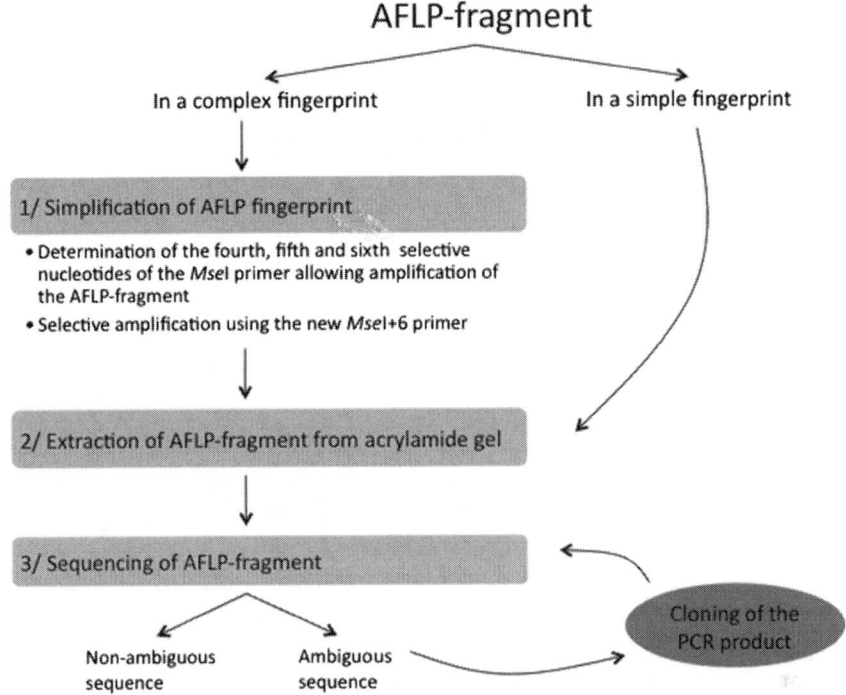

Fig. 1. Different steps of method 1 aiming to determine the sequence of an AFLP fragment using classical molecular tools.

Simplification of the fingerprint is a crucial step for successful excision of the target fragment from the gel, purification, and sequencing. It depends on the density of the initial fingerprint and on the selective bases amplifying the AFLP fragment. We observed that excision of a single fragment is generally possible when the number of fragments in the simplified fingerprint is not higher than ten. In *A. halleri* and *C. intybus*, this value is generally reached with an initial 60- to 70-fragment fingerprint. The fragment density found with the *Mse*I + 6 primers is on average six times lower than the one obtained with the *Mse*I + 3 primers. In some cases, the fingerprint remains dense despite the *Mse*I + 6 primer. It is yet possible to identify and sequence the fragment of interest using an additional cloning step (see Fig. 1).

The second method we describe here takes advantage of the recent development of high-throughput sequencing technologies to combine the classical AFLP protocol with parallel 454 pyrosequencing (Fig. 2). This method was first proposed by van Orsouw et al. (22) to discover SNPs in complex genomes such as maize or crop species, but its potential for population genetic studies remains largely unexplored. Indeed, this approach offers news perspectives for genomics studies by providing sequence information for all fragments present in any AFLP sample. The strategy is to cluster obtained sequences according to their length, and to compare with

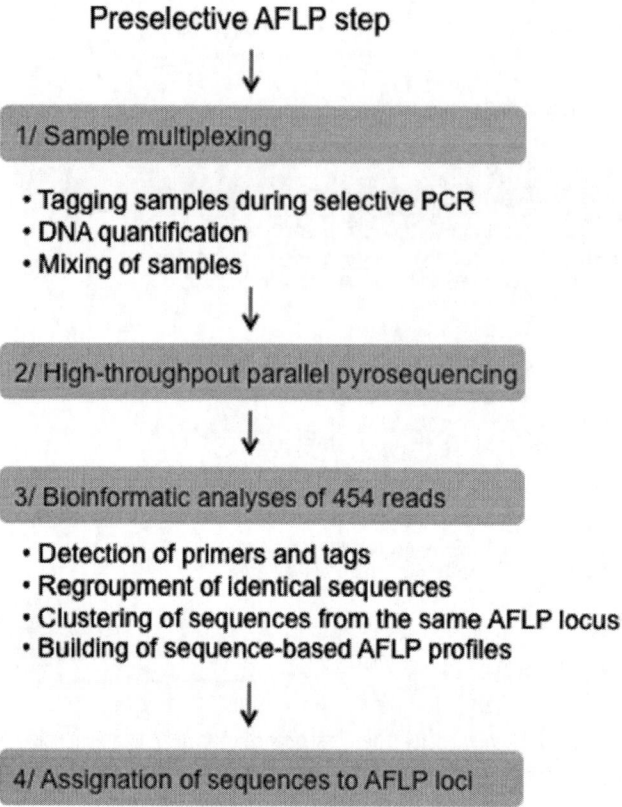

Fig. 2. Different steps of method 2 combining classical AFLP procedure with parallel pyrosequencing to determine the sequence of AFLP loci of interest.

classical AFLP profiles in order to assign each sequence of a given length to one AFLP anonymous peak of the same length. Furthermore, the ingenious use of tagged selective primers by a six-base extension in 5′ allows sequencing hundreds of samples in only one pyrosequencing run, considerably reducing the cost per sample. The reliability of this second method was validated using in silico AFLPs on the model species *Arabidopsis thaliana* (6). Furthermore, its efficiency in non-model species was demonstrated by sequencing AFLP profiles and identifying outlier loci linked to insecticide resistance in the mosquito species *Aedes aegypti* (23) and *Aedes rusticus* (Paris et al. unpublished data), and to adaptation to altitude in *Arabis alpina* ((8), Poncet et al. unpublished data).

The first step of this second method consists in carrying out the AFLP-selective PCR with tagged selective primers in order to recognize the sequences obtained with parallel pyrosequencing and to allow the multiplexing of several individuals in the same 454 run. The primers are tagged with a six-base extension in 5′ composed of a CC at the 5′ extremity followed by four variable nucleotides that identify each individual with a "barcode" (24). As 454 pyrosequencing reads are often truncated in the 5′ extremity, the first CC bases in the composition of the primers prevent losing one of the four variable nucleotides. Several AFLP samples are then pooled within the same 454 run. The crucial step for multiplexing success relies on the precise DNA quantification of each AFLP sample. Indeed, each AFLP sample contains several fragments that differ in length and classical quantifications cannot provide a satisfactory estimation of the number of molecules present in a sample. For this reason, DNA quantification is carried out using a Bioanalyzer (Agilent Technologies) that provides independent quantification of each peak present in the AFLP sample, taking into account the peak length. Then, a mix of AFLP samples is made by taking into account DNA concentration measured for each peak in order to ultimately obtain about 4,000 reads for each sample. A number of 4,000 reads represent an average coverage of 40–50 reads per fragment if we consider that a classical AFLP profile contains usually about 100 peaks (because more peaks would result in profiles with high homoplasy (6)). This coverage value of 40–50 reads per fragments is targeted because it ultimately allows obtaining at least 3 reads for all AFLP fragments, especially for less often amplified fragments of more than 300 bp. The number of reads per sample can be modified according to the number of peaks present in AFLP profiles. For example, only 1,000 reads are needed to obtain on average 40–50 reads per fragment for profiles containing 20 peaks; and about 8,000 reads are needed for species with large genomes that give AFLP profiles containing 200 peaks.

The second and third steps consist in carrying out the 454 parallel pyrosequencing and analyzing the 454 output using Python scripts named ObiTools (freely available at http://www.grenoble.prabi.fr/trac/OBITools) and the clustering program PHRAP (available upon request, home page of developers: http://www.phrap.org/index.html). We have decided to present only freely available programs in this chapter, but any clustering program can be used instead of the PHRAP program. During this step, (a) tags and primers are detected in order to assign all reads to the corresponding samples, (b) identical reads are grouped and counted, (c) reads with less than three counts are deleted in order to limit the impact of sequencing errors (24), and (d) sequences are clustered. Each cluster contains allelic variants for a unique AFLP locus, including heterozygous fragments and fragments with sequencing errors. Sequence-based AFLP profiles can then be built for each sample using cluster length information. Finally, the last step of this second method consists in comparing sequence-based AFLP profiles with classical AFLP profiles to associate a sequence to each AFLP locus.

2. Materials

Enzymes and reagents should be stored frozen at –20°C, and buffers at 4°C.

2.1. Method 1: Extraction from Acrylamide Gel and Direct Sequencing

2.1.1. Simplification of the AFLP Fingerprint

1. LI-COR® DNA analysis system.
2. 41 cm LI-COR® glasses and 0.2-mm spacer.
3. Comb (48 wells, 0.2 mm) for LI-COR® DNA analysis system.
4. Deionized water (18 MΩ-cm or greater).
5. 5× TBE buffer: 0.44 M Tris base, 0.44 M boric acid, 9.9 mM EDTA.
6. Urea (ultrapure).
7. 50% Long Ranger™ acrylamide.
8. TEMED.
9. APS 10%.
10. 700-bp Size Standards IRDye® 700 or 800.
11. Deionized formamide.
12. Congo Red: 25 mg/mL Congo Red, 0.125 M EDTA.
13. Product of the preselective amplification (according to ref. 1 or your own protocol; see Note 1).
14. Primer *Eco*RI + 3 (8×10^{-6} µM, labeled IRDye® 700 or 800) and *Mse*I + 3 (8×10^{-6} µM) allowing amplification of the AFLP fragment of interest.
15. Set of 12 degenerated primers: *Mse*I3N (i.e., *Mse*I + NNN) + A, C, G, or T, *Mse*I4N + A, C, G, or T, and *Mse*I5N + A, C, G, or T (8×10^{-6} µM).
16. 10× PCR buffer.
17. MgCl$_2$ (25 mM).
18. dNTPs mix (2.5 mM each).
19. Taq polymerase (5 U/µL).

2.1.2. Extraction of the AFLP Fragment from Acrylamide Gel

1. Same materials as for Subheading 2.1.1 (but see Note 2).
2. Primer *Mse*I + 6 (8×10^{-6} µM) designed in the previous step.
3. Comb (9 wells, 0.2 mm) for LI-COR® DNA analysis system (see Note 3).
4. Whiteboard marker.
5. Permanent marker.
6. Whatman® Paper 3MMChr 41 cm × 41 cm.
7. TE: 10 mM Tris–HCl, 1 mM EDTA.
8. Transparency film, 41 × 41 cm.
9. Plastic wedge for LI-COR® DNA analysis system.

2.1.3. Direct Sequencing

1. Supernatant containing the AFLP fragment.

2. Deionized water.

3. Primer *EcoR*I + 3 (8×10^{-6} μM; non-labeled) and *Mse*I + 3 (8×10^{-6} μM) allowing amplification of the AFLP fragment.

4. 10× PCR buffer.

5. $MgCl_2$ (25 mM).

6. dNTPs mix (2.5 mM).

7. Taq polymerase (5 U/μL).

8. NucleoSpin® ExtractII (Macherey-Nagel).

9. BigDye® Terminator v3.1 Cycle Sequencing Kit (Applied Biosystems).

10. EDTA (125 mM).

11. Absolute ethanol.

12. Deionized formamide.

2.1.4. Optional Additional Step: Cloning of the AFLP Product

1. Sterile water.

2. TA cloning® Kits (Invitrogen™, Life technologies).

3. Competent *Escherichia coli* suitable for transformation.

4. S.O.C. medium.

5. LB and LB plates containing 50 μg/mL kanamycin.

6. 42°C water bath.

7. 37°C shaking and non-shaking incubator.

8. 10× PCR buffer.

9. $MgCl_2$ (25 mM).

10. dNTPs mix (2.5 μM each).

11. BSA (10 mg/mL).

12. M13F primer (10 μM): 5′-cacgacgttgtaaaacgac-3′.

13. M13R primer (10 μM): 5′-ggataacaatttcacacagg-3′.

14. Taq polymerase (5 U/μL).

15. Deionized water.

16. Mix glycerol 65%, LB.

2.2. Method 2: Combining Classical AFLP and 454 Pyrosequencing

1. Deionized water (18 MΩ-cm or greater).

2. Product of the preselective amplification (but see Note 4).

3. Tagged primers *EcoR*I (10 μM) and *Mse*I (10 μM) (see Notes 5 and 6).

2.2.1. Multiplexing of AFLP Samples

4. 10× PCR buffer.

5. $MgCl_2$ (25 mM).

6. dNTPs mix (2.5 mM each).

7. Taq polymerase (5 U/μL).

8. Qiaquik PCR purification kit (Qiagen GmbH).

9. Agilent 2100 Bioanalyzer with DNA 1000 kit (Agilent Technologies).

2.2.2. High-Throughput Parallel Pyrosequencing

1. 454 Life Science sequencer (Roche).

2. GS FLX Titanium kit (or GS FLX Kit, or GS 20 Kit).

2.2.3. Bioinformatic Analyses of 454 Reads

1. Computer with unix environment.

2. Python scripts named ObiTools (freely available at http://www.grenoble.prabi.fr/trac/OBITools).

3. PHRAP program (freely available upon request, home page of developers: http://www.phrap.org/index.html).

4. BioEdit program v 7.0.9 (freely available at http://www.mbio.ncsu.edu/bioedit/bioedit.html).

3. Methods

Volumes are indicated for one sample.

3.1. Method 1: Extraction from Acrylamide Gel and Direct Sequencing

3.1.1. Simplification of the AFLP Fingerprint

1. Dilute the preselective product with deionized water according to your AFLP protocol (for example, a 50-time preselective dilution is used for the plant species *Arabidopsis halleri* (15) or a 20-time preselective dilution is used for the mosquito species *Aedes rusticus* (14)). Store diluted DNA at –20°C.

2. Prepare the selective mix with the following components: 3 μL of diluted preselective product, 1 μL of 10× PCR buffer, 0.8 μL of 25 mM $MgCl_2$, 0.8 μL of 2.5 μM dNTPs, 0.3 μL of 8×10^{-6} μM labeled *Eco*RI + 3 primer, 0.3 μL of 8×10^{-6} μM *Mse*I + 3 primer, 0.25 unit of Taq polymerase, and deionized water to 10 μL. Amplify using the following program: 13 cycles of 10 s at 94°C, 30 s at 65°C (–0.7°C per cycle), and 1 min at 72°C; then 25 cycles of 10 s at 94°C, 30 s at 56°C, and 1 min at 72°C; followed by a 2-min extension step at 72°C; and a final 5-min step at 4°C. Dilute the *Eco*RI + 3/*Mse*I + 3 selective product 100 times and store at –20°C.

3. Prepare 12 selective mixes with the following components (but see Note 7): 3 μL of diluted *Eco*RI + 3/*Mse*I + 3 product, 1 μL of PCR buffer, 0.8 μL of 25 mM $MgCl_2$, 0.8 μL of 2.5 μM dNTPs, 0.3 μL of 8×10^{-6} μM labeled *Eco*RI + 3 primer, 0.25 unit of Taq polymerase, and deionized water to 9.7 μL. Add to each mix 0.3 μL of one of the following *Mse*I selective primers (8×10^{-6} μM): (1) *Mse*I + 3NA, (2) *Mse*I + 3NT,

(3) *Mse*I + 3NG, (4) *Mse*I + 3NC, (5) *Mse*I + 4NA, (6) *Mse*I + 4NT, (7) *Mse*I + 4NG, (8) *Mse*I + 4NC, (9) *Mse*I + 5NA, (10) *Mse*I + 5NT, (11) *Mse*I + 5NG, and (12) *Mse*I + 5NC. Amplify using the program described in step 2.

4. Prepare a 6% Long Ranger™ acrylamide gel (41 cm, 0.2 mm) according to the manufacturer's instructions (http://biosupport. licor.com/docs/Applications_Manual_4300.pdf).

5. Prepare the formamide loading buffer as follows: 80 µL of CongoRed in 950 µL of deionized formamide. Store the loading buffer at –20°C. Add 8 µL of loading buffer to 10 µL of selective product. Denature the sample at 95°C for 5 min, and then chill immediately on ice.

6. Load 1 µL of the denatured selective products in independent wells and 1 µL of 700 bp Size Standards IRDye® in the first or last well (it is possible to load each product three times with the 48-well comb). Run the electrophoresis according to the manufacturer's instructions (http://biosupport.licor.com/docs/Applications_Manual_4300.pdf). Check on the acrylamide gel the *Mse*I selective primers allowing amplification of the target AFLP fragment and according to these results, design the new *Mse*I + 6 selective primer (see Note 8).

3.1.2. Extraction of the AFLP Fragment from the Acrylamide Gel

1. Re-amplify the diluted preselective product with the *Eco*RI + 3 / *Mse*I + 6 primers as described in Subheading 3.1.1, step 2.

2. Prepare a 6% Long Ranger™ acrylamide gel (41 cm, 0.2 mm).

3. Mount the gel on the sequencer against the heater plate and draw the limits of the scanning window using a whiteboard marker. Load 4 µL of the denatured *Eco*RI + 3 / *Mse*I + 6-amplified product in the first four wells. As control, load 4 µL of the denatured *Eco*RI + 3 / *Mse*I + 3 product (see Subheading 3.1.1, step 2) and 1 µL of 700 bp Size Standards IRDye in the last two wells. Run the electrophoresis.

4. When the fragment is visible at the beginning of the scanning window, stop the electrophoresis and remove the gel from the sequencer.

5. Remove the rails of the gel and put the transparency sheet on the gel. Using the permanent marker, draw on the sheet all elements of the gel (i.e., the top and the bottom, the spacers, the scanning window, and the wells loaded with the *Eco*RI + 3 / *Mse*I + 6 product). Separate carefully the gel plates using the plastic wedge. The acrylamide gel has to adhere to only one plate. Place the Whatman® Paper on the acrylamide gel and unstick the Whatman® Paper plus the acrylamide gel from the plate. Place the transparency sheet on the acrylamide gel according to the limits previously drawn. Cut the scanning

window of the sandwich Whatman® Paper/Acrylamide gel/ Transparency sheet which corresponds to the first four wells. Separate the transparency sheet from the sandwich and transfer the Whatman® Paper plus the acrylamide gel in a 1-mL tube containing 600 µL of TE. Crush the Whatman® Paper and incubate for 1 h at room temperature. Add 500 µL of TE and centrifuge quickly. Store the supernatant at –20°C (see Notes 9 and 10).

3.1.3. Sequencing of the Excised Fragment

1. Prepare the selective mix with the following components: 20–100 ng of DNA from the supernatant, 5 µL of 10× PCR buffer, 4 µL of 25 mM $MgCl_2$, 4 µL of 2.5 µM dNTPs, 1.5 µL of 8×10^{-6} µM *Eco*RI + 3 primer, 1.5 µL of 8×10^{-6} µM *Mse*I + 3 primer, 1.25 unit of Taq polymerase, and deionized water to 50 µL. Amplify using the program described in Subheading 3.1.1, step 2.

2. Purify the PCR product using the NucleoSpin® ExtractII kit.

3. Perform the sequencing reaction with the BigDye® Terminator v3.1 Cycle Sequencing kit and the following amplification program: initial incubation 1 min at 96°C; 25 cycles of 10 s at 96°C, 5 s at 50°C, and 4 min at 60°C, followed by a 5-min step at 4°C.

4. Add to the product 5 µL of EDTA (125 mM) and 60 µL of absolute ethanol. Mix and incubate for 15 min at room temperature. Centrifuge for 30 min at $3,000 \times g$ and then throw supernatant. Centrifuge twice for 1 min at $185 \times g$ and throw supernatant. Add 20 µL of deionized formamide. Denature the sample at 95°C for 5 min and analyze the fragments on a capillary Applied Biosystems DNA analyzer.

3.1.4. Optional Additional Step: Cloning of the Product Extracted from the Acrylamide Gel

When the fingerprint remains dense despite the *Mse*I + 6 primer, the products extracted from the acrylamide gel are likely to contain several AFLP fragments. To separate the fragments and identify the sequence of interest, it is then possible to clone the fragments and sequence the inserts.

1. Perform the PCR reaction described in Subheading 3.1.3, step 1.

2. Clone the AFLP fragments present in the PCR product using the TA cloning® Kits.

3. Prepare the PCR mix with the following components: 30 µL of 10× PCR buffer, 36 µL of 25 mM $MgCl_2$, 24 µL of 2.5 µM dNTPs, 6 µL of 10 mg/mL BSA, 12 µL of 10 µM M13F primer, 12 µL of 10 µM M13F primer, 7.5 units of Taq polymerase, and deionized water to 390 µL. Distribute 13 µL in 30 wells of a PCR plate (96-wells) and chill immediately on ice. Add in each well 100 µL of LB containing 50 µg/mL kanamycin.

Pick 30 transformed colonies in the wells. Amplify using the following program: initial incubation 5 min at 94°C; 30 cycles of 40 s at 94°C, 50 s at 55°C, and 1 min at 72°C, followed by a 10-min extension step at 72°C, and a final 1-min step at 16°C.

4. Sequence directly the PCR products as described in Subheading 3.1.3, steps 3 and 4.

5. Identify the sequence of interest according to its length.

6. Make a glycerol stock of plasmid DNA for long-term storage as follows: growth in a 37°C shaking incubator for at least 12 h. Add one drop of the mix glycerol 65% plus LB and store at –20°C.

3.2. Method 2: Combining the Classical AFLP Protocol and 454 Pyrosequencing

3.2.1. Multiplexing of AFLP Samples

1. Dilute the preselective product with deionized water according to your AFLP protocol (for example, a 50-time preselective dilution is used for the plant species *Arabidopsis halleri* (15) or a 20-time preselective dilution is used for the mosquito species *Aedes rusticus* (14)). Store diluted DNA at –20°C.

2. Prepare the 25 μL selective mix with the following components: 5 μL of diluted preselective product, 11.6 μL of deionized water 2.5 μL of 10× PCR buffer, 2.5 μL of 25 mM MgCl$_2$, 2 μL of 2.5 μM dNTPs, 0.5 μL of 10 μM tagged *Eco*RI primer, 0.5 μL of 10 μM tagged *Mse*I primer, 0.2 μL of BSA (1 mg/mL), and 0.2 μL of Taq polymerase (5 U/μL). Amplify using the following program: 10 min at 95°C; 13 cycles of 30 s at 94°C, 60 s at 65°C to 56°C (–0.7°C per cycle), and 60 s at 72°C; 23 cycles of 30 s at 94°C, 60 s at 56°C, and 60 s at 72°C, followed by a final extension step of 10 min at 72°C (see Note 11). Store the selective PCR product at 4°C.

3. Purify the selective PCR product using the MinElute PCR purification kit (Qiagen GmbH) following the manufacturer's instructions.

4. Quantify the selective PCR product using the Agilent 2100 Bioanalyzer and DNA 1000 chips and kit (Agilent Technologies) following the manufacturer's instructions. This kit allows the detection and quantification of all PCR fragments from 25 to 1,000 bp with a 5-bp resolution. Quantify separately the number of molecules present in each peak of the AFLP profile. For clustered AFLP peaks (separated by less than 4 bp), consider the length indicating the higher fluorescence as the reference fragment length for the quantification of the number of molecules.

5. Pool the tagged selective PCR products so as to obtain about 4,000 reads for each sample. The 454 pyrosequencing is a semiquantitative method, i.e., the number of reads obtained for one sample is dependant on the relative number of molecules from this sample compared to the total number of sequences

in the pooled DNAs. For example, using the GS FLX Titanium Kit generating about 450,000 sequences per region (one classical 454 GS FLX Titanium run contains two different regions), one AFLP sample should represent about 1/112 (or 0.9%) of the total amount of pooled DNAs in order to generate 4,000 reads.

6. Store the pooled DNAs at –20°C.

3.2.2. High-Throughput Parallel Pyrosequencing

1. Sequence the pooled DNAs using a 454 Life Science sequencer (Roche) following the manufacturer's instructions. Use preferentially a GS FLX Titanium kit that provides a read length average of 400 bases and allows the complete sequencing of most of the AFLP fragments present in a profile.

3.2.3. Bioinformatic Analyses of 454 Reads: Complete AFLP Fragments (Less than 350 bp)

1. Sort the sequences according to the primers and tags in order to assign each 454 read to the corresponding sample using the ObiTool python program "fasta454filter.py". For this step, do not include the selective bases in the primer sequences. Create one file per sample using the program "fastaSplit.py". During this step, only reads with full-length primers and tags on the two extremities of the sequence are kept. Put the reads that do not satisfy these constraints in a specific file (named "remain. fasta" in this chapter) using the command "-u". This file can be used for the analysis of long AFLP fragments (described in Subheading 3.2.4). To obtain both *Eco*RI-*Mse*I and *Eco*RI-*Eco*RI fragments present in the classical AFLP profile, search and create an individual file for (a) reads starting by tag + *Eco*RI and finishing by tag + *Mse*I, (b) reads starting by tag + *Mse*I and finishing by tag + *Eco*RI, and then (c) reads starting by tag + *Eco*RI and finishing by tag + *Eco*RI. The script "fasta454filter.py" has the advantage to delete the tag and the primer sequences from the selected reads, a required step before performing the sequence clustering (Subheading 3.2.3, step 7).

2. For each sample, reverse complement the *Mse*I-*Eco*RI reads using the ObiTool python program fastaComplement.py. Reads from the three files (*Eco*RI-*Mse*I, *Mse*I-*Eco*RI, and *Eco*RI-*Eco*RI) now start by the selective bases linked to the primer *Eco*RI.

3. Group together in one file all the reads assigned to the same sample.

4. For each sample, cluster together all the strictly identical reads using the ObiTool python program "fastaUniq.py". A sequence attribute named "count" indicating the number of identical reads is added to the sequence description during this step.

5. Remove the sequences repeated less than three times using the script "fastaGrep.py" and filtering by the "count" attribute (see Note 12).

6. Remove the sequences smaller than 18 bases using the script "fastaGrep.py" and filtering by the "length" attribute. At this step, 18-base sequences correspond to 50-base peaks in a classical AFLP profile (16 bases for each primer and 18 bases of DNA restriction fragment).

7. For each sample, cluster the sequences corresponding to the same AFLP locus (see Note 13) using the program PHRAP. Use a penalty value for substitution of –9 in order to tend to find alignments that are 90% identical for each cluster (see Note 14). Use a vector_bound value of 0 to indicate the absence of potential bases at the beginning of each read. Use the high gap_init value of –15 to limit the false alignment of different sequences using gaps. Finally, create a ".ace" file that contains all the information about reads (presence and location) grouped in each cluster. Each cluster contains allelic variants for a unique AFLP locus, including heterozygous fragments and fragments with sequencing errors. The PHRAP program provides a FASTA file named ".contigs" containing the consensus sequences of all the clusters.

8. Sort the ".contigs" file consensus sequences by their length and create a sequence-based AFLP profile for each sample.

9. The ".ace" file can be used to verify the clustering process, especially recommended (a) for clusters whose length is higher than expected and (b) for species genomes containing a high rate of repeated sequences, such as transposable elements. Convert the global ".ace" file in a ".fasta" file using the ObiTool python program "ace2fasta.py". One FASTA file with aligned sequences is created for each cluster, i.e., for each AFLP fragment. The sequence alignment within each cluster can now be verified using the Bioedit program, for example.

3.2.4. Bioinformatic Analyses of 454 Reads: Incomplete AFLP Fragments (Longer than 350 bp)

The AFLP protocol allows the genotyping of fragments from 50 to 500 bp, and AFLP loci longer than 350 bp are often detected as loci of interest (less homoplasy, less biases due to adjacent fragments of similar length, and higher probability of insertion/deletion within the fragment). The sequence assignation for AFLP loci longer than 350 bp is possible using the file "remain.fasta" obtained during step 1 in Subheading 3.2.3.

1. Sort the reads of the file "remain.fasta" by the 5′ primers and tags only using the ObiTool python program "fasta454filter.py". As for step 1 in Subheading 3.2.3, search and create an individual file for (a) reads starting by tag + *Eco*RI and then (b) reads starting by tag + *Mse*I to obtain both *Eco*RI-*Mse*I and *Eco*RI-*Eco*RI fragments present in the classical AFLP profile.

2. Keep only unfinished sequences of long AFLP fragments and discard the reads containing a primer in 3′ by using the script

"fastaGrep.py" and filtering by the attribute "unidentified = no reverse match" (all primer and tag detections can be found in the sequence attributes). This step allows discarding chimeric fragments containing incorrect primer combinations.

3. The rest of the protocol remains unchanged from steps 2 to 9 in Subheading 3.2.3.

3.2.5. Assignation of 454 Sequences to AFLP Loci

1. For each sample, make a direct comparison between the classical AFLP profile and the sequence-based AFLP profile to assign sequences to peaks (see Note 15). Do not forget to add the primer length (without the tags) to the length of the 454 sequences obtained before the comparison with classical AFLP profile.

2. Compare the sequence results for the loci of interest between at least two samples that share this locus and two samples for which the locus is absent. The sequence should not be present in these later samples.

3. A reduction of the classical AFLP fingerprint (as described in Subheading 3.1.1) can be made to confirm the sequence assignation by verifying the first three bases of the locus of interest (see Note 8).

4. Notes

1. The AFLP fragment length distribution was classically described as L shaped with a very high density of fragments under 120 bp (15, 25, 26). Consequently, simplification of the fingerprint and single excision from the gel, as exposed in this chapter, is very difficult for small fragments. We recommend using this protocol for fragments larger than 120 bp, the success of excision increasing with the fragment length. Furthermore, small fragments are more likely to comigrate (6) and thus the risk to obtain ambiguous sequences is very high.

2. When the fingerprint remains dense, two parameters may be changed to increase the probability to isolate only one AFLP fragment. First, we recommend choosing a 66-cm acrylamide gel rather than a 41-cm one. The running time will be longer and the separation of fragments more efficient (conversely, when the simplification of the fingerprint is very efficient, the 41-cm gel may be changed to a 25-cm one for faster migration). Secondly, the scanning window cut from the sandwich Whatman Paper/Acrylamide gel/Transparency sheet can be separated in at least four longitudinal bands. Each band will be transferred to a 1-mL tube containing 500 μL of TE and

directly sequenced. The sequence of interest will be then identified according to its length.

3. The 9-well comb is a classical comb for LI-COR® DNA analysis system, where some teeth are absent to obtain 9 wells of similar length. It may be very useful when the migration of fragments is not in a straight line. We then recommend drawing each well on the transparency sheet and excising the wells separately.

4. Simplification of the AFLP fingerprint (described in Subheading 3.1.1) can be used before the second method to decrease the number of peaks per profile while keeping the fragment of interest. This has the advantages to (a) decrease the number of reads needed per sample and/or increase the number of reads per AFLP peak, (b) facilitate the bioinformatic analyses and profiles comparisons, and (c) reduce the homoplasy rate and the ambiguous assignation of several sequences with the same length to the locus of interest.

5. The four variable nucleotides of the primer tags should not start by a C in order to distinguish initial CC bases from the tags itself.

6. Any two tags must differ by at least two nucleotide polymorphisms to ensure the good assignation of the 454 reads to the samples.

7. Genotyping errors due to sample mixing or pipetting errors are common when the volume of the PCR mix is small. Consequently, to avoid this bias and misinterpretation for the determination of the *Mse*I + 6 primer (performed on 10 μL), we recommend using several replicates for each reaction.

8. In some rare cases, the fragment of interest is not amplified using the degenerated *Mse*I4N + A, T, G, or C and/or the *Mse*I5N + A, T, G, or C primers. Determination of the fifth and sixth selective bases is then only possible with more specific primers. According to the amplification with the *Mse*I3N + A, T, G, or C, we recommend to design new non-degenerated primers as follows: *Mse*I + 4 selective bases + A, T, G, or C and *Mse*I + 4 selective bases + N + A, T, G, or C.

9. We recommend controlling the success of excision before sequencing. On a 41-cm acrylamide gel, load the product of the extraction re-amplified using the *Eco*RI + 3/*Mse*I + 6 primers and both the *Eco*RI + 3/*Mse*I + 3 product and the 700-bp Size Standards IRDye® as controls. Run the electrophoresis and check that the fragment amplified with the *Eco*RI + 3/*Mse*I + 6 primers has the correct size.

10. As an alternative for direct excision, the acrylamide gel may be scanned on an Odyssey® Infrared Imaging System (LICOR® Biosciences) and gel plugs containing the target fragment may

be excised using a scalpel. Successful fragment extraction will be then verified by re-scanning the gel on the Odyssey® Infrared Imaging System.

11. A modification of the selective PCR parameters is not needed, even with the addition of six nucleotides to AFLP selective primers.

12. The threshold of three counts for the deletion of unique sequences can be modified taking into account the average number of reads per AFLP fragment. A smaller threshold can allow retaining longer fragments that are usually less amplified than smaller ones. An upper threshold limits the presence of allele variants due to sequencing errors in the final clusters.

13. The 454 reads of individuals amplified with the same primer combinations can be pooled before the clustering. With such pooled 454 reads, a cluster is created for each AFLP fragment: some fragments can be present in only one of the individuals, some of them can be shared by several individuals, and monomorphic AFLP fragments are shared by all the individuals. For an AFLP locus shared by several individuals, the cluster contains all allelic variants for each individual, i.e., heterozygous fragments and fragments with sequencing errors. As each 454 read is assigned to an individual, this method allows detecting rapidly which individuals share an AFLP locus of interest using the FASTA files containing aligned sequences for each cluster (created during step 9 in Subheading 3.2.3).

14. The identity value of 90% used for the clustering of the sequences corresponding to the same AFLP locus can be modified depending of the species used. For species with low-sequence polymorphism or with high rates of repeated sequences like transposable elements, a higher identity value will improve the quality of the clustering by limiting the clustering of several different fragments with high homology of sequence. When the 454 reads from several individuals are pooled before the clustering (see Note 13), a lower identity value can be used to ensure the clustering of all the sequences corresponding to the same AFLP locus. For example, a clustering value of 90% was successfully used for the clustering of pooled 454 reads from several individuals from the same population of the mosquito *Aedes aegypti* (Paris et al. unpublished data) and for that of pooled 454 reads from several individuals from different populations of the plant *Arabis alpina* (Poncet et al. unpublished data).

15. A difference from −3 bases to +3 bases is sometimes observed between the estimated length of classical AFLP fragments and the real sequence. For this reason, it is recommended to check carefully the sequences obtained in a 6-bp range around the length of the fragment of interest.

Acknowledgments

We are grateful to Cécile Godé and Pierre Saumitou-Laprade for technical assistance and to Martin C. Fischer and Aurélie Bonin for helpful comments on an earlier version of this work. M.P. and L.D. were supported by a grant from the French Rhône-Alpes region (grant 501545401) and by the French National Research Agency (project ANR-08-CES-006-01 DIBBECO). C.L.M. was funded by the French Ministry of Research and Technology.

References

1. Vos P, Hogers R, Bleeker M, Reijans M et al (1995) AFLP – a new technique for DNA-fingerprinting. Nucleic Acids Res 23: 4407–4414

2. Bensch S, Akesson M (2005) Ten years of AFLP in ecology and evolution: why so few animals? Mol Ecol 14:2899–2914

3. Meudt HM, Clarke AC (2007) Almost forgotten or latest practice? AFLP applications, analyses and advances. Trends Plant Sci 12:106–117

4. Bonin A, Bellemain E, Eidesen PB et al (2004) How to track and assess genotyping errors in population genetics studies. Mol Ecol 13:3261–3273

5. Pompanon F, Bonin A, Bellemain E et al (2005) Genotyping errors: causes, consequences and solutions. Nat Rev Genet 6:847–859

6. Paris M, Bonnes B, Ficetola GF et al (2010) Amplified fragment length homoplasy: *in silico* analysis for model and non-model species. BMC Genomics 11:287

7. Bonin A, Taberlet P, Miaud C et al (2006) Explorative genome scan to detect candidate loci for adaptation along a gradient of altitude in the common frog (*Rana temporaria*). Mol Biol Evol 23:773–783

8. Poncet BN, Herrmann D, Gugerli F et al (2010) Tracking genes of ecological relevance using a genome scan in two independent regional population samples of *Arabis alpina*. Mol Ecol 19:2896–2907

9. Conord C, Lemperiere G, Taberlet P et al (2006) Genetic structure of the forest pest *Hylobius abietis* on conifer plantations at different spatial scales in Europe. Heredity 97: 46–55

10. Egan SP, Nosil P, Funk DJ (2008) Selection and genomic differentiation during ecological speciation: isolating the contributions of host association via a comparative genome scan of *Neochlamisus bebbianae* leaf beetles. Evolution 62:1162–1181

11. Nosil P, Egan SP, Funk DJ (2008) Heterogeneous genomic differentiation between walking-stick ecotypes: "isolation by adaptation" and multiple roles for divergent selection. Evolution 62:316–336

12. Manel S, Conord C, Despres L (2009) Genome scan to assess the respective role of host-plant and environmental constraints on the adaptation of a widespread insect. BMC Evol Biol 9:288

13. Herrera CM, Bazaga P (2008) Population-genomic approach reveals adaptive floral divergence in discrete populations of a hawk moth-pollinated violet. Mol Ecol 17:5378–5390

14. Paris M, Boyer S, Bonin A et al (2010) Genome scan in the mosquito *Aedes rusticus*: population structure and detection of positive selection after insecticide treatment. Mol Ecol 19:325–337

15. Meyer CL, Vitalis R, Saumitou-Laprade P, Castric V (2009) Genomic pattern of adaptive divergence in *Arabidopsis halleri*, a model species for tolerance to heavy metal. Mol Ecol 18:2050–2062

16. Rossi M, Bitocchi E, Bellucci E et al (2009) Linkage disequilibrium and population structure in wild and domesticated populations of *Phaseolus vulgaris* L. Evol Appl 2:504–522

17. Wilding CS, Butlin RK, Grahame J (2001) Differential gene exchange between parapatric morphs of *Littorina saxatilis* detected using AFLP markers. J Evol Biol 14:611–619

18. Campbell D, Bernatchez L (2004) Generic scan using AFLP markers as a means to assess the role of directional selection in the divergence of sympatric whitefish ecotypes. Mol Biol Evol 21:945–956

19. Savolainen V, Anstett MC, Lexer C et al (2006) Sympatric speciation in palms on an oceanic island. Nature 441:210–213

20. Nosil P, Funk DJ, Ortiz-Barrientos D (2009) Divergent selection and heterogeneous genomic divergence. Mol Ecol 18:375–402

21. Brugmans B, van der Hulst RG, Visser RG et al (2003) A new and versatile method for the successful conversion of AFLP markers into simple single locus markers. Nucleic Acids Res 31:e55

22. van Orsouw NJ, Hogers RCJ, Janssen A et al (2007) Complexity reduction of polymorphic sequences (CRoPS): a novel approach for large-scale polymorphism discovery in complex genomes. PLoS One 2:11

23. Paris M, Despres L (2012) Identifying insecticide resistance genes in mosquito by combining AFLP genome scans and 454 pyrosequencing. Mol Ecol 21:1672–1686

24. Valentini A, Miquel C, Nawaz MA et al (2009) New perspectives in diet analysis based on DNA barcoding and parallel pyrosequencing: the *trn*L approach. Mol Ecol Resour 9:51–60

25. Innan H, Terauchi R, Kahl G et al (1999) A method for estimating nucleotide diversity from AFLP data. Genetics 151:1157–1164

26. Vekemans X, Beauwens T, Lemaire M et al (2002) Data from amplified fragment length polymorphism (AFLP) markers show indication of size homoplasy and of a relationship between degree of homoplasy and fragment size. Mol Ecol 11:139–151

Chapter 7

Roche Genome Sequencer FLX Based High-Throughput Sequencing of Ancient DNA

David E. Alquezar-Planas and Sarah L. Fordyce

Abstract

Since the development of so-called "next generation" high-throughput sequencing in 2005, this technology has been applied to a variety of fields. Such applications include disease studies, evolutionary investigations, and ancient DNA. Each application requires a specialized protocol to ensure that the data produced is optimal. Although much of the procedure can be followed directly from the manufacturer's protocols, the key differences lie in the library preparation steps. This chapter presents an optimized protocol for the sequencing of fossil remains and museum specimens, commonly referred to as "ancient DNA," using the Roche GS FLX 454 platform.

Key words: Ancient DNA, Degraded DNA, High-throughput, Sequencing, GS FLX, Roche 454

1. Introduction

Since the inception of the GS20, the first high-throughput sequencer by Roche/454 Life Sciences in 2005 (1), several other next generation sequencers have been developed, including the adapted Roche/454 GS FLX, the Illumina Genome Analyzer (Solexa/GA/GAII/Hi-Seq 2000), and the Applied Biosystems SOLiD platforms. Whereas both other competing platforms have a much higher sequencing throughput, the main advantage of the GS FLX lies in its ability to sequence substantially longer reads than other next generation technologies. This serves as an advantage in many modern and ancient DNA (aDNA) studies by making easier contig assembly, which may be a limiting step to sequence de novo genomes. In 2009, the Roche GS FLX chemistry was improved and optimized to sequence reads of up to ~400 bp. This chemistry, subsequently named Titanium, is not discussed in detail

François Pompanon and Aurélie Bonin (eds.), *Data Production and Analysis in Population Genomics: Methods and Protocols*, Methods in Molecular Biology, vol. 888, DOI 10.1007/978-1-61779-870-2_7, © Springer Science+Business Media New York 2012

in this chapter, as this protocol was developed for the GS FLX kits. However, the principles mentioned here remain applicable to any form of high-throughput sequencing.

Studies involving aDNA face several problems that do not apply to fresh/intact DNA samples due to the postmortem instability of nucleic acids (2, 3). Hence, intrinsic problems including fragmented template, low template quantity, chemically modified template, and the presence of PCR inhibitors need to be considered when sampling ancient specimens. We do not go into the details of these processes, as there are several review papers covering these issues (4–7). Work done on aDNA has shown that, if an ancient specimen does contain DNA, one can generally expect fragments shorter than 500 bp (2, 8, 9). As a result, the emergence of high-throughput sequencing technologies has been seen as a breakthrough in aDNA analyses, as it has suddenly enabled to undergo massively parallel sequencing of these relatively short molecules to generate previously untargettable datasets (10). Examples include enabling the recovery of mitochondrial genomes sequences from the Neandertal, woolly mammoth, and cave bear, and the first ancient human genome (11–14).

The fact that the DNA is a priori fragmented has important implications for the sequencing procedure as no nebulization or other types of fragmentation steps are required during the library steps (Fig. 1). This, in fact, carries another benefit because, in general, fragmentation using techniques such as nebulization can result in a loss of large amounts of the starting material (>90%). Hence, without fragmentation, one is able to start with relatively lower amounts of initial DNA than in conventional studies.

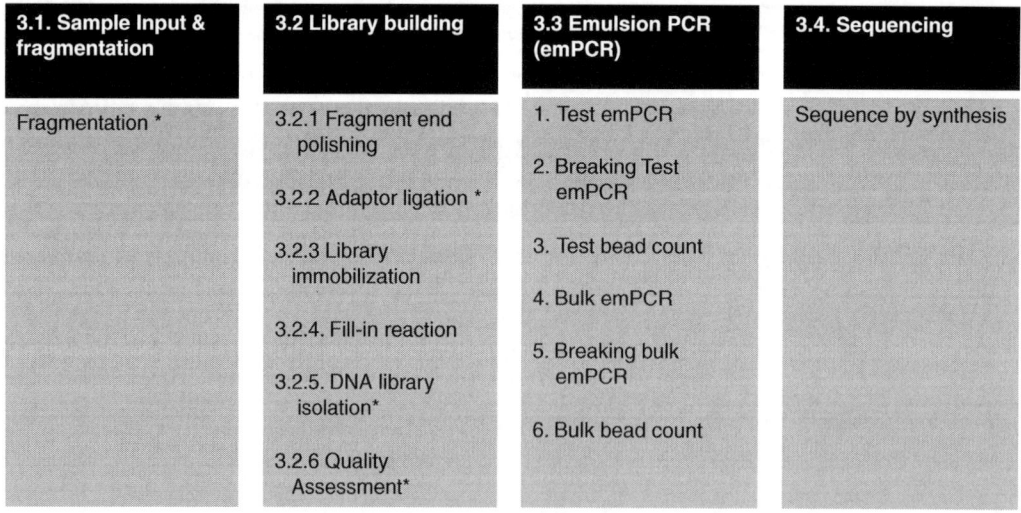

Fig. 1. Schematic overview of GS FLX sequencing. *Asterisks* denote steps which have been altered for a DNA optimized protocol.

Another commonplace concern in any aDNA study is the risk of contamination both within the collected specimen and throughout all downstream experimental applications. At the point of collection, ancient samples can contain very high levels of exogenous DNA, mainly from bacterial/environmental contaminants but also because the specimens themselves may be contaminated with modern DNA pertaining to the same species. Human DNA is another common source of contamination, both from handling of the sample during the excavation/sampling process or during laboratory practices, although control from this source can be minimized with the intervention of appropriate precautions and standard aDNA procedures. The typical levels of human contamination in ancient samples can be seen in the initial Neandertal and mammoth genomic publications (10, 11). Although the issue of coamplification of contaminates during conventional PCR is no longer a problem for high-throughput sequencing techniques due to the lack of predetermined target-specific primer-based amplification, the introduction of highly concentrated exogenous modern DNA, especially from humans, remains a constant concern. Methods of reducing sample contamination during laboratory processes and cleansing of the sample before sampling are discussed in (6, 7).

For certain sequencing applications where more than 16 samples are being run per plate, each run pertaining to a subdivided region of the plate, the use of sample specific molecular identifiers (MID) tags is highly advantageous. This multiplexing capability permits both a larger number of samples to be run simultaneously and reduces the offset cost associated with running numerous expensive sequencing runs. In this protocol, we use the MID Roche protocols that appear in the *GS FLX Shotgun DNA Library Preparation Method Manual (version December 2007)*. The MIDs are unique identifying sequences 10 bp in length that are incorporated into the Roche adaptors and can be identified using the Roche software after sequencing. Once the MIDs have been added to the samples in the Library Preparation steps, subsequent contamination can be identified. If a sequence does not contain a correct MID, then it will be excluded from the sequencing output.

Here, we present an optimized protocol for the sequencing of ancient DNA, based on the Roche *GS FLX Method Manual (version December 2007)*.

2. Materials

For convenience one may consider using Roche's own library build kits. However, there is no constraint to use manufacturer provided reagents, given that library construction involves relatively common molecular biology techniques. In the following protocol, based on methods used by Maricic and Pääbo (15), we provide examples using individually purchased reagents, which we find to be more cost effective. For postlibrary steps, we use dedicated Roche-provided reagents for convenience, available at http://454.com/products-solutions/product-list.asp.

1. T4 DNA Polymerase (Fermentas).

2. T4 Polynucleotide Kinase (Fermentas).

3. T4 Ligase (Fermentas) including 50% PEG-4000 solution.

4. Bst DNA Polymerase, Large Fragment (NEB) including 10× ThermoPol buffer.

5. 10× Buffer Tango (Fermentas).

6. ATP (Fermentas), 100 mM stock solution.

7. dNTPs (GE Healthcare), 25 mM each.

8. Water, HPLC-grade (Sigma).

9. TE buffer (many suppliers or self-made): 10 mM Tris–HCl, 0.1 mM EDTA, pH 8.0.

10. EB buffer (supplied with MinElute PCR Purification kit): 10 mM Tris–HCl, pH 8.5.

11. MinElute PCR Purification Kit (Qiagen).

12. GS FLX DNA Library Preparation Kit (Roche), only the adapters are needed.

13. BindWash buffer (2×): 2 M NaCl, 10 mM Tris–Cl, 1 mM EDTA, 0.05% Tween-20, pH 8.0.

14. BindWash buffer (1×): 1 M NaCl, 10 mM Tris–Cl, 1 mM EDTA, 0.05% Tween-20, pH 8.0.

15. M-270 Dynabeads or MyOne C1 beads (Invitrogen).

16. Melt solution (For 10 ml preparation: 0.125 ml NaOH (10 M) to 9.875 ml Molecular Biology Grade water).

17. Neutralization solution (20% glacial acetic acid; for a 10 ml preparation add 2 ml of glacial acetic acid to 8 ml Molecular Biology Grade water, in that order).

3. Methods

An overview of the method for GS FLX sequencing can be seen in Fig. 1. When preparing ancient DNA material for high-throughput sequencing using the GS FLX platform, sample material is generally unamplified. However, if amplification is a requirement due to expected low DNA yields in the sample, or a specific region of DNA is being targeted for sequencing, then preamplified material can be used as an appropriate starting template. Regardless of the intention to use amplified or nonamplified ancient DNA material for sequencing, follow the Roche *GS FLX Shotgun DNA Library Preparation Method Manual (December 2007)* with the use of MID tags.

3.1. Sample Input and Fragmentation

Fragmentation of ancient DNA material and small fragment removal are not required (see Note 1).

3.2. Library Building

3.2.1. Fragment End Polishing

Fragmented aDNA will likely contain both frayed ends and overhangs, which present restrictions to efficient adapter ligation. In order to maximize the ligation of adapters to the double stranded molecule the reaction requires blunt ended and 5′ phosphorylated DNA, achieved through the action of T4 DNA polymerase and T4 polynucleotide kinase. The reaction (see Table 1) is as follows:

1. Prepare a master mix (Table 1) for the required number of reactions. Mix carefully by pipetting up and down or flicking the tube with a finger. Do not vortex after adding enzymes. Keep the master mix on ice if not immediately used to maintain full enzyme activity.

2. Add 20 μl of master mix to 20 μl sample (dissolved in EB, TE or water) to obtain a total reaction volume of 40 μl and mix. Incubate in a thermal cycler for 15 min at 25°C.

3. Purify the reaction in a Qiagen MinElute silica spin column according to the manufacturer's instructions. Elute in 20 μl EB.

3.2.2. Adaptor Ligation

1. When performing *section 3.3* of the *GS FLX Shotgun DNA Library Preparation Method Manual (December 2007)*, use MID tags as adaptors as this will barcode each of the working samples with a known 10 bp DNA sequence (see Note 2).

2. Prepare a ligation master mix for the required number of reactions (see Table 2). Since PEG is highly viscous, vortex the

Table 1
Master mix for fragment end polishing reaction

Reagent	Volume (μl) per sample	Final concentration in reaction
Water (ad 20 μl)	7.6	
Buffer Tango (10×)	4	1×
dNTPs (2.5 mM each)	1.6	100 μM each
ATP (10 mM)	4	1 mM
T4 Polynucleotide Kinase (10 U/μl)	2	0.5 U/μl
T4 Polymerase (5 U/μl)	0.8	0.1 U/μl

Table 2
Master mix for adaptor ligation reaction

Reagent	Volume (µl) per sample	Final concentration in reaction
Water (ad 20 µl)	6	
T4 Ligase buffer (10×)	4	1×
PEG-4000 (50%)	4	5%
454 adapter mix (50 µM each)	5	1.25 µM each
T4 Ligase (5 U/µl)	1	0.125 U/µl

Table 3
Master mix for fill-in reaction

Reagent	Volume (µl) per sample	Final concentration in reaction
Water (add 50 µl)	37	
Thermopol buffer (10×)	5	1×
dNTPs (2.5 mM each)	5	250 µM each
Bst Polymerase (8 U/µl)	3	0.5 U/µl

master mix before adding T4 ligase and mix gently thereafter (see Note 3).

3. Add 20 µl of master mix to each eluate from Subheading 3.2.3 to obtain a total reaction volume of 40 µl and mix. Visually verify that all reaction components are mixed well. Incubate for 1 h at 22°C in a thermal cycler.

4. Purify the reaction over a Qiagen MinElute silica spin column according to the manufacturer's instructions. Elute in 25 µl EB.

3.2.3. Library Immobilization

Proceed with *section 3.4. "Library Immobilization"* in the *GS FLX Shotgun DNA Library Preparation Method Manual (December 2007)*, for this step.

3.2.4. Fill-In Reaction

Gaps present at the 3' end, due to the adaptors not being phosphorylated, need to be filled-in. These gaps can be repaired using strand-displacing DNA polymerase, making the library full-length dsDNA. The reaction is as follows:

1. Prepare a fill-in master mix (see Table 3) for the required number of samples.

2. Resuspend the bead pellet in 50 µl fill-in master mix. Incubate at 37°C for 20 min in a thermal cycler.

3. Pellet the beads using the magnet and discard supernatant.

4. Wash the beads twice with 100 μl 1× BW buffer.

3.2.5. DNA Library Isolation   For DNA Library Isolation, do not use the NaOH melt step recommended in the Roche protocol, rather follow the heat/boiling protocol from Maricic and Paabo (15). By this approach, DNA library recovery is increased to up to 98% (see Note 4).

1. Resuspend the pellet in 50ul of 0.05% TE + Tween-20 solution.

2. Place the resuspended pellet at 90°C for 2 min.

3. Place on ice for at least 1 min.

4. Put on magnetic particle collector (MPC) for 1 min to separate beads from solution.

5. Collect the solution containing the dsDNA and discard the beads.

3.2.6. Quality Assessment   The Roche protocol suggests using the Agilent Bioanalyzer to assess the average size of the DNA library. However, given the already low DNA template yields used in ancient DNA and the further loss of DNA through the library building procedure a decreased sensitivity may be encountered using this technology. It is recommended as an alternative that a quantitative PCR (qPCR) be run with primers based on the 5′ end of the MID adaptor sequences (sequence is shared by all the MIDs) at different dilution series (typically 1:100 to $1:10^6$) to determine the quantity of starting material needed for the emulsion PCR (emPCR) (16) (see Note 4).

3.3. Emulsion PCR (emPCR)

1. Perform the Test emPCR by following the Roche *GS FLX emPCR Method Manual (December 2007)*, sections 3.1–3.4, using the "*Shotgun sstDNA*" instructions (see Note 5).

2. Break the Test emPCR. Proceed with the Roche *GS FLX emPCR Method Manual (December 2007)*, sections 3.5–3.7, following the "*Shotgun sstDNA*" instructions (see Note 6).

3. Perform a bead count in duplicate using a coulter counter. Section 6.5 of the Roche *GS FLX emPCR Method Manual (December 2007)* contains a list of Roche approved coulter counters and instructions to establish the total amount of enriched DNA beads (see Notes 7 and 8).

4. Use the determined CPB ratio from the above step to continue with a bulk emPCR required to fill a PTP. Follow the *GS FLX Sequencing Method Manual (December 2007)* sections 4.1–4.4.

5. Break the Bulk emPCR. Refer to the Roche *GS FLX Sequencing Method Manual (December 2007)* sections 4.5–4.7.

6. Perform a bead count using the coulter counter to determine if the CPB ratio achieved during the bulk emPCR will produce optimal sequencing results. Refer to step 3 of this section (Subheading 3.2).

3.4. Sequencing   Proceed with the Roche *GS FLX Sequencing Method Manual (December 2007)* protocol in this step.

4. Notes

1. Fragmentation of ancient DNA material is not required nor recommended due to the inherent characteristics of DNA degradation already mentioned. Given the intended nature to sequence already fragmented aDNA the use of the small fragment removal step is unnecessary. If a sample is expected to be heavily degraded, we advise against doing these steps. Given the variability of quality within sample material it is recommended that the DNA integrity be assessed prior to undergoing expensive library building and sequencing. This can be performed by running purified ancient DNA on the Agilent Bioanalyzer with the use of a High Sensitivity Chip. This will allow simultaneous assessment of the quality and the average size of already fragmented DNA material.

2. MID tags can be used to discriminate true aDNA from other contaminant sequence that might arise from other previously built libraries. This simple and effective measure will allow filtering of any irrelevant sequences and will identify early contamination so that appropriate preventative measures can be enforced.

3. White precipitate may be present in the ligation buffer after thawing. Heat the buffer vial briefly to 37°C and vortex until the precipitate has dissolved. When starting from low template quantities (<10 ng) it is safe to use less adapter mix (0.5 or 0.1 μl per reaction).

4. When using the heat step described by Maricic and Paabo (15) remember that this will yield double strand DNA and not single strand DNA as is the case with the NaOH melt step described by Roche; this is vital for subsequent quantification steps.

5. The ultimate goal of the clonal based emPCR is to immobilize one molecule of DNA per bead and one bead per aqueous microreactor (oil micelle). If too little DNA is added the emPCR will result in a large proportion of null beads (without immobilized DNA) that, if carried over to the sequencing reaction, will take up a well in the Picotitre Plate (PTP) and subsequently decrease the amount of sequences produced from the sequencing run. Conversely, if too much DNA is used in preparation of the emPCR some beads will contain multiple amplifiable DNA library templates which will register as either a "mixed" or a "dot" read by the *Genome Sequencer FLX Data Analysis Software Manual (December 2007)*, which is then consequently

filtered out, again resulting in a reduced amount of sequence reads from the run. Refer to the *GS FLX Shotgun DNA library Method Manual (December 2007) section 3.7.3 "Functional Quantitation of the sstDNA Library"* for more details.

6. Optimal counts of enriched DNA beads are best achieved by performing a "test" emPCR or titration to determine the optimal DNA concentration, resulting in a single DNA molecule immobilized on to one bead. This is performed by preparing a small number of standard reactions (one reaction being 1/32 of the total amount of beads required to fill a PTP) with varying amounts of DNA copies per bead (CPB). Typically, between 1 and 5 CPB ratios are sufficient to enrich for the required amount of beads. This titration reaction is especially important when working with both low quality and quantity DNA, such as ancient DNA, in order to maximize sequence throughput and minimize sample consumption. To establish your optimal DNA concentration to your DNA beads, perform a test emPCRs with at least two different CPB ratios.

7. We use the Beckman Coulter Z2 model, but most if not all Coulter counters should yield similar results, as the beads are much larger than a cell. However, in the latest *GS FLX Titanium series (revised version 2010)*, the use of six devices has been validated for the quantitation of amplified DNA library beads.

8. After establishing an average bead count, confirm which count per beads (CPB) ratio is best for recovery of optimal DNA bead quantities. As a general rule approximately **50,000 beads** should be recovered for one reaction when performing both a shotgun library and/or an amplicon library with MID tags for adapters. However, the optimal amount of beads enriched and recovered will depend on how many regions of the PTP you intend to sequence. Refer to the *GS FLX Sequencing Method Manual (December 2007) section 3.2.3.2, Preparation of the DNA Beads, p. 46*. If insufficient beads are recovered during your bulk emPCR, you can subsequently pool any enriched test emPCR DNA beads with your enriched bulk emPCR DNA beads to augment your total bead count or alternatively, reassess the CPB ratio and if necessary repeat the test emPCR step at modified ratios.

References

1. Margulies M, Egholm M, Altman WE et al (2005) Genome sequencing in microfabricated high-density picolitre reactors. Nature 437: 376–380

2. Pääbo S (1989) Ancient DNA: extraction, characterization, molecular cloning and enzymatic amplification. Proc Natl Acad Sci USA 86: 1939–1943

3. Lindahl T (1993) Instability and decay of the primary structure of DNA. Nature 362:709–715

4. Hofreiter M, Jaenicke V, Serre D et al (2001) Ancient DNA. Nat Rev Genet 2:353–359

5. Pääbo S, Poinar H, Serre D et al (2004) Genetic analyses from ancient DNA. Annu Rev Genet 38:645–679

6. Willerslev E, Cooper A (2005) Ancient DNA. Proc Biol Sci 272:3–16

7. Gilbert MTP, Bandelt HJ, Hofreiter M et al (2005) Assessing ancient DNA studies. Trends Ecol Evol 20:541–544

8. Handt O, Richards M, Trommsdorff M et al (1994) Genetic analyses of the Tyrolean Ice Man. Science 264:1775–1778

9. Höss M, Jaruga P, Zastawny TH et al (1996) DNA damage and DNA sequence retrieval from ancient tissues. Nucleic Acids Res 24:1304–1307

10. Poinar NH, Schwarz C, Qi J et al (2006) Metagenomics to paleogenomics: large-scale sequencing of mammoth DNA. Science 311: 392–394

11. Green RE, Krause J, Ptak SE et al (2006) Analysis of one million base pairs of Neanderthal DNA. Nature 444:330–336

12. Miller W, Drautz DI, Ratan A et al (2008) Sequencing the nuclear genome of the extinct woolly mammoth. Nature 456:387–390

13. Noonan JP, Hofreiter M, Smith D et al (2005) Genomic sequencing of Pleistocene cave bears. Science 309:597–599

14. Rasmussen M, Li Y, Lindgreen S et al (2010) Ancient human genome sequence of an extinct Palaeo-Eskimo. Nature 463:757–762

15. Maricic T, Pääbo S (2009) Optimization of 454 sequencing library preparation from small amounts of DNA permits sequence determination of both DNA strands. Biotechniques 45:51–57

16. Meyer M, Briggs AW, Maricic T et al (2008) From micrograms to picograms: quantitative PCR reduces the material demands of high-throughput sequencing. Nucleic Acids Res 36:e5

Chapter 8

Preparation of Normalized cDNA Libraries for 454 Titanium Transcriptome Sequencing

Zhao Lai, Yi Zou, Nolan C. Kane, Jeong-Hyeon Choi, Xinguo Wang, and Loren H. Rieseberg

Abstract

Transcriptome sequencing from cDNA libraries has been extensively and efficiently used to analyze sequence variation in protein-coding genes (Expressed Sequence Tags) in eukaryote species. Rapid advances in next-generation sequencing (NGS) technology, in terms of cost, speed, and throughput, allow us to address previously unanswerable questions in the fields of ecology, evolution, and systematics using these genomic tools. Transcriptome sequencing from individuals across different populations and species enables researchers to study the evolution of gene sequence variation at a population genomics level. In this chapter, we describe a customized protocol that has been successfully optimized for the development of normalized cDNA libraries in eukaryote systems suitable for Roche 454 GS FLX sequencing, requiring only small quantities of starting material.

Key words: EST, Total RNA isolation, SMART™ cDNA synthesis, cDNA normalization, DSN, Roche 454 GS FLX, Transcriptome sequencing

1. Introduction

Expressed Sequence Tag (EST) sequencing of cDNA libraries has become a powerful tool for dissecting the protein-coding fraction of a genome. Due to the considerable cost and time required for deep sequencing of ESTs with Sanger methods, until very recently EST sequencing was limited to model species or key representatives of other taxa. In the past few years, however, with the rapid development of next-generation sequencing (NGS) techniques and applications, whole transcriptome sequencing has been successfully utilized as the basis for novel gene characterization, marker development, and population genomics studies in many non-model species (e.g., ref. 1).

François Pompanon and Aurélie Bonin (eds.), *Data Production and Analysis in Population Genomics: Methods and Protocols,*
Methods in Molecular Biology, vol. 888, DOI 10.1007/978-1-61779-870-2_8, © Springer Science+Business Media New York 2012

Among the currently available NGS platforms (e.g., Roche 454, Illumina Genome Analyzer and HiSeq2000, Life Sciences SOLID), the Roche 454 Genome Sequencer FLX Titanium platform is particularly suitable for large-scale de novo EST sequencing in emerging and non-model organisms. 454 Titanium sequencing can generate the long read lengths (approximately 400–500 bp) in the current NGS platforms, and the read lengths are comparable to traditional Sanger sequencing. Long reads are extremely useful for de novo transcriptome assembly for non-model species lacking whole genome sequences or Sanger EST collections (e.g., ref. 2). Indeed, comparative analyses of the age distribution of duplicate genes based on Sanger, 454 Titanium, and Illumina sequence data indicate that the shorter Illumina reads (paired end or not) fail to resolve close paralogs (3). In contrast, assemblies of Sanger and 454 Titanium reads resolve similar proportions of close paralogs ($Ks < 0.1$).

454 EST sequencing requires generation of cDNA from RNA samples. Due to the advantageous characteristic of largely full-length cDNA synthesis, SMART™ (Clontech) technology has become the most widely adapted approach for cDNA synthesis. The method is based on synthesis of the first cDNA strand with an anchored oligo-dT using total RNA from eukaryote as template. The terminal C addition and template switching features of M-MLV reverse transcriptase, such as Superscript II Reverse Transcriptase, allow the synthesis of the first strand of cDNA (ss-cDNA) based on full-length mRNA templates (see Fig. 1). Then, long-distance PCR is performed to amplify ss-cDNA to double-stranded cDNA (ds-cDNA) using high-fidelity DNA polymerase.

In order to maximize the rate of gene discovery, it is necessary to employ a normalization strategy to remove high-abundance cDNA transcripts and generate a low-redundancy cDNA library. We use a simple and efficient duplex-specific nuclease (DSN) normalization strategy using the TRIMMER cDNA normalization kit (Evrogen), which is based on the denaturing and re-association of cDNAs, followed by digestion with DSN. The enzymatic degradation occurs primarily on the highly abundant cDNA fraction, and thus significantly and efficiently reduces the redundancy of a cDNA library. The single-strand cDNA fraction is then amplified twice by sequential PCR reactions according to the manufacturer's protocol.

However, owing to the nature of the pyrosequencing procedure adapted by Roche GS FLX 454, cDNA sequencing is especially challenging for Roche's 454 platform. Sequencing is sensitive to homopolymers, such as the polyA/Ts existing in the sequences, which not only result in excessive light production and cross talk between neighboring cells, but also lead to polymerase delay and incomplete extension of template in the next cycle (4). We have, thus, developed a customized strategy to overcome homopolymer

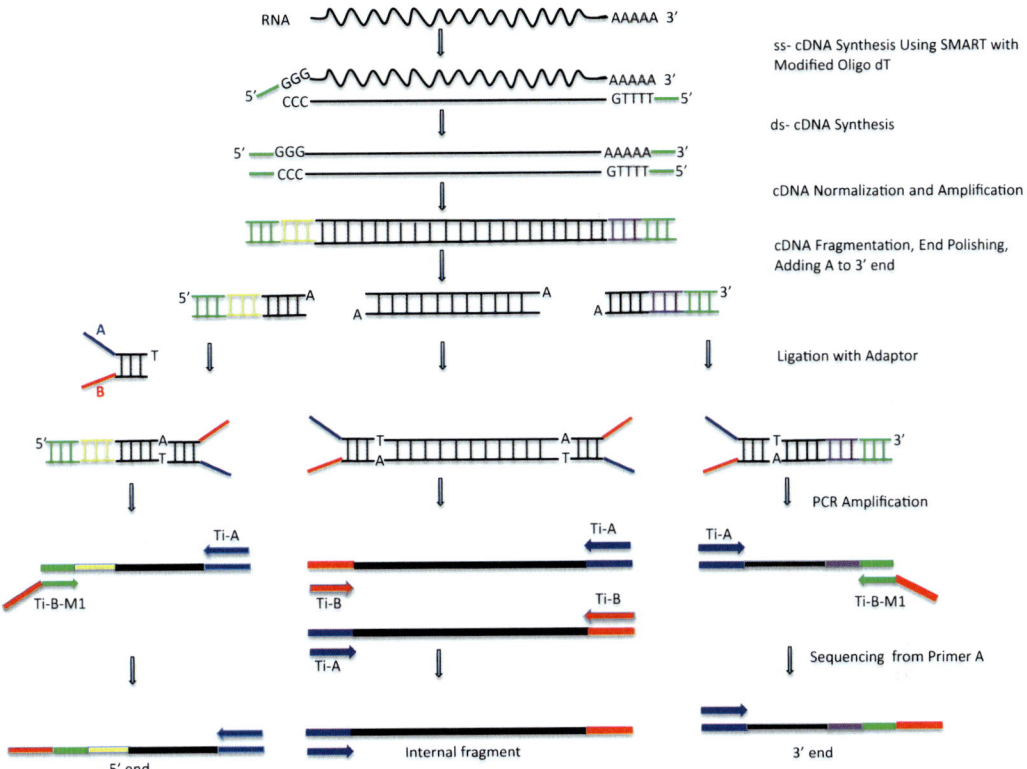

Fig. 1. Overview of the procedures. See text for details.

problems in 454 runs within cDNA samples. Firstly, we use a modified oligo-dT primer to prime the polyA+ tail of mRNA during ss-cDNA synthesis (5). We further break down the polyA tail using another modified oligo-dT primer during the ds-cDNA synthesis (5). These modified oligonucleotides are designed to effectively convert the long run of adenosine residues into a sequence that causes fewer problems for pyrosequencing. Secondly, we employ a different end polishing procedure so that only physically fragmented ends are polished (6). We design a unique adaptor modified from Illumina technology (7, 8) but using the 454 primer A and primer B sequences. The ligation of cDNA fragments with adaptors results in the unique placement of ligation products (see Fig. 1). After PCR amplification, only those fragments with the A primer at one end and the B primer at another end will be amplified. Lastly, because the tag sequence (M1 primer sequence, used for amplification of normalized cDNA) is in both ends, it is possible to separate the internal fragments from the 5′ and 3′ end fragments. The resulting library is a double-stranded DNA library instead of the standard single-strand DNA library, making size selection much easier and more efficient (see Fig. 1).

The overall workflow is described as follows (see Fig. 2). Firstly, total RNA is extracted from a biological source with the use of *TRIzol®* reagent and purified with the Qiagen RNeasy mini column, coupled with on-column DNase I treatment. Then, ss-cDNA synthesis is performed using SMART™ cDNA synthesis technique with modified oligo dT primer to prime polyA+ of total RNA. ds-cDNA synthesis converts ss-cDNA to ds-cDNA using PCR amplification following the Trimmer cDNA normalization manual. Normalized cDNA is sonicated to 500–800-bp fragments and the fragmented ends are polished and ligated with adaptors. The optimal ligation products are selectively amplified and subjected to two rounds of size selection, including gel electrophoresis and AMPure SPRI bead purification. With this novel cDNA library preparation strategy, we have successfully obtained approximately 250–330 Mb total output per half plate with a median read length of 480 bp, which is substantial improvement over standard ss-cDNA libraries.

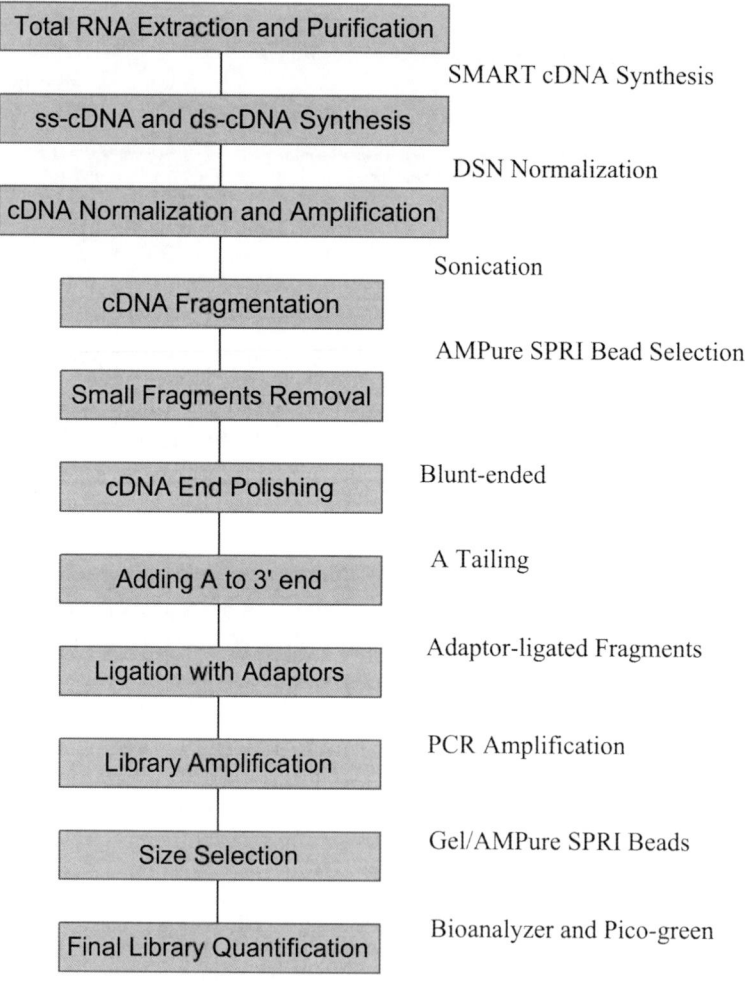

Fig. 2. Overview of the workflow. The *words in the boxes* indicate the experiment steps and the *text descriptions outside of boxes* indicate the experimental approaches.

2. Materials

All reagents, including buffers, dNTPs, primers, and enzymes, should be appropriately stored at the conditions recommended by manufacturers (see Note 1). RNA needs to be stored at –80°C. cDNA and all intermediate-stage samples generated during the process can be stored at either 4°C or 20°C. Essential equipment items include the Eppendorf PCR thermo-cycler, heat block, centrifuge, Nanodrop, and Agilent Bioanalyzer.

2.1. Total RNA Extraction and Purification

1. *TRIzol*® reagent (Invitrogen) (see Note 2).
2. Chloroform.
3. 100% Ethanol.
4. RNase Mini Kit (Qiagen).
5. RNase-free DNase set (Qiagen).

2.2. First-Strand and Second-Stranded cDNA Synthesis

1. 10 mM dNTPs.
2. Nuclease-free water.
3. Superscript® II Reverse Transcriptase (200 U/µL) (Invitrogen).
4. Phusion® Hot Start High Fidelity DNA Polymerase (NEB).
5. Primer SMART IIA (use for ss-cDNA synthesis, see Note 3):
   5′-AAGCAGTGGTATCAACGCAGAGTACGCrGrGrG-3′.
6. Primer CDSIII-1M (use for ss-cDNA synthesis):
   5′-AAGCAGTGGTATCAACGCAGAGT(T)4G(T)9C(T)10 VN-3′.
7. Primer SMART IIA_DNA (use for ds-cDNA synthesis, see Note 4):
   5′-AAGCAGTGGTATCAACGCAGAGTACGCGGG-3′.
8. Primer CDSIII-2M (used for ds-cDNA synthesis):
   5′-AAGCAGTGGTATCAACGCAGAGT(T)4GTC(T)4GTT CTG(T)3C(T)4VN-3′.

2.3. cDNA Normalization and Amplification

1. Trimmer cDNA normalization kit (Evrogen).
2. 10 mM dNTPs.
3. Nuclease-free water.
4. Phusion® Hot Start High Fidelity DNA Polymerase.
5. Primer M1: 5′-AAGCAGTGGTATCAACGCAGAGT-3′.
6. Primer M2: 5′-AAGCAGTGGTATCAACGCAG-3′.

2.4. cDNA Fragmentation, End Polishing, Adding A to 3′ End

1. Misonix Ultrasonic Liquid Processors sonicator.
2. AMPure SPRI beads (Beckman).
3. 10 mM dNTPs.
4. Nuclease-free water.
5. T4 DNA polymerase (3 U/μL, NEB).
6. Klenow fragment of DNA polymerase I (5 U/μL, NEB).
7. Klenow Fragment (3′–5′ exo-) (50 U/μL, NEB).

2.5. Ligation with Adaptors and Amplification

1. 10 mM dNTPs.
2. Nuclease-free water.
3. T4 DNA ligase (2,000 U/μL, NEB).
4. Adaptor 454-1 (* indicates phosphorothioate bond, see Note 5):

   5′-CCATCTCATCCCTGCGTGTCTCCGACTCAGGCTCT TCCGATC*T-3′.

5. Adaptor 454-2 (p indicates phosphate, see Note 5):

   5′-pGATCGGAAGAGCCTGAGACTGCCAAGGCA CACAGGGGATAGG-3′.

6. Primer Ti-A: 5′-CCATCTCATCCCTGCGTGTCTCCGAC TCAG-3′.

7. Primer Ti-B: 5′-CCTATCCCCTGTGTGCCTTGGCAGTCT CAG-3′.

8. Primer TiB-M1: 5′-CCTATCCCCTGTGTGCCTTGGCAGT CTCAGAAGCAGTGGTATCAACGCAGAGT-3′.

2.6. Final Size Selection and Library Quantification

1. GTG® SeaKem® agarose.
2. SYBR-safe dye.
3. 100 bp DNA ladder (NEB).
4. AMPure SPRI beads.
5. QIAquick Gel extraction kit (Qiagen).
6. Bioanalyzer DNA 7500 chip.

3. Methods

3.1. Total RNA Extraction and Purification

1. For plant tissues, we recommend using mortar and pestle to grind the tissues with liquid nitrogen. Add 1 mL of *TRIzol*® Reagent directly to the tube containing the frozen tissue powder (approximately 500 μL). Vortex mixture thoroughly to homogenize tissues. Incubate mixture for 5 min at room temperature.

2. Add 1/5 volume (200 µL) of chloroform to each sample. Immediately shake the mixture vigorously for 15 s. Centrifuge the mixture for 15 min at 4°C at 12,000×*g*.

3. Transfer upper aqueous phase (approximately 600 µL) to labeled RNase-free 1.7-mL tube. Precipitate RNA by adding 0.53× volume (approximately 320 µL) of 100% ethanol. Gently pipette to mix.

4. Transfer mixture, including any precipitate that may have formed, to Qiagen RNeasy mini-spin column. Proceed immediately. Follow manufacturer's instructions in the next few steps.

5. Centrifuge for 30 s at 12,000×*g*. Discard flow through.

6. On-Column DNase Treatment: Wash the column with 350 µL of Buffer RW1. Spin for 30 s. Discard the flow through. Pipette the DNase I/RDD Buffer mixture 80 µL (add 70 µL of RDD Buffer to 10 µL of DNase I stock solution; gently mix) directly onto the RNeasy silica-gel membrane of the RNeasy mini-spin column. Incubate for 15 min at room temperature. Wash the column with another 350 µL of Buffer RW1. Spin for 30 s. Discard the flow through.

7. Wash the column twice with 500 µL of Buffer RPE each. Spin for 30 s. Discard flow through.

8. Centrifuge for 1 min to dry the column. Transfer column to a new 1.5-mL tube.

9. Apply 30 µL of RNase-free water directly onto silica-gel membrane. Incubate for 1 min at room temperature. Spin for 1 min.

10. The elution now contains purified total RNA. Determine RNA concentration with NanoDrop. Determine integrity using Bioanalyzer/RNA 6000 Nano kit.

3.2. First-Strand and Double-Stranded cDNA Synthesis

3.2.1. First-Strand cDNA Synthesis

1. Prepare total RNA/primer mixture in a 0.2-mL PCR tube. The components per reaction are as follows: 1 µL CDSIII-1M primer, 1–1.5 µg total RNA (maximum volume of 3 µL), and 1 µl SMART™ IIA primer. Add DNase and RNase-free water to bring a final volume of 5 µL. Mix gently.

2. Incubate for 3 min at 65°C and quick chill for 2 min on ice. Spin briefly. Keep on ice.

3. At room temperature, prepare first-strand master mix. The component per reaction is as follows: 2 µL 5× First-Strand Buffer, 1 µL 10 mM dNTPs, 1 µL 100 mM DTT, and 1 µL SuperScript II. Gently pipette to mix. Spin briefly.

4. Incubate the reaction in PCR thermo-cycler for 1 h at 42°C. Then, inactivate the enzyme activity by incubating the reaction for 15 min at 70°C.

5. Remove from thermo-cycler. Spin briefly. Place samples on ice. Immediately proceed to ds-cDNA synthesis.

1. Prepare second-strand master mix on ice. The components per reaction are as follows: 69.5 µL Nuclease-free water, 20 µL 5× HF PCR Buffer, 2 µL 10 mM dNTP Mix, 2 µL SMART™ IIA_DNA primer, 2 µL CDSIII-2M primer, 1.5 µL MgCl₂, and 1 µL Phusion® Hot Start DNA Polymerase. Usually, two reactions per sample are required.

2. Add 2 µL ss-cDNA template to each tube. Add 98 µL second-strand master mix to make 100-µL total volume. Mix gently. Centrifuge briefly.

3. PCR program: 98°C for 1 min, and then 98°C for 7 s; 66°C for 20 s; and 72°C for 5 min for 18–20 cycles.

4. When the PCR cycling program is completed, check 5 µL of each PCR product alongside the 1 kb DNA ladder on regular 1% TBE agarose gel. A moderately strong smear of PCR amplification ranging from 500 bp to 4 kb indicates the good example of ds-cDNA for most plant species.

5. Purify the amplified ds-cDNA using the Qiagen PCR purification Kit following the manufacturer's instructions. The final elution volume is 20 µL EB buffer.

6. Ascertain the concentration of purified ds-cDNA by Nanodrop. 800 ng cDNA are needed for the normalization step. If the resulting concentration is less than 1 µg in 20 µL, concentrate using a Microcon® YM-30 column.

3.3. cDNA Normalization and Amplification

3.3.1. cDNA Normalization

1. The following procedures are based on the Evrogen Trimmer cDNA normalization kit manual with a few modifications.

2. Prepare the cDNA hybridization step. The components per reaction for three reactions are as follows: 800 µg (≤9 µL) cDNA from the previous step, 3 µL 4× hybridization buffer, and add H₂O to a final volume of 12 µL.

3. Aliquot 4 µL cDNA hybridization mix into three individual PCR tubes and overlay each with a drop of sterile mineral oil; centrifuge briefly to collect liquid and separate phases.

4. Incubate the tubes at 98°C for 2 min in a thermal-cycler to denature the ds-cDNA, and then at 68°C for 5 h. Proceed immediately to the next step.

5. Dilution of the DSN and checking of the DSN activity before starting a new tube are based on the manual.

6. DSN treatment: Near the end of the hybridization period, preheat the DSN master buffer at 68°C for 5 min. Prepare ½ and ¼ strength dilutions of the DSN using DSN storage buffer as the diluents; store on ice until ready to use. At the end of the hybridization period, add 5 µL preheated master buffer to each tube. Spin briefly in a benchtop centrifuge 2× and return immediately to the thermal-cycler. To three individual PCR tubes,

add 1 μL ½ dilution DSN enzyme, 1 μL ¼ dilution DSN enzyme, and 1 μL DSN storage buffer (as control) separately. Incubate at 68°C for 25 min. Add 10 μL of DSN stop solution to each tube, mix well, and spin briefly. Add 20 μL H_2O to each tube and then store at −20°C or proceed with the next step.

3.3.2. First Amplification of Normalized cDNA

1. Prepare the master mix for first amplification of normalized cDNA: Set up three separate PCR reactions for each reaction above; components per reaction are as follows: 34.85 μL H_2O, 10 μL 5× HF PCR buffer, 1 μL 10 mM dNTPs, 1.5 μL Primer M1 (10 μM), 0.75 μL 50 mM $MgCl_2$, 0.4 μL Phusion® Hot start Polymerase, and 1.5 μL Normalized cDNA from above. Amplify using the following thermal profile: 98°C for 1 min; 98°C for 7 s, 66°C for 20 s, and 72°C for 4 min for seven cycles.

2. Take out all tubes from thermal-cycler. Put the experimental tubes (½ and ¼ DSN-treated tubes) on ice. Remove a 5-μL aliquot from the control tube and set this aside.

3. Amplify the control tube (0 DSN, just storage buffer tube) for an additional two cycles (total = 9). Remove another 5-μL aliquot and set this aside.

4. Repeat step 3 twice more, producing aliquots from this tube that correspond to 11 and 13 cycles.

5. Load 5-μL aliquots from each 7th-, 9th-, 11th-, and 13th-cycle PCR products side by side on an agarose gel to evaluate optimum cycle number X as described in the manufacturer's instructions, where X = optimal numbers of cycles required for amplification of each of the control tube (see Note 6).

6. Return experimental tubes to the thermal-cycler and amplify for an additional $N + X$ cycles, where $N = X − 7$ (see Note 6).

7. Load 5 μL on a gel to determine which enzyme dilution treatment (½ or ¼) gives the best results, as described in the manual (see Note 7).

8. Once optimum enzyme treatment has been established, make dilutions of first amplified normalized cDNAs by adding 5-μL cDNA aliquots into 50 μL H_2O. This will be used as the template for the second amplification of normalized cDNA.

3.3.3. Second Amplification of Normalized cDNA

1. Set up two separate 100 μL PCR reactions for each sample. The components per reaction are as follows: 69.7 μL H_2O, 20 μL 5× HF PCR buffer, 2 μL 10 mM dNTPs, 4 μL Primer M2 (10 μM), 1.5 μL 50 mM $MgCl_2$, 0.8 μL Phusion® Hot start Polymerase, and 2 μL first amplified diluted cDNA from above.

2. Amplify using the following PCR program: 98°C for 1 min; 98°C for 7 s, 64°C for 20 s, and 72°C for 4 min for 12 cycles.

3. Pool the products, purify on a Qiagen Qiaquick column, elute in 50 µL EB buffer, and quantify them using a Nanodrop. Normalized cDNA can be stored at –20°C.

3.4. cDNA Fragmentation, End Polishing, Adding A to 3′ End

3.4.1. cDNA Fragmentation by Sonication

1. For each sample, dilute 2.0–2.5 µg second amplified cDNA from above into EB buffer for a final volume of 100 µL.

2. Load the diluted samples into the Misonix Ultrasonic Liquid Processors sonicator, ensuring that the liquid level in the tubes is below the liquid level in the sonicator. Ensure that the tubes are equally spaced around the center of the cup horn, and at equal depths in the water. Close the latch on the sonicator door.

3. Sonicate the tubes at the following intervals at amplitude of 1. Use the following program for the sonicator: 30-s *cycles*, 30-s *rest time*, and 40-s *processing time*.

4. Check the result of sonication by running 2.5 µL of each of sonicated cDNA on regular 1% TBE agarose gel. The target size of fragments ranges from 300 to 1,000 bp.

3.4.2. Small Fragment Removal

1. This process is performed following the AMPure beads manual. Note that each bottle of AMPure beads has to be calibrated before starting the new bottle. The target size removal is the fragments less than 500 bp. So use the ratio of beads that will remove fragments less than 500 bp (usually 0.5× or 0.55×).

2. Measure the volume of the samples using the pipettor. Add EB Buffer to a final volume of 100 µL. Add the needed amount of AMPure beads (volume is relative to the amount of starting sonicated cDNA, usually will be 50 or 55 µL). Vortex for 10–20 s followed by a quick centrifuge. Incubate for 5 min at room temperature.

3. Using a Magnetic Particle Collector (MPC), pellet the beads against the wall of the tube. Let the tubes stand in the MPC for 2 or 3 min.

4. Remove supernatant with a pipette carefully. Wash beads twice with 500 µL of 70% ethanol. Incubate each wash at room temperature for 30 s before removing with a pipette.

5. Remove all the supernatant carefully. To dry completely, place on a preheated 45–50°C heating block for about 5 min.

6. Add 20 µL EB buffer to each tube. Vortex each tube for 10–20 s and then centrifuge it briefly. This step elutes the cDNA from the AMPure beads.

7. Using the MPC, pellet the beads against the wall of the tube once more, and transfer the supernatant containing the purified, sonicated cDNA to a new tube. Quantify cDNA

concentration using a Nanodrop. If the amount of cDNA is greater than 200 ng, proceed to the next step (see Note 8).

3.4.3. cDNA End Polishing

1. Polish the fragmented cDNA to ensure that all ends are blunted. Prepare the end-polishing reaction master mix by combining the following in a tube at room temperature: 20 µL water, 5 µL 10× NEB2 buffer, 5 µL 10× BSA, 2 µL 10 mM dNTPs, 1.4 µL T4 DNA polymerase (4 U), and 0.8 µL Klenow fragment of DNA polymerase I (4 U). Aliquot 34.2 µL master mix into each sample tube. Add 200–500 ng fragmented cDNA from the previous step. Mix gently (see Note 9).

2. Incubate the reaction at room temperature for one and a half hours, and then heat to inactivate the enzyme activity at 70°C for 15 min.

3. The reaction products are purified by Qiaquick column. Elution volume is 32 µL.

3.4.4. Adding A to 3′ End

1. Prepare the adding A master mix: 2.7 µL water, 5 µL 10× NEB2 buffer, 10 µL 1 mM dATP, and 0.3 µL Klenow 3′–5′ exo (15 U). Aliquot 18 µL master mix into each sample tube. Add 32 µL end-polished cDNA from the previous step to make the reaction total volume of 50 µL. Mix gently.

2. Incubate the reaction at 37°C for 30 min.

3. The reaction products are purified by Qiaquick column. Elute with 30 µL EB buffer.

3.5. Ligation with Adaptors and Amplification

3.5.1. cDNA Ligation with Adaptors

1. Prepare the ligation master mix: 13.6 µL water, 5 µL T4 DNA ligase buffer, 1 µL of mix adaptor of 454-1 and 454-2, and 0.4 µL T4 ligase. Add 20 µL master mix into each sample tube. Add 30 µL 3′ A-added cDNA from the previous step to make the reaction total volume of 50 µL. Mix gently.

2. Incubate the reaction at 16°C for 30 min.

3. The reaction products are purified by Qiaquick column. Elute with 30 µL EB buffer.

3.5.2. PCR Test of Ligation

1. Set up four PCR test tubes for each sample and label them as 1, 2, 3, and 4. Add 1 µL ligation product from the previous step into each tube. Add 2 µL Ti-A primer into the tube 1, 2 µL Ti-B primer into the tube 2, 1 µL Ti-A primer and 1 µL Ti-B primer into the tube 3, and 1 µL Ti-A primer and 1 µL Ti-B-M1 primer into the tube 4.

2. Prepare the PCR master mix. Assemble the component per reaction as follows: 34.75 µL water, 10 µL 5× HF PCR buffer, 1 µL 10 mM dNTPs, 0.75 µL 50 mM $MgCl_2$, and 0.5 µL Phusion® Hot Start Polymerase. Aliquot 47 µL master mix into each sample tube. Gently mix.

3. Amplify using the following PCR program: 98°C for 30 s; and then 98°C for 7 s, 65°C for 30 s, and 72°C for 1 min for 16 cycles.

4. Check the result of the PCR test by running 5 μL of each PCR product on an agarose gel. The expected results are no products from the tube 1 and 2 and a nice smear of PCR products ranging from 500 to 1,000 bp from tubes 3 and 4.

5. If the PCR test result is as expected, then proceed to the next step for final library amplification.

3.5.3. Final Library Amplification

1. Set up two PCR tubes for each sample, and label them as 1 and 2. Add 2 μL ligation product from the previous step into each tube. Add 2 μL Ti-A primer and 2 μL Ti-B primer into the tube 1 for amplifying cDNA internal fragments. Add 2 μL Ti-A primer and 2 μL Ti-B-M1 primer into the tube 2 for amplifying 5′ end and 3′ end of cDNA fragments (see Note 10).

2. Prepare the final library amplification PCR master mix. Assemble the component per reaction as follows: 69.5 μL water, 20 μL 5× HF PCR buffer, 2 μL 10 mM dNTPs, 1.5 μL 50 mM $MgCl_2$, and 1 μL Phusion® Hot Start Polymerase. Aliquot 94 μL master mix into each sample tube. Mix gently.

3. Amplify using the following PCR program: 98°C for 30 s; 98°C for 7 s, 65°C for 30 s, and 72°C for 1 min for 16 cycles.

4. The reaction products are purified by Qiaquick column. Elute with 14 μL EB buffer. Quantify the purified products using a Nanodrop.

5. If the total cDNA amount of tube 1 and tube 2 from each sample is in the range of 3–6 μg, then proceed to the next step for final size selection.

3.6. Final Size Selection and Library Quantification

1. Prepare a 0.8% agarose gel using SeaKem® GTG® agarose/1× TAE buffer with SYBR-SAFE Dye.

2. Prepare samples for gel loading. For each sample, mix up to 24 μL samples (3–6 μg of cDNA) with 8 μL loading dye. Prepare one aliquot of ladder per sample, according to the following: 5 μL water, 3 μL Tit-gel Dye, and 2 μL 100 bp DNA Ladder.

3. Loading samples: Leave at least one well between each ladder and sample, placing a ladder before each sample. Run the gel at 100 V for 2 h.

4. Place the gel on the DarkReader. Carefully cut the PCR products ranging from 500 to 800 bp. Put the gel slice into a 2-mL tube.

5. Isolate the cDNA from the gel slice according to the Qiagen Qiaquick Gel Isolation Kit instructions for the next few steps.

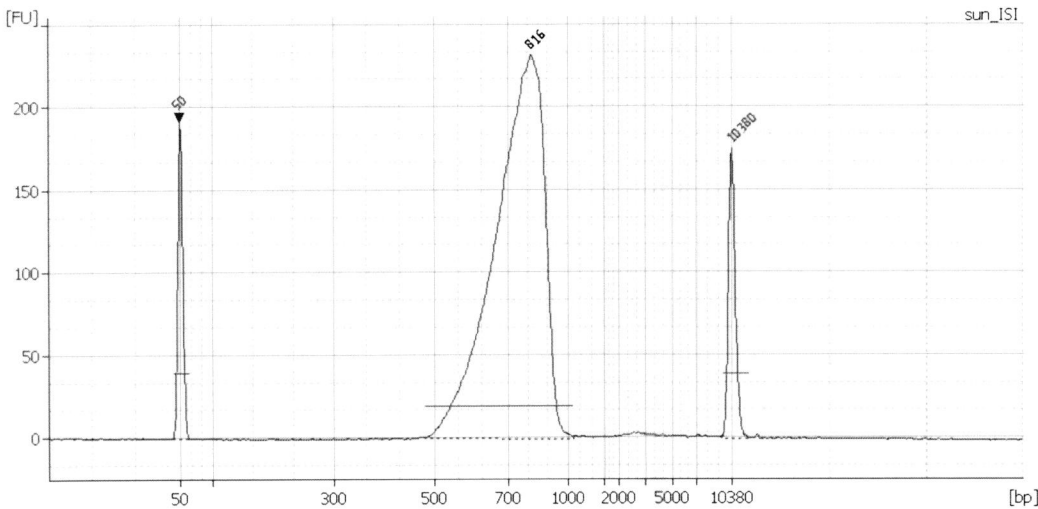

Fig. 3. Electrophrogram of the final cDNA library assessed by a Bioanalyzer DNA7500 chip. The electrophrogram shows the high quality of the final library with the average size of 816 bp and no noise before 500 bp and after 1,000 bp.

6. Add 3× volume of Buffer QG to the tube with gel slice. Put into the turboMix to shake for 5 min until the gel dissolves completely.

7. Add 1× volume of isopropanol and mix. Apply the sample to QIAquick column and spin for 1 min. Discard the flow through. Add 500 µL Buffer QG to the column and spin for 1 min.

8. Wash the column twice with 500 µL PE buffer. Incubate PE buffer for 2–5 min before spinning. Discard the flow through. Spin for additional 1 min to dry the column.

9. Elute the column twice with 25 µL EB buffer each.

10. Add 25 µL (0.5×) AMPure SPRI beads to the eluted sample. This step will remove any fragments less than 500 bp. Wash the beads with 12 µL EB buffer.

11. Assess the size distribution of the completed library using Bioanalyzer with an Agilent DNA 7500 chip (see Fig. 3).

12. Quantify the final library concentration using Quan-It PicoGreen.

4. Notes

1. It is assumed that users have knowledge of molecular biology techniques and safe laboratory practices. Before undertaking a new protocol or using unfamiliar reagents, users should review relevant Material Safety Data Sheets to identify potential hazards and recommended precautions. For background in general molecular biology, please see ref. 9.

2. For all steps involved in RNA isolation, it is important to guard against sources of dust and nucleases. All reagents must be prepared with RNase-free water using RNase-free chemicals and DNase- and RNase-free plastics. In addition, decontaminate both workspaces and pipettes with RNase Zap, according to manufacturer's instructions. Among the reagents used in the whole process, *TRIzol*® contains phenol, which is toxic in contact with skin and if swallowed. Thus, it should be handled with caution and should therefore be manipulated under a fume hood using gloves.

3. SMART IIA primer is used for ss-cDNA synthesis. This is the RNA-based primer and the last three Gs need to be RNA bases if you synthesize your own. Therefore, this primer needs to be diluted with RNase-free water, aliquoted, and stored at –80°C.

4. SMART IIA_DNA primer is used for ds-cDNA synthesis. It is used together with CDSIII-2M primer to further break down the polyA/T homopolymer sequences. This primer is a regular DNA-based primer.

5. The design of the Adaptor 454-1 and Adaptor 454-2 used for the ligation is based on the Illumina technology. So when synthesizing your own adaptors, a phosphorothioate bond is needed for the Adaptor 454-1 and a 5′ end phosphate is needed for the Adaptor 454-2. These adaptors are modified for further 454 sequencing and thus are based on the 454 primer A and primer B sequences. The part of 454 sequence from Adaptor 454-1 is in the same orientation as the original 454 primer A: 5′-CCATCTCATCCCTGCGTGTCTCCGACTCAG-3′. The part of 454 sequence from Adaptor 454-2 is the reverse-complementary sequence of the original 454 primer B: 5′-CTGAGACTGCCAAGGCACACAGGGGATAGG-3′. Prepare the mix of adaptor by mixing the primers at the final concentration of 10 μM each. All primers should be purified by high-performance liquid chromatography (HPLC) grade.

6. For each sample, determine optimal cycle number, "X." The optimal cycle should show a fairly dark smear in the 500 bp to 4 kb range, and no nonspecific amplification above the 8 kb range. In most of our cases, PCR product from 9th cycle's control tube yielded the optimum results, so $X=9$. Then, the number of additional cycles to run is $N+X$, where $N=X-7$. For instance, if $X=9$, $N=9-7=2$. Then, $2+9=11$; more cycles are needed for these 7-cycle experimental tubes and 18 cycles in total for experimental tubes.

7. During the DSN normalization step, most of the time, both the ½ and ¼ DNS dilution tubes contain good amplification products and can be combined for the next cDNA fragmentation step.

8. Small fragment removal using AMPure beads is performed twice in the protocol. The removal of the small fragments after cDNA fragmentation ensures that fragments of the right size are ligated to the adaptor. If small fragments remain in the final cDNA library, they will be preferably amplified during the following emPCR amplification and this will result in short sequence reads.

9. For the end-polishing step, it is important that no kinase is added (7) so that the "new" ends resulting from the fragmentation/polishing will have 5′ phosphates to ligate with the adaptor. Original 5′ end and 3′ end with the tag sequence (corresponding to the M1 sequence) will not ligate with the adaptor because they do not bear 5′ phosphates.

10. With a separate PCR amplification of ligation products, cDNA internal fragments and end fragments are generated. The primer A sequence, which corresponds to the 454 Titanium sequencing primer, will always correspond to internal cDNA sites, thus minimizing the chance of hitting polyA/T homopolymer sequences.

Acknowledgments

The authors thank James Ford, Jade Buchanan-Carter, Zach Smith, and Keithanne Mockaitis for technical support for sequencing, and Jie Huang for assistance with manuscript preparation. This work was supported in part by a Roche Applied Science/454 Life Sciences 1 GB sequencing grant program to Z.L. and L.H.R., a Natural Sciences and Engineering Research Council of Canada (NSERC) grant #353026 to L.H.R, and the Indiana METACyt Initiative of Indiana University, funded in part through a major grant from the Lilly Endowment, Inc.

References

1. Vera JC, Wheat CW, Fescemyer HW et al (2008) Rapid transcriptome characterization for a non-model organism using 454 pyrosequencing. Mol Ecol 17:1636–1647

2. Jarvie T, Harkins T (2008) Transcriptome sequencing with the genome sequencer FLX system. Nat Methods Application Notes September Issue

3. Lai Z, Kane NC, Kozik A, Hodgins, K et al (2012) Genomics of Compositae weeds: EST libraries, microarrays and evidence of introgression. 2011 Nov. 4 [Epub ahead of print], Am J Bot

4. Metzker M (2010) Sequencing technologies – the next generation. Nat Rev Genet 11:31–46

5. Beldade P, Rudd S, Gruber JD, Long AD (2006) A wing expressed sequence tag resource for *Bicyclus anynana* butterflies, an evo-devo model. BMC Genomics 7:130

6. Meyer E, Aglyamova GV, Wang S et al (2009) Sequencing and *de novo* analysis of a coral larval transcriptome using 454 GSFlx. BMC Genomics 10:219

7. Bentley DR, Balasubramanian S, Swerdlow HP et al (2008) Accurate whole human genome sequencing using reversible terminator chemistry. Nature 456:53–59

8. Quail MA, Kozarewa I, Smith F et al (2008) A large genome center's improvements to the Illumina sequencing system. Nat Methods 5:1005–1010

9. Sambrook J, Fritsch E, Maniatis T (1989) Molecular cloning: a laboratory manual. Cold Spring Harbor, New York

Chapter 9

RAD Paired-End Sequencing for Local De Novo Assembly and SNP Discovery in Non-model Organisms

Paul D. Etter and Eric Johnson

Abstract

Restriction-site Associated DNA (RAD) markers are rapidly becoming a standard for SNP discovery and genotyping studies even in organisms without a sequenced reference genome. It is difficult, however, to identify genes nearby RAD markers of interest or move from SNPs identified by RAD to a high-throughput genotyping assay. Paired-end sequencing of RAD fragments can alleviate these problems by generating a set of paired sequences that can be locally assembled into high-quality contigs up to 1 kb in length. These contigs can then be used for SNP identification, homology searching, or high-throughput assay primer design. In this chapter, we offer suggestions on how to design a RAD paired-end (RAD-PE) sequencing project and the protocol for creating paired-end RAD libraries suitable for Illumina sequencers.

Key words: Genetic mapping, Population genomics, Genotyping, Single nucleotide polymorphisms, Next-generation sequencing, RAD-seq, Non-model species

1. Introduction

Although the cost of sequencing a genome has drastically decreased, full-genome sequencing remains out of reach for most genetic studies of populations, which may have tens to thousands of individuals. Thus, complexity reduction sequencing approaches such as *Restriction-site Associated DNA* (RAD) have been used to sample genomes at discrete loci (1–3). RAD tags are short DNA fragments flanking a particular restriction site and defined by the restriction site on one extremity and randomly sheared DNA on the other. Sequencing of RAD tags (RAD-seq) focuses next-generation sequencers on a small portion of a genome, allowing high coverage of each RAD tag to be rapidly achieved (4–7). Single (or simple)

François Pompanon and Aurélie Bonin (eds.), *Data Production and Analysis in Population Genomics: Methods and Protocols*,
Methods in Molecular Biology, vol. 888, DOI 10.1007/978-1-61779-870-2_9, © Springer Science+Business Media New York 2012

nucleotide polymorphisms can then be discovered by comparing RAD tags between samples.

RAD-seq can be used in organisms both with or lacking a reference genome, as the sequence comparison is between RAD sequences. However, the short reads of current sequencers impose some bottlenecks in the use of RAD-seq for studying the genetics of a population. First, the short reads remain somewhat anonymous without the help of a reference genome. That is, it is difficult to move from a RAD sequence of interest to finding nearby genes. Second, some researchers desire to move SNPs identified by RAD to a high-throughput genotyping assay, such as GoldenGate or Sequenom (8–11). These assays require flanking genomic sequence for primer design, and short read lengths are not sufficient for this step. Third, high-density markers are sometimes required and RAD complexity reduction generally does not provide sufficient sequence to find the number of markers needed.

These problems can be avoided by using a RAD paired-end (RAD-PE) contig protocol. Paired-end sequencing of a library sequences both ends of each RAD fragment, and explicitly links one sequence with the other. Paired-end sequencing of RAD fragments generates a set of paired sequences in which the first sequence corresponds to the restriction site end and the second sequence to the random sheared end. Thus, the second reads randomly sample genomic DNA some distance from the restriction site (see Fig. 1). This allows a computational step of local assembly of the second reads, performed sequentially on each restriction cut site sequence

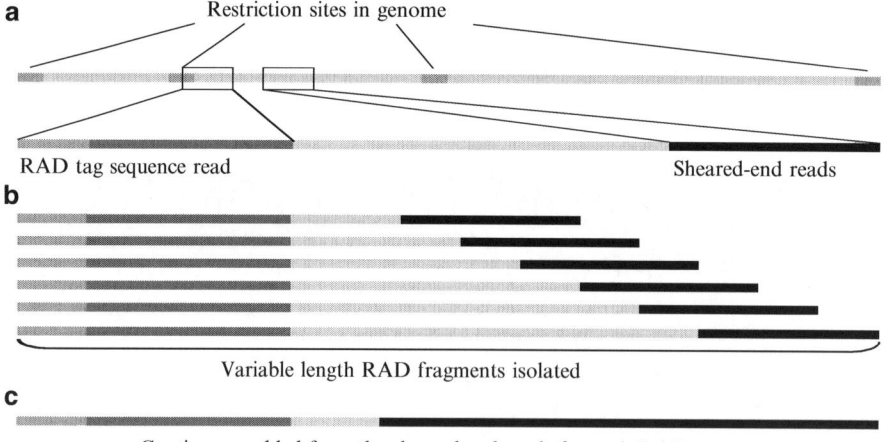

Fig. 1. Local assembly with RAD paired-end contig libraries. (a) DNA fragments created by RAD tag library preparation have a restriction site (*light gray*) and associated sequence (*dark gray*) at one end, and a random sheared end sequence (*black*) at the other. (b) Paired-end sequencing of RAD tag libraries allows the assembly of the sheared end sequences into contigs (c), one RAD site sequence at a time. The distance at which the random-end sequence lies, and hence the length of the contigs assembled, is dictated by the distribution of fragments isolated during the gel extraction step in the protocol.

in turn. The small number of second reads sharing a restriction cut site sequence can be gathered, sent to a de novo assembler, and the resulting high-quality contigs used for SNP identification, homology searching of related genomes, or high-throughput assay primer design. Here, we provide some guidelines for designing a RAD-PE project and the protocol for creating RAD-PE libraries suitable for sequencing on Illumina Genome Analyzer IIx and HiSeq 2000 sequencers (7).

The first step in project design is to determine the number of markers needed to attain the project goals. Once the number of markers needed is decided, there are two choices to be made that will determine the number of SNP markers identified for a given genome size and polymorphism rate. First, the choice of restriction enzyme used to digest the genome will determine the number of cut sites, which will in turn produce two RAD sequences each. The number of restriction enzyme cut sites in a genome is dependent on the length of the restriction site sequence (typically four to eight nucleotides), the GC content of the restriction site sequence and of the genome of interest, and if the enzyme is sensitive to DNA methylation. The second choice is the desired average contig length. Longer contigs will increase the amount of sequence in which to identify SNPs, and provide more information for finding genes or searching related reference genomes. Of course, the cut site frequency and contig length must be balanced by the cost of the sequencing needed to create the contigs at the coverage needed to make accurate SNP calls. In the absence of precise information about the genome size, GC content of the genome, or polymorphism rate between samples, a trial project might be necessary to test several enzyme choices. Because one lane of Illumina sequencing typically produces far more reads than is needed for a single sample, RAD adapters contain barcodes that allow multiplexing of samples and the association of reads back to a particular sample after the run is completed.

Conversion of the raw sequence reads into long contigs is straightforward, although there are many areas of possible further optimization. Generally, a lane of multiplexed samples is first split into sample-by-sample sequence files by the barcode sequence in the adapter. For each sample, the RAD site reads are then stacked (see Fig. 1); that is, the number of instances of each unique sequence is tracked. These reads may be filtered to eliminate sequences with too many instances (likely repetitive genomic sequences) or too few (unlikely to create good contigs or have sufficient depth for SNP detection). Then, the paired-end reads associated with a single RAD read are collected and sent to a de novo assembly tool, such as Velvet (12). The resulting contig or contigs are then collected from the output, stored with the contigs assembled from the previous RAD site sequences, and then the process repeated until all the RAD sites have had their associated paired-end sequences assembled. The contigs created can then be used as a reference sequence for alignment and SNP detection. Note that contigs can be compared between samples that are highly inbred, haploid, or otherwise known to be highly homozygous. But if heterozygous sequences are expected, the contigs will reflect only a single majority state at any particular nucleotide. In these circumstances, the original paired-end reads are aligned to the contigs using any short-read aligner (Novoalign (13) or BWA (14), for example) and the alignments used to find mismatches that meet quality and coverage thresholds and may be true polymorphisms in the sample.

As an example, a researcher might desire to create a high-throughput SNP assay of 1,536 markers for a plant species with a large genome (e.g., a 2 Gb diploid). Although the genome size is not precisely known, most plant genomes are largely made up of methylated repetitive DNA, with the remaining non-methylated gene space sequence typically being several hundred million base pairs. The mapping crosses planned do not involve varieties with a high polymorphism rate between them, which will reduce the number of polymorphic RAD markers found from the RAD tags. However, the low polymorphism rate increases the rate of SNP conversion to a high-throughput assay, which requires 50–100 bp of flanking sequence free of other polymorphisms, resulting in more usable markers. The enzyme used should be methylation sensitive in order to only create RAD tags in the gene space, and relatively infrequent given the small number of markers needed. The restriction enzyme *SbfI* is a reasonable choice, as the CCTGCAGG restriction site sequence is likely to be found every 30–70 kb, and *SbfI* is sensitive to the CpNpG methylation commonly found in plants. The number of RAD tags produced is likely to be around 10,000, and fewer than one million paired-end reads per sample would give good coverage over a contig of 300 nucleotides. When calculating the reads needed (10 k tags × 20× coverage × 300 bp contig length/100 bp read length = 600,000 reads), it is a good practice to add 50% more reads to overcome uneven read distribution over the RAD sites and between samples. If the polymorphism rate is 1 in 300, there would be 10,000 SNPs expected to be found distributed in the 10,000 RAD tags. However, some contigs will have no SNPs (perhaps ~3,500), and others will have 2 or more. Because the SNPs found need to be in the center 200 bp to have the flanking DNA required for conversion to a high-throughput assay, only two-thirds of the SNPs found will be in the right location. Finally, some of these SNPs will have additional SNPs in the flanking sequence, so the 10,000 SNPs first found will likely be reduced to less than 3,000 after filtering.

2. Materials

2.1. DNA Extraction, RNase A Treatment, and Restriction Endonuclease Digestion

1. DNeasy Blood and Tissue Kit (Qiagen).
2. RNase A (Qiagen).
3. High-quality genomic DNA from 2.1: 25 ng/μl (see Notes 1 and 2).
4. Restriction enzyme (NEB; see Note 3).

2.2. P1 Adapter Ligation, Sample Multiplexing, and DNA Shearing

1. NEB buffer 2.
2. rATP (Promega): 100 mM.
3. Barcoded P1 adapter: 100 nM stocks in 1× Annealing Buffer (AB; 10× AB: 500 mM NaCl, 100 mM Tris–Cl, pH 7.5–8.0). Prepare 100 μM stocks for each single-stranded oligonucleotide in 1× Elution Buffer (EB, Qiagen: 10 mM Tris–Cl, pH 8.5). Combine complementary adapter oligos at 10 μM each in 1× AB. Place in a beaker of water just off boil and cool slowly to room temperature (RT) to anneal. Alternatively, use a boil and gradual cool program in a PCR machine. Dilute to desired concentration in 1× AB (see Notes 4–6).
4. Concentrated T4 DNA Ligase (NEB): 2,000,000 U/ml.
5. QIAquick or MinElute PCR Purification Kit (Qiagen).
6. Bioruptor, nebulizer, or Branson sonicator 450.

2.3. Size Selection/ Agarose Gel Extraction and End Repair

1. Agarose.
2. 5× TBE: 0.45 M Tris–Borate, 0.01 M EDTA, pH 8.3.
3. 6× Orange Loading Dye Solution (Fermentas).
4. GeneRuler 100 bp DNA Ladder Plus (Fermentas).
5. Razor blades.
6. MinElute Gel Purification Kit (Qiagen).
7. Quick Blunting Kit (NEB).

2.4. 3′ dA Overhang Addition and PE-P2 Adapter Ligation

1. NEB buffer 2.
2. dATP (Fermentas): 10 mM.
3. Klenow Fragment (3′ to 5′ exo⁻, NEB): 5,000 U/ml.
4. rATP: 100 mM.
5. PE-P2 adapter: 10 μM stock in 1× AB prepared as P1 adapter described above (see Notes 4 and 5).
6. Concentrated T4 DNA Ligase.

2.5. RAD Tag Amplification/ Enrichment

1. Phusion High-Fidelity PCR Master Mix with HF buffer (NEB).
2. RAD amplification primer mix: 10 μM. Prepare 100 μM stocks for each oligonucleotide in 1× EB. Mix together at 10 μM in buffer EB (see Note 7).

3. Methods

The DNA fragments created by RAD tag library preparation have a restriction site at one end and are randomly sheared at the other. This arrangement, when combined with Illumina paired-end sequencing, results in each instance of a restriction site sequence being sampled many times by the first reads and the genomic DNA sequence in the nearby region being randomly sampled at a lower coverage by the second reads. The explicit linking of second reads that sample a genomic region with a common first read RAD sequence allows the second reads to be assembled on a local basis, one RAD site at a time (see Fig. 1).

We have modified the sequenced RAD tag protocol (4) in order to create paired-end compatible libraries by altering two key aspects of the protocol (7). First, as the size of contigs assembled from the paired-end reads is dependent on the size range of fragments selected during library construction, a wider distribution of fragments is isolated after shearing. Second, a longer, divergent PE-P2 adapter that possesses the reverse sequencing primer sequence is ligated to the variable end of the RAD tags before amplification, allowing the randomly sheared end of the RAD fragments to be sequenced by the second read.

The protocol for creating RAD paired-end libraries for contig assembly and SNP discovery in stickleback, reported in Etter et al. (7), is described here in detail as an example of RAD paired-end library preparation (see Note 8). We prepared barcoded *SbfI* libraries from 2 three-spine stickleback individuals from a phenotypically polymorphic population (High Ridge Lake, Alaska). This study showed that RAD-PE contigs are a robust way to generate long local assemblies at consistent regions of the genome between samples, making it a useful approach for comparative genomics, including marker discovery. Both homozygous and heterozygous alleles were successfully identified.

3.1. DNA Extraction, RNase A Treatment, and Restriction Endonuclease Digestion

1. We recommend extracting genomic DNA samples using the DNeasy Blood and Tissue Kit (Qiagen) or a similar product that produces very pure, high-molecular-weight, RNA-free DNA. Follow the manufacturer's instructions for extraction from your tissue type. Be sure to treat samples with RNase A following manufacturer's instructions to remove residual RNA. The optimal concentration after elution into buffer EB is 25 ng/µl or greater (see Notes 1 and 2).

2. Digest equal quantities of genomic DNA (0.1–1.0 µg) for each sample with the appropriate restriction enzyme in a 50-µl reaction volume, following the manufacturer's instructions. For example, for *SbfI* digestion, combine the following in a microcentrifuge

tube: 5.0 μl 10× NEB buffer 4, 0.5 μl *SbfI*-HF, DNA and H₂O to 50 μl (see Notes 9–11). Incubate reaction at 37°C for 60 min to overnight (follow manufacturer's recommendations).

3. Heat-inactivate the restriction enzyme for 20 min at 65°C (or following manufacturer's instructions). If the enzyme cannot be heat-inactivated, purify with a QIAquick column following manufacturer's instructions prior to ligation. QIAquick purification should work equally well, although conceivably the representation of fragments from distantly separated sites could be reduced when using an infrequent cutter. Allow reaction to cool to ambient temperature before proceeding to ligation reaction or cleanup (see Note 4).

3.2. P1 Adapter Ligation, Sample Multiplexing, and DNA Shearing

1. Ligate barcoded, restriction site overhang-specific P1 adapters onto complementary compatible ends on the genomic DNA created in the previous step. If barcoding only a few individuals or samples, choose barcodes whose sequences differ as much as possible from one another to avoid causing the Genome Analyzer software to lose cluster registry, as it is prone to do when it encounters a nonrandom assortment of nucleotides. For example, in Etter et al. (7), DNA samples from two stickleback individuals were uniquely barcoded using the barcodes AGAGT and CAGTC and were sequenced in a single lane on the GAII (see Note 6).

2. To each inactivated digest, add 1.0 μl 10× NEB buffer 4, 4.0 μl barcoded P1 adapter (100 nM), 0.6 μl rATP (100 mM), 0.5 μl concentrated T4 DNA Ligase (2,000,000 U/ml), and 3.9 μl H₂O, 60 μl total volume (see Note 12). Be sure to add P1 adapters to the reaction before the ligase to avoid re-ligation of the genomic DNA. Incubate reaction at room temperature for 30 min to overnight. Reduce the amount of P1 used in the ligation reaction if starting with less than 1 μg genomic DNA or if cutting with an enzyme that cuts less frequently than *SbfI* in stickleback (stickleback has close to 23,000 *SbfI* restriction sites in its 460-Mb genome, which are more than expected by chance). It is critical to optimize the amount of P1 adapter added when a given restriction enzyme is used for the first time in an organism (see Note 13).

3. Heat-inactivate T4 DNA Ligase for 10 min at 65°C. Allow reaction to cool slowly to ambient temperature before shearing.

4. Combine barcoded samples at an equal or otherwise desired ratio (see Note 14). Use a 100- to 300-μl aliquot containing 1–2 μg DNA total to complete the protocol and freeze the rest at –20°C in case you need to optimize shearing in the next step. For example, in Etter et al. (7), the two 60-μl ligation reactions were combined before shearing.

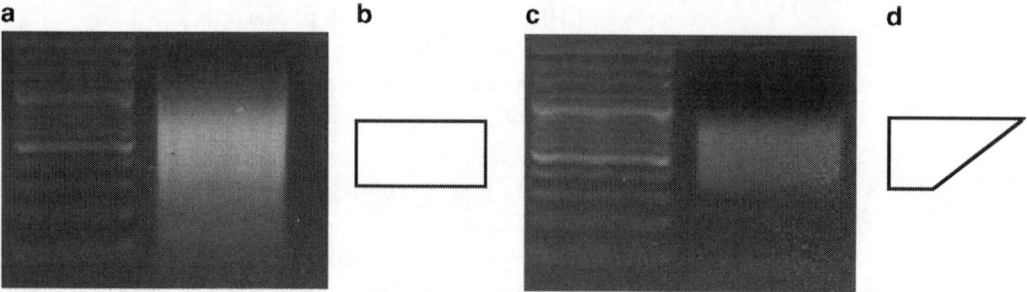

Fig. 2. RAD-PE library isolation steps. (**a**) Agarose gel image with the barcoded, multiplexed, and sheared library described in step 1 of Subheading 3.3 in the protocol (*lane 2*) and 2.0 µl GeneRuler 100 bp DNA Ladder Plus (*lane 1*, Fermentas—bright bands correspond to 500 bp and 1 kb). (**b**) Shape of fragments isolated from the gel in (**a**) that were carried through the rest of the library preparation protocol. (**c**) Agarose gel image with the final amplified RAD-PE contig library described in step 4 of Subheading 3.5 in the protocol (*lane 2*) and 2.0 µl Ladder (*lane 1*). (**d**) Trapezoidal shape of fragments isolated from the gel in (**c**) that were sequenced on the paired-end module of a Genome Analyzer II.

5. Shear DNA samples to an average size of 500 bp to create a pool of P1-ligated molecules with random, variable ends. This step requires some optimization for different DNA concentrations and for each type of restriction endonuclease. The following protocol has been optimized to shear stickleback DNA digested with either *Eco*RI or *Sbf*I using the Bioruptor and is a good starting point for any study (see Note 15). The goal is to create sheared product that is predominantly smaller than 1 kb in size (see Fig. 2a).

6. Dilute ligation reaction to 100 µl in water (or take a 100- to 300-µl aliquot from multiplexed samples) and shear in the Bioruptor 10× for 30 s on high following manufacturer's instructions. Make sure that the tank water in the Bioruptor is cold (4°C) before starting. All other positions in the Bioruptor holder not filled by your sample/s should be filled with balance tubes containing an equal volume of water.

7. Clean up the sheared DNA using a MinElute column following manufacturer's instructions. This purification is performed in order to remove the ligase and restriction enzyme and to concentrate the DNA so that the entire sample can be loaded in a single lane on an agarose gel. Elute in 20 µl EB.

3.3. Size Selection/
Agarose Gel Extraction
and End Repair

1. This step in the protocol removes free un-ligated or concatemerized P1 adapters and restricts the size range of tags to those that can be sequenced efficiently on an Illumina Genome Analyzer flow cell. Run the entire sheared sample in 1× Orange Loading Dye on a 1.25% agarose, 0.5× TBE gel for 45 min at 100 V, next to 2.0 µl GeneRuler 100 bp DNA Ladder Plus for size reference; run the ladder in lanes flanking the samples until the 300- and 500-bp ladder bands are sufficiently resolved from 200- and 600-bp bands (see Fig. 2a and Note 16).

2. Being careful to exclude any free P1 adapters and P1 dimers running at ~130 bp and below, use a fresh razor blade to cut a slice of the gel spanning 300–800 bp (see Fig. 2b and Note 17). Extract DNA using MinElute Gel Purification Kit following manufacturer's instructions with the following modifications: to improve representation of AT-rich sequences, melt agarose gel slices in the supplied buffer at room temperature (18–22°C) with agitation until dissolved (usually less than 30 min) (15); let column stand for 5 min after addition of buffer PE, before centrifugation. Elute in 20 μl EB into a microcentrifuge tube containing 2.5 μl 10× Blunting Buffer from the Quick Blunting Kit used in the following step (see Note 18).

3. The Quick Blunting Kit protocol converts 5′ or 3′ overhangs, created by shearing, into phosphorylated blunt ends using T4 DNA Polymerase and T4 Polynucleotide Kinase.

4. To the eluate from the previous step, add 2.5 μl dNTP mix (1 mM) and 1.0 μl Blunt Enzyme Mix. Incubate at RT for 30 min.

5. Purify with QIAquick column. Elute in 43 μl EB into a microcentrifuge tube containing 5.0 μl 10× NEB buffer 2.

3.4. 3′ dA Overhang Addition and PE-P2 Adapter Ligation

1. This step in the protocol adds an "A" base to the 3′ ends of the blunt phosphorylated DNA fragments using the polymerase activity of Klenow Fragment (3′ to 5′ exo⁻) and prepares the DNA fragments for ligation to the PE-P2 adapter, a "Y" adapter with divergent ends and a complementary 3′ dT overhang. This creates RAD-seq library template ready for amplification.

2. To the eluate from the previous step, add 1.0 μl dATP (10 mM) and 3.0 μl Klenow (exo⁻). Incubate at 37°C for 30 min. Allow the reaction to cool to ambient temperature.

3. Purify with a QIAquick column. Elute in 45 μl EB into a microcentrifuge tube containing 5.0 μl 10× NEB buffer 2.

4. To the eluate from the previous step, add 1.0 μl PE-P2 adapter (10 μM), 0.5 μl rATP (100 mM), and 0.5 μl concentrated T4 DNA Ligase. Incubate reaction at room temperature for 30 min to overnight.

5. Purify with a QIAquick column. Elute in 50 μl EB and quantify (see Notes 17 and 19).

3.5. RAD Tag Amplification/ Enrichment

1. In this step, high-fidelity PCR amplification is performed on P1 and PE-P2 adapter-ligated DNA fragments, enriching for RAD tags that contain both adapters and preparing them to be hybridized to an Illumina Genome Analyzer or HiSeq flow cell.

2. Perform a test amplification to determine library quality. In a thin-walled PCR tube, combine 10.5 μl H_2O, 12.5 μl

Phusion High-Fidelity Master Mix, 1.0 µl RAD amplification primer mix (10 µM), and 5–10 ng RAD library template (see Note 20). Perform 18 cycles of amplification in a thermal cycler: 30 s at 98°C, 18× [10 s at 98°C, 30 s at 65°C, 30 s at 72°C], 5 min at 72°C, and hold at 4°C. Run the entire PCR reaction in 1× Orange Loading Dye out on a 1.0% agarose gel next to the same amount of RAD library template used in the amplification and 2.0 µl GeneRuler 100 bp DNA Ladder Plus.

3. If the amplified product is at least twice as bright as the template (the template should appear dim, yet visible on the gel), perform a larger volume amplification (typically 50–100 µl), but with fewer cycles (12–14, to minimize bias), to create enough product to retrieve a large amount of the RAD tag library from a final gel extraction (see Note 21). Purify the large volume reaction with a MinElute column. Elute in 20 µl EB. If amplification looks poor, use more library template in a second test PCR.

4. The following purification step is performed to eliminate any artifactual bands that may appear due to an improper ratio of P1 adapter to restriction-site compatible ends (see Note 13) and to eliminate residual unligated PE-P2 adapters (see Note 4). Load the entire sample in 1× Orange Loading Dye on a 1.25% agarose, 0.5× TBE gel and run for 45 min at 100 V, next to 2.0 µl GeneRuler 100 bp DNA Ladder Plus for size reference (Fig. 2c). Being careful to exclude any free adapters or adapter dimers running at ~130 bp and below, use a fresh razor blade to cut a slice of the gel spanning ~350–850 bp in an inverted trapezoidal shape (see Fig. 2d and Note 22). Extract DNA using MinElute Gel Purification Kit following manufacturer's instructions, but melt agarose gel slices in the supplied buffer at room temperature and let the PE buffer wash stand for 5 min before centrifugation. Elute in 10–20 µl EB (see Note 23).

5. Quantify the DNA using a fluorometer to accurately measure the concentration. Concentrations will range from 1 to 20 ng/µl. Determine the molar concentration of the library by examining the gel image and estimating the median size of the library smear, which should be around 600 bp. Multiply this size by 650 (the average molecular mass of a base pair) to get the molecular weight of the library. Use this number to calculate the molar concentration of the library (see Note 24).

6. Sequence libraries on a paired-end flow cell on the Illumina Genome Analyzer or HiSeq following manufacturer's instructions.

4. Notes

1. Clean, intact, high-quality DNA is required for optimal restriction endonuclease digestion and is important for the overall success of the protocol. We have found that lower quality DNA can be used, but the starting amount will likely need to be increased because a large number of DNA fragments that have a correctly ligated P1 adapter (see the following steps of the protocol) may not end up in the proper size range when the starting DNA is partially degraded. When working with heavily degraded DNA samples is the only option, we have found that parameters of the protocol can be optimized (such as using more input DNA to start with and shearing less) to create usable libraries. These libraries often do not amplify as well as ones made with intact, high-molecular-weight genomic DNA. The "Best Practices" sections of the most recent Illumina Sample Prep Guides are a good resource for quantification, handling, and temperature considerations.

2. We recommend using a fluorescence-based method for DNA quantification in order to get the most accurate concentration readings. Since they bind specifically to double-stranded DNA, the dyes used in fluorometric assays are not as affected by RNA, free nucleotides, or other contaminants commonly found in DNA preparations (which can lead to inaccurate concentration predictions when using absorbance). If using another form of DNA quantification, such as UV spectrometer 260/280 absorbance readings, be sure to confirm the concentration by comparing a known calibration sample or running the sample on an agarose gel and comparing to a known quantity of DNA or ladder. We recommend checking the integrity of samples on a gel prior to embarking on this protocol regardless of the quantification method. Genomic DNA should consist of a fairly tight high-molecular-weight band without any visible degradation products or smears.

3. The choice of the particular restriction enzyme to use for a study is based upon several parameters, such as the desired frequency of RAD sites throughout the genome, GC content, depth of coverage necessary, and size of the genome. An average restriction endonuclease with an 8-bp recognition sequence would produce one tag every 64 kb in an organism with equal and random frequency of cut sites. Of course, the predicted and actual number of restriction sites in a genome can be quite different. For example, the predicted number of *Sbf*I sites (CCTGCAGG) in the stickleback genome is 7,069, whereas we have identified 22,829 sites found at an average distance of

20.2 kb between sites. Yet *Sgr*DI (CGTCGACG), with the same nucleotides as *Sbf*I, but in a different order, has 2,892 sites found at an average distance of 160.2 kb between sites, nearly an eightfold difference. In addition, the depth of coverage for calling a genotype in an outbred sample is quite a bit higher than what is necessary for an isogenic, recombinant inbred line. In general, RAD-seq experimental design is a challenge of optimizing the number of individuals, markers, and coverage given a fixed sequencing effort due to budgetary constraints.

4. The P1 and PE-P2 adapters are modified Solexa© adapters (2006 Illumina, Inc., all rights reserved). The presence of some salt is necessary for the double-stranded adapters used in this protocol to hybridize and remain stable at ambient temperatures and above. Care should be taken to allow reagents to cool to ambient temperature before double-stranded adapters are ligated to digested fragments so as not to denature them. In addition, heating short AT-rich RAD fragments may result in their denaturation. Since single-stranded RAD fragments will not ligate to the double-stranded adapter overhangs, these RAD tags may be underrepresented in the final library. The PE-P2 adapter used here and in Hohenlohe et al. (5) is longer than the P2 adapter we used in the first few published RAD studies. The new, longer PE-P2 adapter is paired-end compatible and has a higher T_m closer to that of the P1 adapters. Thus, this salt concentration consideration in the ligation becomes less important. However, the longer PE-P2 does not get effectively eliminated with a column cleanup, necessitating a gel extraction after amplification no matter how well P1 adapter titration has gone unless quantitative PCR is going to performed prior to sequencing (see Note 13).

5. Below are examples of barcoded *Sbf*I P1 and PE-P2 adapter sequences. "P" denotes a phosphate group and "x" refers to barcode nucleotides (see Note 6), and an asterisk denotes a phosphorothioate bond introduced to confer nuclease resistance to the double-stranded oligo (15). Phosphorothioate bonds should be added to any 3′ overhangs on P1 adapters (e.g., *Sbf*I adapters).

P1 top: 5′-AATGATACGGCGACCACCGAGATCTACACTCT TTCCCTACACGACGCTCTTCCGATCTxxxxxTGC*A-3′
P1 bot: 5′-P-xxxxxAGATCGGAAGAGCGTCGTGTAGGGAA AGAGTGTAGATCTCGGTGGTCGCCGTATCATT-3′
PE-P2 top: 5′-P-GATCGGAAGAGCGGTTCAGCAGGAATG CCGAGACCGATCAGAACAA-3′
PE-P2 bot: 5′-CAAGCAGAAGACGGCATACGAGATCGGT CTCGGCATTCCTGCTGAACCGCTCTTCCGATC*T-3′

6. Three mismatch barcodes are optimal because although a significant number of reads will have a sequencing error in the

barcode, it is very unlikely that a single read will have two errors in the same 5-bp sequence. Therefore, most of the reads that have an error in the barcode can still be assigned to the correct sample when the adapters are designed to have three mismatches, whereas with only two mismatches a read with a sequencing error in the barcode may have come from one of the two samples.

7. RAD amplification primers are Modified Solexa Amplification primers (2006 Illumina, Inc., all rights reserved). An asterisk denotes a phosphorothioate bond.

P1-forward primer: 5′-AATGATACGGCGACCACCG*A-3′
P2-reverse primer: 5′-CAAGCAGAAGACGGCATACG*A-3′

We recommend using aerosol-resistant filter tips for setting up the amplification and downstream steps in the protocol to avoid library contamination.

8. This protocol has been modified from that used in Baird et al. (4) to make paired-end compatible RAD libraries and incorporates critical improvements made since that publication, including the ones adopted from Quail et al. (15) and Illumina library preparation protocols. Although we recommend following the protocol as described, other companies may offer superior (or cheaper) versions of reagents that come at different enzyme concentrations or activities, which should work just as well. Use of them may require additional optimization, including different incubation times or reaction volumes for efficient RAD-seq library preparation. Many other brands of DNA cleanup and gel extraction columns could instead be used also, but an important consideration is the minimum required elution volume. We have successfully substituted Zymo's DNA Clean and Concentrator for Qiagen's MinElute kit and ten Weiss units/μl Epicentre T4 ligase instead of NEB's 2000 cohesive end units/μl ligase.

9. "H$_2$O" in this text refers to water that has a resistivity of 18.2 MΩ cm and total organic content of less than 5 ppb.

10. Lower quantities of starting DNA are allowable when multiplexing samples. If working with dilute DNA concentrations, set up larger reaction volumes and then concentrate the samples with a column before proceeding to ligation. This will cut down on throughput of the protocol, of course depending on the application.

11. In general, when making master mixes, using multichannel pipettes and working with samples in 96- or 384-well plates will speed up the restriction digest and P1 ligation steps when multiplexing multiple barcoded individuals.

12. NEB buffers 2 and 4 are used in the ligation reactions instead of ligase buffer because the salt they contain (50 mM NaCl and KAc, respectively) ensures that the double-stranded adapters

remain annealed during the reactions (see Note 4). T4 DNA Ligase is active in all four NEB buffers if supplemented with 1 mM rATP, but does not work at maximum efficiency in NEB buffer 3 because of the high levels of salt in that buffer. The presence of some salt is necessary for the double-stranded adapters to remain stable at ambient temperature during the ligation, but too much may cause a problem. Be aware of the amount of salt put into the ligation reactions and adjust the concentration of the adapters accordingly (for instance, if working with a frequent cutter, use lower volumes of P1 at 1 μM instead of higher volumes at 100 nM to cut down on salt added to the reaction). If the restriction buffer used for digestion does not contain 50 mM potassium or sodium ions or if the restriction endonuclease cannot be heat-inactivated, purify the reaction in a column prior to P1 ligation and add 6.0 μl NEB buffer 2. This will negate the benefits of multiplexing somewhat, but may be useful in certain RAD library applications involving one or only a few individuals.

13. *Sbf*I has been shown to work robustly in multiple organisms in our experience. Restriction enzymes that cut less frequently create fewer RAD tags, and thus require more input DNA and less P1 adapter to keep the molar ratio approximately equal. Libraries produced with less frequent cutters are more difficult to amplify in general and protocol parameters may take some optimization for favorable results. It is critical to perform preliminary studies to optimize the appropriate amount of P1 adapter for a given restriction enzyme that is used for the first time in an organism, unless the actual number of sites is known (i.e., if a genome sequence is available). A range of P1 adapter-to-DNA ratios can be used in a preliminary study, and the efficiency of ligation can then be assayed via gel visualization. If the ratio of P1 adapter overhangs to available genomic compatible ends is too low, you can get insufficient amplification and/or biased representation of some RAD tags. However, if the ratio of P1 to genomic overhangs is too high, a contaminant band that runs around 130 bp will appear after the final PCR. If this contaminant overwhelms the amplification reaction, it can lead to significant adapter sequence reads in the final sequencing output (even after gel extraction following the final PCR). This phenomenon is completely dependent upon the number of available cut sites present in that genome and the corresponding amount of P1 adapter used. Our *Sbf*I study in stickleback (4) used 2.5 μl of 100 nM P1 per microgram of starting material and performed very well for library construction; however, this is likely due to the fact that there are actually more *Sbf*I sites than expected by chance (see Note 3).

Therefore, it may be preferable to start with less P1 when working on genomes with closer to the expected number of sites. In our hands, RAD-seq libraries created using adapters that have the phosphorothioate modifications described here amplify better and appear much less prone to adapter contamination than those without that were used previously.

14. This step allows multiple individually barcoded samples to be combined and processed as one to cut down on cost, work time, and differences in amplification efficiency that may arise between different library preparations when processing many at once.

15. Although we have optimized our protocol for shearing via sonication, other forms of shearing should work (the "Alternate Fragmentation Methods" sections of the Illumina Sample Prep Guides are a good resource for important considerations when choosing a shearing method).

16. We have found that it is unwise to run more than one library sample on the same agarose gel, unless they will be combined and sequenced in the same lane on the flow cell, because it can lead to contamination between samples.

17. A wider size range can be isolated (see Subheading 3.4, step 2) in order to have more RAD fragments to carry through the protocol if low template amounts are evident after the PE-P2 ligation (see Note 20). In addition, you can isolate a wider distribution of fragments in order to make longer contigs (we have isolated from 200 bp up to 1,200 bp (7) and successfully sequenced and assembled contigs off the fragments); however, it will require greater coverage per site in order to assemble the sheared end sequences and the Illumina Genome Analyzer and HiSeq have difficulty sequencing fragment lengths longer than 1,000 bp.

18. Use MinElute columns and not QIAquick columns, which require a larger elution volume.

19. An optional step used in Etter et al. (7) is to re-digest the RAD template with *Sbf*I-HF again before amplification to remove rare genomic DNA concatemers formed from re-ligation of short fragments with two *Sbf*I sites within 800 bp of each other. This step is truly only necessary if using a more frequent cutter and should only be performed if the adapters used do not have barcodes ending with the nucleotide CC, which would recapitulate the *Sbf*I recognition site when the adapter was ligated to genomic overhangs and cause RAD tags with that barcode to be digested out of the final library. For instance, the barcode AGAGT would be fine, but AGACC would not.

20. The optimal amount of template used for amplification is dependent on the restriction enzyme used and its occurrence

throughout the particular genome; however, a good starting point is 5–10 ng per 25-μl reaction. Lower template input concentrations are preferable, as you can be more confident of the true concentration of amplified RAD tag molecules in your final sample, which have both P1 and P2 sequences, and are therefore able to bind the adapter oligonucleotides present on the Illumina flow cell. Poorly amplified libraries will contain a greater number of background sheared genomic DNA fragments with only PE-P2 adapters attached, which cannot bind to the flow cell and will effectively lower the concentration of readable fragments in the library and throw off your molarity calculation (see Note 24).

21. Libraries that amplify robustly can be amplified with only 14 or fewer cycles of amplification to avoid skewing the representation of the library (15). The goal is to use as few PCR cycles as possible to obtain robustly amplified libraries with a minimum of amplification bias. For the library described here, 14 cycles of amplification from 36 ng of template in a 100-μl reaction yielded 180 ng of product before gel extraction and 82 ng after the trapezoidal slice was excised and purified.

22. PCR amplification of a wide range of fragment sizes often results in biased representation of amplified products with an increased number of short fragments. We found this to be true in our current protocol, but reduced the effects by selecting a trapezoidal slice during gel extraction to reduce the level of short fragment lengths from the PCR.

23. For long-term storage of DNA samples, Illumina recommends a concentration of 10 nM and adding Tween-20 to the sample to a final concentration of 0.1%. This helps to prevent adsorption of the template to plastic tubes upon repeated freeze–thaw cycles, which would decrease the effective DNA concentration and therefore the number of sequencing clusters the library will produce.

24. For example, a measured DNA concentration of 10 ng/μl expressed in grams per liter is 0.01 g/L. 650 g/mol/bp × 600 bp = 390,000 g/mol = 0.00039 g/nmol. To calculate the nanomolarity, then, divide 0.01 g/L by 0.00039 g/nmol to get 25.6 nmol/L or 25.6 nM.

Acknowledgments

The authors thank the University of Oregon researchers who have helped troubleshoot preliminary versions of this protocol. The project described was supported by grant R21HG003834 from the National Human Genome Research Institute (E.A.J.).

References

1. Miller MR, Dunham JP, Amores A, Cresko WA, Johnson EA (2007) Rapid and cost-effective polymorphism identification and genotyping using restriction site associated DNA (RAD) markers. Genome Res 17:240–248

2. Miller MR, Atwood TS, Eames BF et al (2007) RAD marker microarrays enable rapid mapping of zebrafish mutations. Genome Biol 8:R105

3. Lewis ZA, Shiver AL, Stiffler N et al (2007) High-density detection of restriction-site-associated DNA markers for rapid mapping of mutated loci in *Neurospora*. Genetics 177: 1163–1171

4. Baird NA, Etter PD, Atwood TS et al (2008) Rapid SNP discovery and genetic mapping using sequenced RAD markers. PLoS One 3:e3376. doi:10.1371/journal.pone.0003376

5. Hohenlohe P, Bassham S, Stiffler N, Johnson EA, Cresko WA (2010) Population genomics of parallel adaptation in threespine stickleback using sequenced RAD tags. PLoS Genet 6:e1000862

6. Emerson KJ, Merz CR, Catchen JM et al (2010) Resolving post-glacial phylogeography using high throughput sequencing. Proc Natl Acad Sci U S A 107:16196–16200

7. Etter PD, Preston JL, Bassham S, Cresko WA, Johnson EA (2011) Local *de novo* assembly of RAD paired-end contigs using short sequencing reads. PLoS One 6(4):e18561

8. Fan JB, Oliphant A, Shen R et al (2003) Highly parallel SNP genotyping. Cold Spring Harb Symp Quant Biol 68:69–78

9. Fan JB, Chee MS, Gunderson KL (2006) Highly parallel genomic assays. Nat Rev Genet 7:632–644

10. Cox A, Dunning AM, Garcia-Closas M et al (2007) A common coding variant in CASP8 is associated with breast cancer risk. Nat Genet 39:352–358

11. Gabriel S, Ziaugra L, Tabbaa D (2009) SNP genotyping using the Sequenom MassARRAY iPLEX platform. Curr Protoc Hum Genet Chapter 2:Unit 2.12

12. Zerbino DR, Birney E (2008) Velvet: algorithms for de novo short read assembly using de Bruijn graphs. Genome Res 18:821–829

13. Hercus C (2009) www.novocraft.com. Accessed Nov 2009

14. Li H, Durbin R (2010) Fast and accurate long read alignment with Burrows-Wheeler transform. Bioinformatics 26:589–595

15. Quail MA, Kozarewa I, Smith F et al (2008) A large genome center's improvements to the Illumina sequencing system. Nat Methods 5:1005–1010

Part III

Analyzing Data

Chapter 10

Automated Scoring of AFLPs Using RawGeno v 2.0, a Free R CRAN Library

Nils Arrigo, Rolf Holderegger, and Nadir Alvarez

Abstract

Amplified Fragment Length Polymorphisms (AFLPs) are a cheap and efficient protocol for generating large sets of genetic markers. This technique has become increasingly used during the last decade in various fields of biology, including population genomics, phylogeography, and genome mapping. Here, we present RawGeno, an R library dedicated to the automated scoring of AFLPs (i.e., the coding of electropherogram signals into ready-to-use datasets). Our program includes a complete suite of tools for binning, editing, visualizing, and exporting results obtained from AFLP experiments. RawGeno can either be used with command lines and program analysis routines or through a user-friendly graphical user interface. We describe the whole RawGeno pipeline along with recommendations for (a) setting the analysis of electropherograms in combination with PeakScanner, a program freely distributed by Applied Biosystems; (b) performing quality checks; (c) defining bins and proceeding to scoring; (d) filtering nonoptimal bins; and (e) exporting results in different formats.

Key words: Scoring optimization, Command lines, Graphical user interface, Data mining, Bin editing

---

1. Introduction

The Amplified Fragment Length Polymorphism (AFLP) technique is increasingly used in phylogeographic and population genomics studies, particularly in non-model organisms for which no prior DNA sequence information is available (1). This relatively cheap technique is based on complete endonuclease restriction digestion of total genomic DNA followed by selective PCR amplification and electrophoresis of a subset of fragments, resulting in a unique, (theoretically) reproducible fingerprint for each individual. Although the AFLP technique is able to generate a large number of informative markers, the success of this method is compromised by different factors (2). For instance, manual scoring relies on

François Pompanon and Aurélie Bonin (eds.), *Data Production and Analysis in Population Genomics: Methods and Protocols*, Methods in Molecular Biology, vol. 888, DOI 10.1007/978-1-61779-870-2_10, © Springer Science+Business Media New York 2012

visual inspection and subjective interpretation of the electrophoretic profiles during a time-consuming and tedious task. In the last decades, several improvements in automatic scoring have been proposed and implemented in commercial software [see ref. 3 for a review]. However, until recently, no free open-source software was available to process AFLP data from raw data to ready-to-use presence/absence binary matrices. Three years ago, we developed RawGeno 1.0 (4), a library performing automated binning, scoring, and data mining analyses under the R CRAN environment, based on outputs from the freely available electropherogram-analyzing software PeakScanner (Applied Biosystems, Foster City, USA, http://www.appliedbiosystems.com/peakscanner). Implementing RawGeno 1.0 solutions in a free environment has provided an accessible and accurate solution to many users (with 982 downloads from http://sourceforge.net/projects/rawgeno two years after release). Here, we present RawGeno 2.0, an updated version of this software, built around an optimized, less time-consuming algorithm, implementing new features such as binning and editing. We provide examples both in a user-friendly and in a command-line interface. In order to allow users to customize queries (which may vary depending on dataset quality/size), we further provide tips for setting analyses and improving output robustness. All stages of a whole analysis are detailed according to the following five sections: (a) importing features, (b) quality check, (c) binning and scoring algorithms, (d) bin filtering, and (e) exporting options.

2. Program Utilization

2.1. Overview of the Analysis

Analysis of AFLP electropherograms is achieved using two programs: PeakScanner and RawGeno. Whereas PeakScanner detects AFLP peaks along electropherograms and calculates their intensity and size in base pairs (by relying on an internal size standard included in electrophoresis), RawGeno proceeds to the binning and scoring of AFLP electropherograms.

RawGeno includes several filters to assess the quality of electropherograms and checks the consistency of binning. In addition, several preliminary analyses are available to the user for making biological inferences and/or remove outlier samples. Finally, several functions allow exporting resulting datasets into properly formatted files for further analyses.

2.2. Building Up a Scoring Project

In RawGeno 2.0, the AFLP scoring project should be organized according to the following procedure.

1. Create a folder (hereafter "project folder") from which the project will be managed.

2. In this folder, add a sub-directory including all electrophero-gram files (*.fsa). Also add an R shortcut (Windows users) to conveniently launch RawGeno scoring sessions. Right-clicking on this R shortcut allows defining the default working directory of R by specifying it into the "Start into" addressing field of the shortcut. Copy-pasting the project folder address into this field will set up the working directory of R accordingly.

3. Create a text-tabulated table listing individuals included in the project (hereafter referred to as "info table"). The info table is optional as the minimal RawGeno analysis can proceed without it. However, RawGeno includes several functions relying on this table, for instance to label individuals during preliminary analyses or facilitate the sorting and selection of individuals (for example, according to populations or species) during the production of exports. Therefore, the info table should include any additional relevant information that the user would like to consider. It must contain at least the name of individuals (i.e., in a column named "Tag") and any supplementary information in extra columns.

2.3. Obtaining the Raw Data from *.fsa Files Using PeakScanner

The analysis of AFLPs starts by using PeakScanner in order to detect peaks along electropherograms and to calculate their size. The procedure is highly automated, letting the user set peak detection parameters and check the quality of electropherograms. The peak detection parameters are set up using the so-called Analysis Method, which is available from the graphical interface in PeakScanner (menu "Resources/Manage Analysis Methods"). Typically, a proper peak detection attempts to detect only peaks that are biologically relevant and exclude peaks only reflecting technical background noise.

We advise to set the "Analysis Method" using the following guidelines.

1. Prior to the detection of peaks *per se*, a light smoothing of electropherograms might be desirable in order to eliminate small secondary peaks due to technical background noise.

2. The detection of peaks is achieved through a "sliding window" analysis that inspects electropherograms locally. Within the inspected region, PeakScanner first creates a modeled version of the electropherogram by fitting a polynomial curve to the data. Peaks are detected according to this modeled signal, based on their absolute width. Therefore, the detection sensitivity is adjusted by modifying the width of the sliding window (i.e., in terms of data points, the smaller it is, the more sensitive the procedure becomes), the goodness of fit reachable by the polynomial curve (again, increasing the polynomial degree of fitting increases the detection sensitivity), and the minimal

width above which a peak is recorded as present. We advise to use the default parameters as a starting point, as they have been shown to provide reliable results (4, 5): set 15 points for the sliding window width, use a third-degree polynomial curve, and consider peaks that at least have two "points" (i.e., electrophoresis mobility units that are specific to PeakScanner) of half-width.

3. Downstream to peak detection, PeakScanner filters peaks according to their absolute fluorescence intensity, i.e., the peak height, measured in relative fluorescent units (rfu). Visually checking electropherograms obtained from blank samples generally helps to adjust the fluorescence threshold to the upper limit of the technical background noise. While some applications might benefit from considering only peaks with a strong fluorescence (e.g., greater than 150 rfu to provide conservative estimates for band presence statistics), most users will prefer using a more permissive threshold at this stage and apply a posteriori filtering strategies based on bin quality statistics (6, 7). We advise to use 50 rfu as a minimal fluorescence for considering individual AFLP peaks.

4. Save the customized "Analysis Method" in order to use it during electropherogram analysis.

5. Once the "Analysis Method" is set up, import the electropherograms (stored as *.fsa files) into PeakScanner using the "Add Files" button.

6. Define the size standard and the "Analysis Method" to be used for all individuals included in the project (set this information for the first individual, then select the columns "Size Standards" and "Analysis Method," and use the "ctrl + D" keyboard shortcut to apply these settings to the remaining individuals).

The detection and sizing of peaks are processed using the "Analysis" menu, from the graphical interface. Once achieved, electropherograms can be visualized and compared among individuals. This might help identifying AFLP reactions that were not successful (e.g., individuals with a systematically low fluorescence or showing abnormal peaks). Removing such individuals prior to the RawGeno analysis will help to enhance the final quality of scoring.

The PeakScanner analysis ends with a simple export process in which the list of peaks detected throughout the complete set of analyzed individuals is stored in a table. This is achieved using the "Export/Export Combined Table" menu, producing a text-tabulated file containing the size, height, area, and width of all detected peaks (this can be checked using the "Edit Table Settings" menu).

2.4. From Raw Data to Ready-to-Use Matrices Using RawGeno

2.4.1. Installing RawGeno

RawGeno is freely available from http://sourceforge.net/projects/rawgeno as a zip file. In Windows, the installation is achieved either using the graphical user interface of R (menu "Packages/Install package(s) from local zip files") or the following command line in the R console:

```
utils:::menuInstallLocal()
```

Installing RawGeno with Linux requires decompressing RawGeno.zip into the library folder of R. Using the shell command line, this is done as follows:

```
sudo  mv  RawGeno.zip  /usr/lib64/R/library/
  RawGeno.zip

cd /usr/lib64/R/library/RawGeno.zip

unzip RawGeno.zip

rm RawGeno.zip
```

Finally, RawGeno requires the installation of two companion packages: vegan and tkrplot that are available from usual R CRAN repositories (Linux users should see Note 1). Their installation is achieved either using the graphical user interface of R (menu "Packages/Install package(s) from CRAN") or with the following command lines (prompted into the R console):

```
install.packages("vegan")

install.packages("tkrplot")
```

2.4.2. Importing PeakScanner Results

All the following steps are performed in the R CRAN environment using the R shortcut described above (or ensuring that the correct working directory has been selected). Once the RawGeno package has been installed, it can be called applying the following command line:

1. Call RawGeno, vegan, and tkrplot as libraries into R and launch the graphical user interface:

```
require(RawGeno)

require(vegan)

require(tkrplot)

RawGeno()
```

2. Importing the PeakScanner text-tabulated file in RawGeno can then be done using the graphical user interface (menu "RawGeno/1. Files/Electroph./PeakScanner (*.txt)") or using the following command lines:
Choose interactively the PeakScanner file:

```
myfile = tk_choose.files(caption = 'Choose
  PeakScanner File')

OPENAFLP(myfile, pksc = T, dye = "B")
```

Or explicitly specify the path of file of interest:

```
mypath = "C:/MyDocuments/MyPeakScannerFile.
txt"
OPENAFLP(mypath, pksc = T, dye = "B")
```

During importation, RawGeno handles a single dye color at a time, which is user specified and considers the "dye" parameter with the following values: "B" (blue; FAM), "G" (green; HEX), "Y" (yellow; NED), "R" (red; ROX), or "O" (orange; LIZ). If electrophoresis was achieved using several dyes simultaneously (e.g., multiplexing of PCR products), each dye must be analyzed separately in RawGeno. Datasets obtained from several dyes can be merged a posteriori in a final binary table (see below).

2.4.3. Quality Check

Because the detection of AFLP peaks is based on a defined threshold, it is not easy to handle reactions showing electropherograms with varying intensities (see Note 2). When improperly handled, such a situation leads to the inclusion of samples characterized by many false absences in the final dataset. Although the only way to correctly address this issue is a robust wet-lab protocol, RawGeno still attempts limiting the influence of low-quality AFLPs on binning and scoring by filtering individuals that were unsuccessful. Here, the variability in the number of peaks detected per individual is used as a proxy of AFLP reactions' quality. Empirical evidence shows that this statistics is dependent on the specific dataset used and the biological organism studied. The lower bound of this distribution most generally includes individuals with low AFLP intensities, characterized by many AFLP peaks that remain undetected in the electropherograms. Because such individuals usually represent a small fraction of the complete project, we advise removing them from the dataset. The upper bound of the distribution can either reflect a biologically relevant signal (e.g., hybridization) or a technical bias (e.g., contamination, odd PCR reaction). Such individuals should be either discarded or identified as outliers for proper interpretation in further analyses.

RawGeno includes the two following options for checking quality of individuals.

1. Filter individuals according to the number of detected peaks by manually selecting individuals that should be kept for further analysis or using a dedicated device (Fig. 1a). The command line version of this operation relies on percentiles and conserves individuals within 5–95% bounds of the detected peaks' distribution:

   Retrieve imported electropherograms

   ```
   data.electroph = AFLP$all.dat
   ```

   Compute the number of AFLP peaks per individual:

   ```
   pk.per.smp = table(data.electroph$sample.id)
   ```

Compute 5 and 95% quantiles:

```
qtles=quantile(pk.per.smp, probs=c(0.05, 0.95))
```

Determine what samples can be kept:

```
to.keep=which(pk.per.smp>=qtles[1] & pk.per.
    smp<=qtles[2])
```

Filter electropherograms, keep only retained individuals, and update individuals' indexing accordingly:

```
smp.ok=match(data.electroph$sample.id,
    to.keep)
```

```
data.clean=data.electroph[is.na(smp.ok)==F, ]
```

```
data.clean$sample.id=as.factor(data.
    clean$sample.id)
```

```
levels(data.clean$sample.id)=1:length(to.
    keep)
```

```
smp.names=AFLP$samples.names
```

```
smp.clean=smp.names[to.keep]
```

```
AFLP$all.dat=data.clean
```

```
AFLP$samples.names=smp.clean
```

2. Check the quality of individuals by taking into account their position in PCR plates (only available from the graphical interface, Fig. 1b). This display helps to highlight systematic biases having a technical origin (e.g., pipetting errors or thermocycler

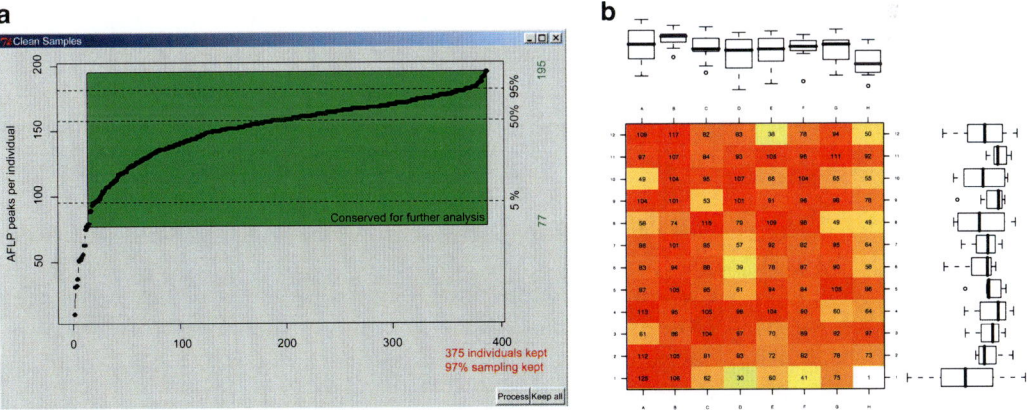

Fig. 1. Checking quality of AFLP reactions. (**a**) Interactive device for filtering samples, before performing binning. Samples are sorted according to the number of AFLP peaks successfully amplified. Summary statistics (i.e., 5, 50, and 95% quantiles) are provided to help users selecting samples to be included in further analyses. (**b**) Visualization of 96-well PCR plates. Four quality statistics are available in RawGeno (number of AFLP peaks per sample, mean and variance of fluorescence intensity, and outlier detection index; see RawGeno manual). *Box plots* provide summary statistics for rows and columns of PCR plates, respectively.

bias) and to identify batches of individuals that were not successful.

2.4.4. Binning

Building a presence/absence matrix requires recognizing which AFLP peaks are homology across individuals. This procedure relies on the size of peaks along electropherograms and assumes homology for peaks sharing identical sizes. Because peak sizes are determined empirically using electrophoresis, measurements generally include technical variations preventing the observation of strictly identical sizes across homology AFLP peaks. Indeed, Holland et al. (8) reported measurement variations ranging between 0 and 0.66 bp (with 0.08 bp in average) for replicated AFLP peaks. Therefore, properly recording the signals of AFLP peaks asks for taking into account size variations by defining size categories (i.e., "bins") into which the presence/absence of AFLP peaks are recorded. Bins are characterized by their position along the electropherogram (i.e., the average size of peaks they include) and their width (the size difference between the longest and shortest peaks included in the bin). RawGeno uses a binning algorithm relying on the size of AFLP peaks over all individuals included in the project and defines bins in a way that respects the two following conditions (Fig. 2a).

1. The first constraint is a maximal bin width. This prevents the definition of too large bins that could lead to homoplasy (i.e., erroneously assigning non-homology AFLP peaks within the same bin). This limit is set using the MaxBin parameter. We advise to use MaxBin values ranging between 1.5 and 2 bp (see Note 3). Using small values (i.e., MaxBin < 0.5 bp) should be avoided as this generally causes oversplitting, a situation where the presence/absence of homology AFLP bands are coded using an exaggerated number of bins.

2. The second constraint prevents the assignment of more than one peak from the same individual within the same bin (i.e., "technical homoplasy," as defined in 4). If such a situation occurs, RawGeno defines two separate bins in which the two peaks are assigned. This constraint can be relaxed by increasing the "MinBin" parameter in order to include two peaks of the same individual in the same bin (when not exceeding the "MinBin" value in size difference). Such a relaxing might be desirable, for instance, when artifactual peaks (i.e., shoulder, stutter, or secondary peaks bordering the authentic peak in a

---

Fig. 2. (continued) and the shortest amplicons included in the considered bin) and the technical homoplasy rates (i.e., HR, the mean number of peaks belonging to the same sample that are included in a same bin) are indicated. (**b**) Binning edition device, allowing users to translate, resize, add, or remove bins interactively, starting from bins that were initially defined by the automated algorithm. Summary statistics (i.e., the average size of bins and the number of AFLP peaks present per bin) are provided as editing guidelines.

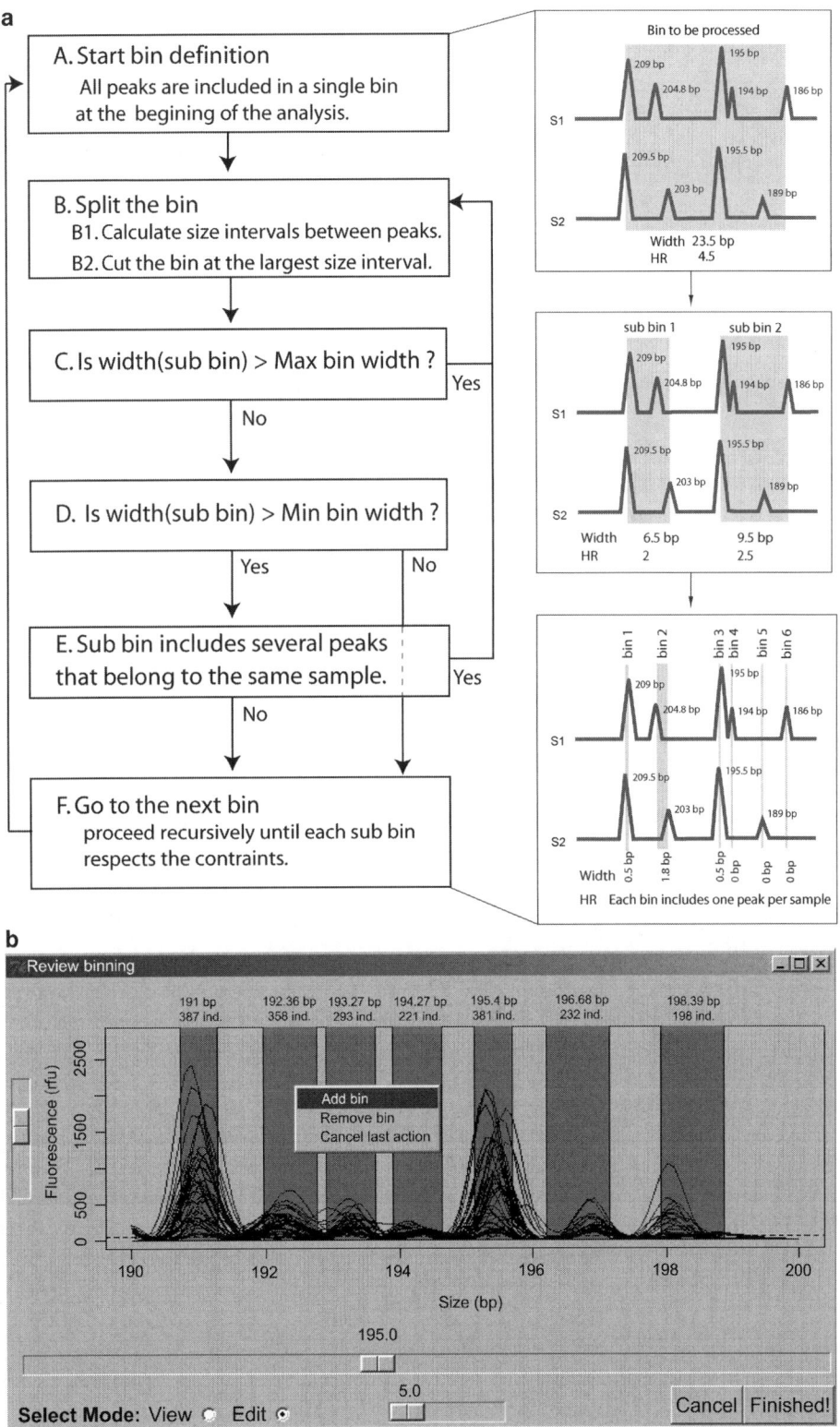

Fig. 2. Binning algorithm implemented in RawGeno. (**a**) *Left panel*: main steps followed by the algorithm to define bins; *right panel*: illustration of binning with two samples (S1 and S2); the bin widths (i.e., the difference in size between the longest

same individual) lead to the definition of numerous extra bins. In such a situation, artifactual peaks can cause the local definition of extra bins into which homology peaks can be inconsistently assigned. We advise to use MinBin values ranging between 1 and 1.5 bp. The binning is launched through the graphical interface (menu "RawGeno/2. Scoring") or using the following command lines:

Proceed to binning:

```
EXTRACTAFLP(all.dat=AFLP$all.dat, samples.names=
    AFLP$samples.names, MAXBIN=2, MINBIN=1)
```

View results (that are assigned into a "data.binary" object):

```
attributes(data.binary)
```

```
data.binary$data.binary
```

Binning is an automated and straightforward analysis step that users might want to review interactively. RawGeno includes a visualization device for manually editing the binning by adding, removing, or modifying the width and position of bins (this tool is only available from the graphical interface, Fig. 2b). This device includes several help-to-decision statistics, such as the average size and the number of presences associated with each bin.

2.4.5. Filtering

Once defined, bins can be filtered according to their properties and/or quality. Note that such filtering strategies require analyzing AFLP reactions with a consistent quality across individuals. Filtering options are accessible from the scoring menu when using the graphical user interface (menu "RawGeno/2. Scoring"). Command line users will set accordingly the EXTRACTAFLP function.

Proceed to binning and filtering simultaneously:

```
EXTRACTAFLP(all.dat=AFLP$all.dat,    samples.
    names=AFLP$samples.names,MAXBIN=2,MINBIN=1,
    RMIN=100,   RMAX=500,   cutRFU=50,   who='B',
    thresh=95)
```

In its current version, RawGeno includes three kinds of filters.

1. The size filter restricts binning to a given portion of the electropherogram (RMIN and RMAX parameters). We advise to limit the binning to peaks included in the range of the size ladder because their size is accurately interpolated by PeakScanner (in contrast to larger peaks, where the size is extrapolated). We recommend discarding peaks with small sizes (i.e., smaller than 100 bp, RMIN = 100) as they are more likely to be homoplasic (9, 10). In addition, large size peaks should as well be considered cautiously because their fluorescence intensity might not always be consistent across individuals. Because this upper limit might vary according to

datasets, we advise to run preliminary analyses and check electropherograms to confidently determine it.

2. The second filter eliminates bins according to their average fluorescence (i.e., the cutRFU parameter). This filter assumes that bins with a high average fluorescence retrieve a more consistent signal than bins with a low fluorescence. The rationale for this strategy is the following. The fluorescence of an AFLP fragment largely determines its detection probability during the PeakScanner analysis of electropherograms. Therefore, fragments that systematically produce low fluorescence are more likely to be erroneously recorded as absent from electropherograms as they might pass the threshold in some reactions but not in others just by chance. The computation starts by normalizing fluorescence intensities across samples using the sum normalization method (6), before computing the average fluorescence of each bin. Hence, keep in mind that the cutRFU parameter applies on normalized values and does not scale with fluorescence measures provided in PeakScanner. Setting this filter is dataset dependent and we recommend running several trials before producing a definitive dataset (see Note 4). Refer to the works of Whitlock et al. (6) and Herrmann et al. (7) for more sophisticated filtering R scripts based on fluorescence intensities. Bridging RawGeno with these algorithms is achieved by exporting fluorescence results instead of a binary matrix. Once binning is achieved, use the following command lines.

Retrieve raw fluorescence results, stored in a matrix corresponding to the usual binary matrix:

```
mat.rfu = t(data.binary$data.height.raw)
```

For normalized fluorescence (sum normalization), use instead:

```
mat.rfu = t(data.binary$data.height)
```

Prepare for export and save as a text-tabulated file (refer to programs' documentation to properly format files):

```
mat.rfu[is.na(mat.rfu) ==T] = 0

write.table(mat.rfu, "MyFluorescenceFile.txt",
  quote = F, sep = "\t")
```

3. The reproducibility filter evaluates bin quality according to their robustness across AFLP reactions by relying on replicated samples. This filter assumes that replicated individuals were selected randomly from the original dataset so as to scan the genetic diversity at best (see Notes 2 and 4). Keep in mind that RawGeno identifies replicated individuals using their names. Replicates must be named using the original individual name plus a suffix letter. The suffix is matched using the "who" parameter of the

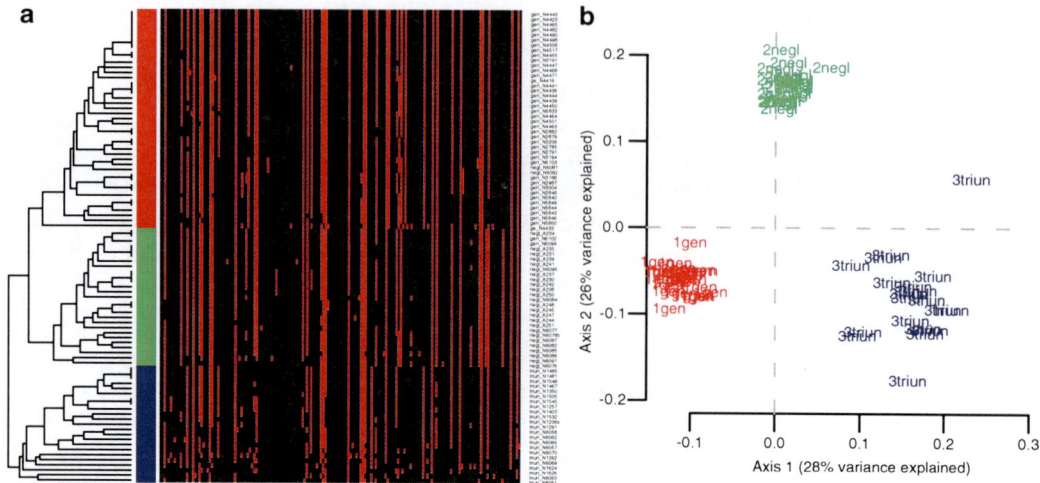

Fig. 3. Reviewing results. RawGeno includes basic visualization devices for performing preliminary data mining. Specifically, results can be reviewed using (**a**) heatmaps of the binary matrix, where samples are sorted according to their genetic similarity and (**b**) principal coordinates analysis of the corresponding matrix. Both devices can compare AFLP results with either quality statistics (i.e., number of AFLP peaks per sample, mean and variance in fluorescence intensity, and outlier detection index) or external information provided by users (e.g., the population from where samples were collected). In addition, both devices are handled through a graphical user interface for sorting and selecting samples to be visualized.

filtering algorithm. As an example, "mysample.fsa" and "mysampleB.fsa" are a pair of original–replicated samples, being identified with a "B" suffix (therefore, set who = "B" when filtering). For each bin, RawGeno compares original to replicated individuals and calculates the percentage of original–replicated pairs for which the AFLP signal is successfully reproduced. Bins, where reproducibility cannot reach a satisfactory rate (i.e., the "thresh" parameter, a user-defined reproducibility percentage), are eliminated from the final dataset.

2.4.6. Review of Results

RawGeno offers two displays for exploring scoring results (menu "RawGeno/3. Quality Check/Samples Checking"). The binary matrix can be directly visualized using a heatmap, showing individuals sorted according to their genetic relatedness. Alternatively, individuals can be examined with a principal coordinates analysis. Both displays allow plotting quality statistics or external information (i.e., picked from the "info table" cited above) against the AFLP results. These displays are only available from the graphical user interface (Fig. 3a, b).

2.4.7. Exporting Files

RawGeno includes functions for producing binary tables and standard exports for Arlequin, Hickory, Popgen, AFLPsurv, STRUCTURE 2.2, Mltr, Spagedi, Dfdist, Treecon, Baps, PAUP, Structurama, MrBayes, and NewHybrids (11). Furthermore, these exports can be sliced according to information provided in the info table and produce ad hoc subsets. These functions are accessible from the graphical interface ("RawGeno/4. Save") or using the following command lines.

Retrieve the binary matrix from RawGeno and import the info table:

```
mat01 = t(data.binary$data.binary)
```

```
matinfo = read.delim("MyInfoTable.txt",
header = T)
```

Cross-reference the AFLP results to the info table:

```
popsA = row.names(mat01)
```

```
popsB = matinfo$Tag
```

```
mat01 = mat01[match(intersect(popsA,   popsB),
popsA), ]
```

```
matinfo = matinfo[match(intersect(popsA,
popsB), popsB), ]
```

Remove monomorphic bins:

```
mat01 = mat01[  , colSums(mat01) > 0  &  colSums
(mat01) < nrow(mat01)]
```

Produce the required outputs, e.g., for STRUCTURE 2.2. (refer to the library documentation for further details regarding exporting functions):

```
Structure.popsD(mat01,   pops = matinfo$MyPops
Column,   path = getwd(),   name = "MyStructure2.
2File.txt")
```

Users willing to analyze AFLP signals as codominant markers (12) should use command lines described above to export fluorescence data (a proxy of allele copy number in genomes) associated with binary matrices.

2.4.8. Handling Data from Previously Scored Projects

Users willing to merge, visualize, and/or produce exports from datasets that were already scored can import binary tables within RawGeno using the "RawGeno/1. Files/Import" menu. From the command line, such an operation is done as follows.

Select files to merge:

```
list.merge = tk_choose.files(caption = 'Choose
    Files to Merge')
```

Or specify a directory in which the binary matrices are stored:

```
mypath = "C:/MyDocuments/MyBinaryMatrices
Directory"
```

```
list.merge = dir(mypath, pattern = '.txt')
```

Proceed to merging:

```
MERGING(transpose="indRows",        exclude=T,
replacewith=NA)
```

The transpose parameter states whether the binary matrices store individuals as lines ("indRows") or columns ("indColumns"), and the exclude parameter defines whether individuals that are not shared by all matrices will be removed from the final merged dataset (exclude="T"). If kept (exclude="F"), individuals with missing AFLP genotypes will be completed using NA values (replacewith=NA) when no data is available. Note that the merged matrix is stored into a "mergedTable" object. Visualization and exports can be performed using the graphical user interface as described above. Command lines for exporting merged matrices are given below:

```
mat01=mergedTable

matinfo = read.table("MyInfoTable.txt",
header=T)

popsA=row.names(mat01)

popsB=matinfo$Tag

mat01=mat01[match(intersect(popsA,   popsB),
popsA), ]

matinfo = matinfo[match(intersect(popsA,
popsB), popsB), ]

mat01=mat01[ , colSums(mat01)&colSums(mat01)
<nrow(mat01)]

Structure.popsD(mat01,   pops=matinfo$MyPops
Column,   path=getwd(),   name="MyStructure2.
2File.txt")
```

3. Conclusions and Perspectives

We present here a complete suite of tools to automate the scoring of AFLP datasets using free software applications. Our program proposes an integrated solution to manage all the components of the analysis pipeline. Accordingly, samples are checked by removing non-satisfactory electropherograms at the very beginning of the analyses and AFLP genotypes are associated with user-specified information while producing ad hoc exports. Bins are managed using an automated algorithm and can be edited manually using a dedicated graphical user interface.

As a next milestone, we plan to develop RawGeno into two complementary directions: incorporating the handling of electropherograms (which is now part of PeakScanner) and enhancement of bin filtering possibilities. Indeed, the RawGeno version currently under development already integrates functions for detecting and calculating the size of AFLP peaks along electropherograms. Finally, achieving connections with the R scripts of Herrmann et al. (6) and Whitlock et al. (7) is another way to improve RawGeno.

4. Notes

1. Linux users might need to run R as "sudo" users to properly install companion packages (i.e., vegan and tkrplot). In addition, troubles might arise because R libraries are downloaded as source code and compiled locally before being installed. This requires that all compilers needed by R (such as gc, gcc, gcc-fortran, and others) have been installed locally, before attempting the installation of external R packages. In OpenSUSE, the necessary compilers can be obtained using YaST2 (into the rpm groups dedicated to development tools). Ubuntu users are more fortunate because Synaptic Manager can install ready-to-use R libraries in addition to usual compilers (refer to http://cran.r-project.org/bin/linux/ubuntu/README for further details regarding repository addresses).

2. Whereas manual scoring allows permanent but subjective adjusting of the criteria defining whether an AFLP peak should be recorded as present or absent, automated peak detection algorithms apply uniform fluorescence thresholds. This requires datasets showing low AFLP quality variation because fluorescence differences between samples will be reflected in the final binary matrix. For instance, we encountered problematic situations with AFLP reactions showing different fluorescence offsets among PCR plates. These "plate effects" can be highlighted with principal coordinate analyses, as samples are clustered according to PCR plates (i.e., due to plate-specific losses of AFLP bands). These situations are especially difficult to handle without repeating wet-lab experiments. Solving this problem computationally remains difficult and asks for setting sample (or plate)-specific detection sensitivities, which is beyond PeakScanner possibilities. The present version of RawGeno conservatively proposes the removal of samples with unusual AFLP profiles by relying on the distribution of the number of AFLP peaks. Future developments of RawGeno will attempt analyzing electropherograms directly. The problem

highlights the crucial importance of using robust and standardized lab protocols. In the following, we list several tips helping to retrieve consistent results from AFLP reactions, when using automated scoring (also consult Bonin et al. (13) and Gugerli et al. (14)).

(a) Randomize reactions on PCR plates in order to properly discriminate technical bias from biological signals.

(b) Standardize all reaction steps: Adjust DNA concentrations after spectrometer quantification, limit impacts of pipetting errors by preparing reaction mixes in batches, run critical reactions in uniform conditions (for instance perform restriction steps in an incubator rather than in a thermocycler), and, importantly, run PCRs on the same thermocycler.

(c) Optimize signal detection during genotyping analysis, for instance, by increasing the injection time of automated sequencers at the beginning of electrophoresis, and prefer primer pairs showing strong and consistent amplifications.

3. Based on empirical case-study datasets (available from the authors upon request), we provide help-to-decision statistics aimed at guiding users in setting their RawGeno analysis. We analyzed 17 AFLP primer datasets: *Aegilops geniculata*, two primer pairs (5); *Arum* spp., two primer pairs (Espindola et al., unpublished data); *Baldellia* spp., two primer pairs (Arrigo et al., unpublished data); *Bupleurum ranunculoides*, three primer pairs (Labhardt et al., unpublished data); *Deschampsia litoralis* and *Deschampsia caespitosa*, each with three primer pairs (15); and *Peucedanum ostruthium*, two primer pairs (Borer et al., unpublished data). These studies included between 87 and 509 individuals (mean: 255) of either intra- and interspecific sampling; 4 to 51% (mean: 23%) of the individuals were replicated. We varied binning parameters (i.e., MinBin and MaxBin) and measured their effects on the width of bins, the datasets' polymorphism (i.e., the proportion of bins with presence frequencies ranging between 5 and 95%), and the bin reproducibility (i.e., the proportion of replicated sample pairs over which the focal bin is successfully reproduced; this measure is independent of the total number of bins in the dataset and is therefore suitable in the context of binning optimization).

MinBin and MaxBin acted as boundaries on the width of bins. They determined how accurately an AFLP signal was reflected into the final presence/absence matrix. Both parameters were explored for values ranging between 0.1 and 5, with 0.1-bp increments, therefore totalizing 1,176 "binning trials" per dataset (Fig. 4a, b).

Using exaggeratedly small MaxBin values forces the binning algorithm to define narrow bins. This situation leads to "oversplitting," a bias where AFLP signals are coded into more

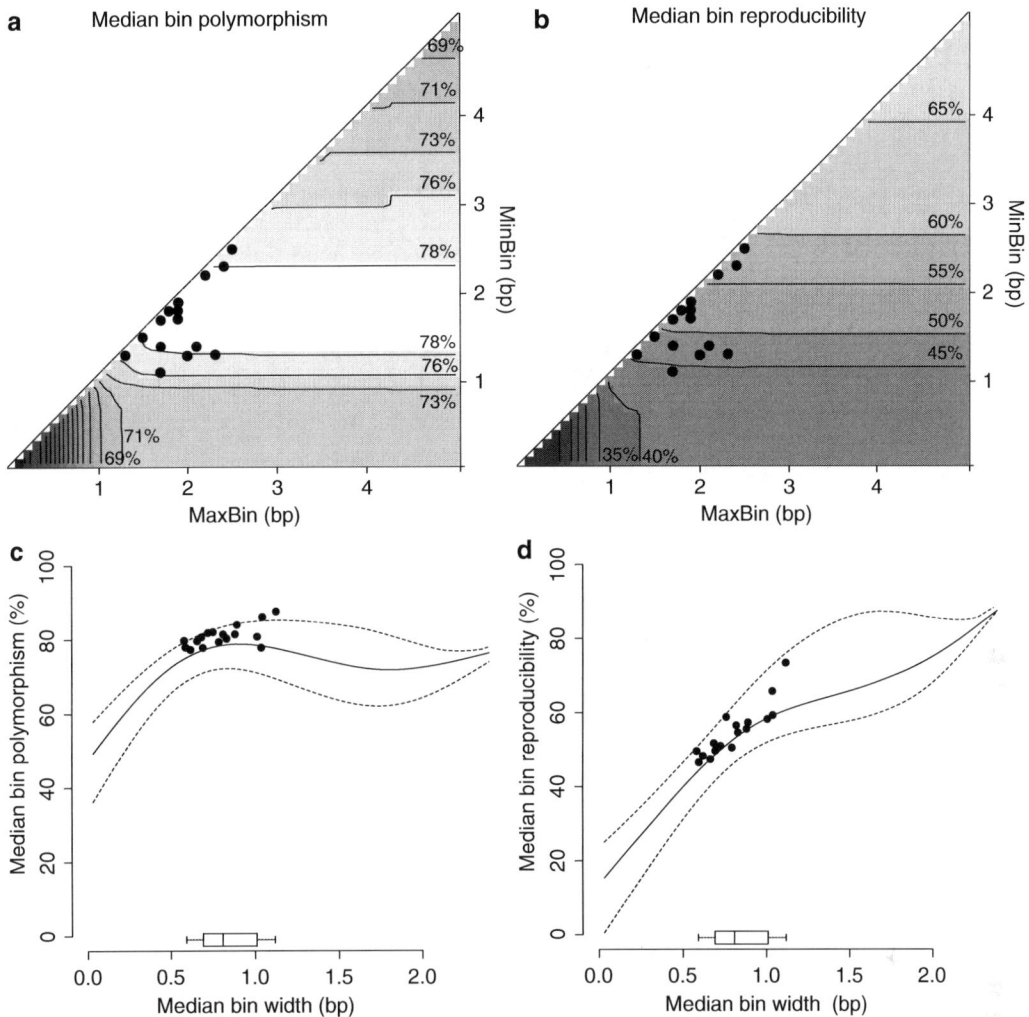

Fig. 4. Binning parameters. We analyzed 17 different datasets (see Note 3) in RawGeno by varying binning parameters that have an effect on the width of bins (i.e., MinBin and MaxBin) and measured the associated effects on bin width, polymorphism, and reproducibility. All results were corrected in order to normalize statistics among datasets (i.e., using nonlinear mixed effect models, with the dataset of origin considered as a covariable). *Upper row*: *Triangle plots* displaying (**a**) bin polymorphism and (**b**) reproducibility averaged for the 17 datasets as a function of MinBin and MaxBin parameters. *Lower row*: *Scatter plots* of corresponding (**c**) bin polymorphism and (**d**) bin reproducibility statistics, displayed according to the median bin width of binning trials. 5 and 95% confidence intervals (*dashed lines*) and the average (*continuous line*) are displayed. (**a**–**d**) Trials optimizing bin polymorphism are represented as dots for each of the 17 datasets (*box plots* indicate bin widths associated to optimized trials).

bins than needed. In this case, an AFLP locus is coded using several adjacent bins appearing as inconsistent when considered independently. Our results showed oversplitting evidence for bin widths below 0.5 bp, with decreased bins' polymorphism and reproducibility (Fig. 4c, d). This situation should be avoided, and we recommend using values larger than 0.5 bp for MaxBin (in contrast, the MinBin parameter had little effects on oversplitting).

Using exaggeratedly large MinBin and MaxBin values introduces "technical homoplasy" (as defined in 4), a bias where AFLP signals are coded using less bins than required. Although merging artifactual secondary peaks with authentic peaks is desirable, a process that increases the consistency of binning (see above), exaggerated merging tends to artificially increase similarity between unrelated samples and has immediate effects on AFLP polymorphism and reproducibility. Our results showed that technical homoplasy was reflected by a decrease in bin polymorphism, when MinBin and MaxBin both exceeded 2 bp (i.e., corresponding to bin widths larger than 0.8 bp). On the other hand, reproducibility increased along with technical homoplasy due to the addition of some level of artifactual similarity among samples.

From these results, we suggest to screen binning parameters by considering bin polymorphism as a main optimization criterion (Fig. 4a, c). Reproducibility statistics should not be considered for binning optimization because of their inability to detect technical homoplasy (Fig. 4b, d); see Holland et al. (7) for further considerations about binning optimization.

4. In its current version, RawGeno includes three filters that can be applied after binning has been achieved. These are a size (i.e., the region of electropherograms over which the analysis must be carried out), a fluorescence, and a reproducibility filters. We applied filters to the 17 datasets explored during binning optimizations (see Note 3). The size filter was not tested and all datasets were analyzed for bins ranging between 100 and 400 bp (i.e., corresponding to the electrophoresis region where size interpolation is accurate). The two remaining filters were explored starting from binned datasets that maximized polymorphism.

The first filter considers bins' quality to vary according to average fluorescence. Indeed, bins with an average fluorescence close to the detection threshold used in PeakScanner are more likely to reflect inconsistent signals (e.g., technical false absences) than bins with strong average fluorescence (see above). The filter removes bins according to a user-defined lower fluorescence limit. Keep in mind that this limit applies on normalized values. We tested limits ranging between 0 (i.e., no filtering) and 400 rfu (with 10 rfu increments) and measured polymorphism and reproducibility changes caused by this filtering (Fig. 5a, b). All but one dataset gained in polymorphism and reproducibility when filtering was optimized. However, while filtering increased dataset reproducibility, it drastically decreased polymorphism when set up improperly. Its use hence requires dataset-depending optimization to limit information reduction in datasets. We recommend testing fluorescence thresholds below 200 rfu.

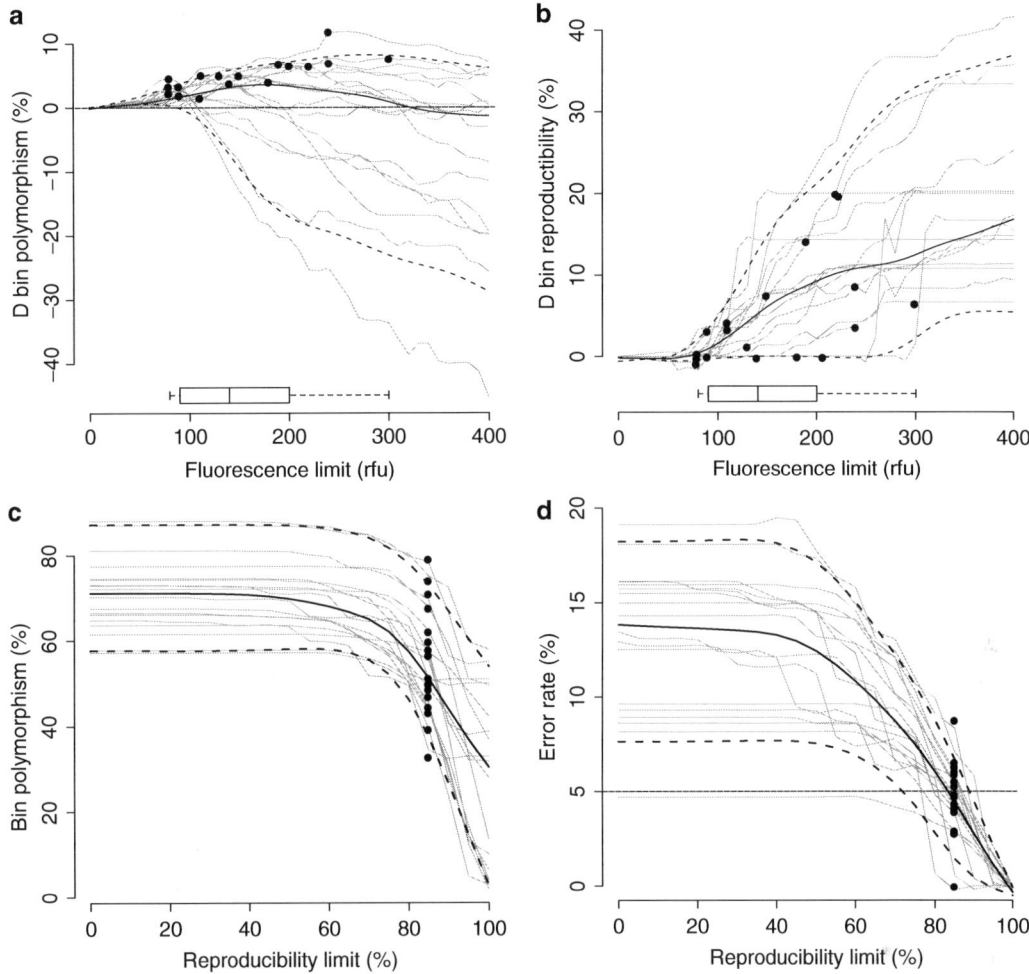

Fig. 5. Filtering parameters. We filtered datasets (using optimal binning parameters as defined in Note 3) by increasing bin fluorescence and reproducibility thresholds. Median bin polymorphism and error rates were measured for the final datasets (dataset-specific offsets were not corrected). *Upper row*: Bins are filtered according to their average fluorescence. Results are displayed as a comparison with non-filtered trials, with "D bin polymorphism" and "D bin reproducibility" being the difference between filtered and non-filtered datasets. Therefore, values greater than zero reflect cases where filtering increased the median (**a**) polymorphism or (**b**) reproducibility of bins in the final dataset. Conversely, values lower than zero indicate decreases in these statistics associated with filtering. Lower row: Filtered bins according to their reproducibility and consequences on median bin (**c**) polymorphism and (**d**) error rate of the final datasets. Results of each dataset (*gray lines*) are displayed along with summary statistics (*dashed lines*: 5 and 95% confidence intervals; *continuous line*: average). (**a–d**) Trials optimizing bin polymorphism after filtering are presented as dots for each of the 17 datasets (*box plots* indicate fluorescence threshold associated to optimized filtering).

The reproducibility filter assesses the robustness of AFLP signals and removes bins that are not satisfactorily reproducible. We tested filtering using bin reproducibility limits ranging between 0 and 100% (with 5% increments) and measured polymorphism and global error rate of datasets (Fig. 5c, d). We present error rates (13) instead of bin reproducibility here because error rates are more commonly reported in AFLP studies.

Filtering effectively reduced the global error rate along with the removal of non-reproducible bins from datasets. As expected, filtering also decreased the polymorphism of datasets. Nevertheless, this filter generally removed numerous bins, and we clearly propose avoiding the use of reproducibility values larger than 95%. Our experience shows that using 85% as a threshold provides satisfactory results by balancing the error rates of datasets (i.e., 5% in average, which we consider reasonable) with polymorphism. A more stringent filtering can be applied (e.g., in studies where peak reproducibility is especially relevant such as in genome scans), but it might affect the information content of datasets and thus requires the inclusion of additional primer pairs to achieve a reasonable level of polymorphism.

Acknowledgments

We are grateful to Anahi Espindola, Matthias Borer, and Aurélien Labhardt for kindly providing the AFLP datasets presented in Notes 2–4. We thank Nicolas Salamin for having provided a working environment to N. Arrigo during the redaction of this manuscript. Felix Gugerli, Sabine Brodbeck, Piya Kittipadakul, Julie B. Hébert, Julie Lee Yaw, Christian Parisod, Roland Dubillard, and Robin Arnoux provided helpful comments during the development of RawGeno. This project was supported by the National Centre of Competence in Research (NCCR) Plant Survival and the PNR-59, two programs of the Swiss National Science Foundation. N. Alvarez and N. Arrigo were funded by the Swiss National Science Foundation (Ambizione fellowship PZ00P3_126624 and prospective researcher fellowship PBNEP3_132747, respectively).

References

1. Vos P, Hogers R, Bleeker M et al (1995) AFLP: a new technique for DNA fingerprinting. Nucleic Acids Res 23:4407–4414

2. Pompanon F, Bonin A, Bellemain E, Taberlet P (2005) Genotyping errors: causes, consequences and solutions. Nat Rev Genet 6: 847–859

3. Meudt HM, Clarke AC (2007) Almost forgotten or latest practice? AFLP applications, analyses and advances. Trends Plant Sci 12(3):106–117

4. Arrigo N, Tuszynski JW, Ehrich D, Gerdes T, Alvarez N (2009) Evaluating the impact of scoring parameters on the structure of intraspecific genetic variation using RawGeno, an R package for automating AFLP scoring. BMC Bioinformatics. doi:10.1186/1471-2105-10-33

5. Arrigo N, Felber F, Parisod C et al (2010) Origin and expansion of the allotetraploid *Aegilops geniculata*, a wild relative of wheat. New Phytol 187(4):1170–1180

6. Whitlock R, Hipperson H, Mannarelli M, Butlin RK, Burke T (2008) An objective, rapid and reproducible method for scoring AFLP peak-height data that minimizes genotyping error. Mol Ecol Resour 8:725–735

7. Herrmann D, Poncet BN, Manel S et al (2010) Selection criteria for scoring amplified fragment length polymorphisms (AFLPs) positively affect the reliability of population genetic parameter estimates. Genome 53(4):302–310

8. Holland BR, Clarke AC, Meudt HM (2008) Optimizing automated AFLP scoring parameters to improve phylogenetic resolution. Syst Biol 57(3):347–366

9. Vekemans X, Beauwens T, Lemaire M, Roldan-Ruiz I (2002) Data from amplified fragment length polymorphism (AFLP) markers show indication of size homoplasy and of a relationship between degree of homoplasy and fragment size. Mol Ecol 11(1):139–151

10. Paris M, Bonnes B, Ficetola GF, Poncet BN, Després L (2010) Amplified fragment length homoplasy: *in silico analysis* for model and non-model species. BMC Genomics. doi:10.1186/1471-2164-11-287

11. Ehrich D (2006) AFLPdat: a collection of R functions for convenient handling of AFLP data. Mol Ecol Notes 6:603–604

12. Gort G, van Eeuwijk F (2010) Codominant scoring of AFLP in association panels. Theor Appl Genet. doi:10.1007/s00122-010-1313-x

13. Bonin A, Bellemain E, Eidesen PB et al (2004) How to track and assess genotyping errors in population genetics studies. Mol Ecol 13:3261–3273

14. Gugerli F, Englisch T, Niklfeld H et al (2008) Relationships among levels of biodiversity and the relevance of intraspecific diversity in conservation—a project synopsis. Perspect Plant Ecol Evol Systemat 10:259–281

15. Peintiger M, Arrigo N, Brodbeck S et al (2010) Genetische und morphologische Differenzierung der endemischen Grasart *Deschampsia littoralis* (Gaudin) Reut.—Wie verschieden sind die Population am Bodensee und am Lac de Joux im Vergleich zu *D. cespitosa* (L.) P. Beauv. Vogelwarte Radolfszell, University of Neuchâtel, WSL

Chapter 11

Haplotype Inference

Olivier Delaneau and Jean-François Zagury

Abstract

The information carried by combination of alleles on the same chromosome, called haplotypes, is of crucial interest in several fields of modern genetics as population genetics or association studies. However, this information is usually lost by sequencing and needs, therefore, to be recovered by inference. In this chapter, we give a brief overview on the methods able to tackle this problem and some practical concerns to apply them on real data.

Key words: Genomic, SNP, Haplotype, Statistical inference, Combinatorial algorithm

1. Introduction

Following the major progress in modern genotyping technologies, allele distributions of single nucleotide polymorphisms (SNP) have become available in high density throughout the genome (millions of markers) for large population samples (thousands of subjects). These bi-allelic polymorphisms are genetic markers with potentially relevant biological properties such as protein coding or gene regulation. They also constitute a useful information in population genomics approaches to characterize past demographic events or for association studies to detect correlations with phenotypes (1–4). However, the mere analysis of single SNPs is not sufficient to fully exploit genetic information since alleles of neighbouring SNPs on the same chromosome are generally inherited together through generations. These units of inheritance, usually called haplotypes, can be used as multi-allelic markers to improve the power of analyses (5, 6).

Basically, haplotypes are created in the course of evolution via mutations and recombinations. Then, selection or demographic

François Pompanon and Aurélie Bonin (eds.), *Data Production and Analysis in Population Genomics: Methods and Protocols*, Methods in Molecular Biology, vol. 888, DOI 10.1007/978-1-61779-870-2_11, © Springer Science+Business Media New York 2012

events such as random genetic drift are responsible for the spread or disappearance of these haplotypes in populations. Unfortunately, standard genotyping (based on PCR/sequencing or on chips) does not allow the direct determination of haplotypes, and current experimental solutions to this problem are still expensive, time-consuming, and not suitable at a genome-wide scale (7). Moreover, the strategy of collecting small family pedigrees such as trios to determine the transmitted haplotypes via Mendel inheritance laws remains often infeasible since parental genotypes are rarely available. These constraints have motivated the development of computational alternatives for the last 20 years to determine haplotypes in samples of unrelated individuals (8). This field is one of the most challenging and dynamic of statistical genetics as shown by the number and the variety of methods that have already been developed (more than 50 in 2004) (9).

Owing to the large number of methods developed so far, this chapter is not a comprehensive review of all these methods, but rather an introduction to the haplotype reconstruction problem and to the main strategies developed to face it in the context of diploid organisms. We first describe briefly the genetic processes implied in the creation of haplotypes via a toy example (see Subheading 2.1). The readers acquainted with the concept of linkage disequilibrium and haplotypes may skip this part to go directly to the following one that deals with the haplotype reconstruction problem (Subheading 2.2). Once these basic points are clarified, the main topic of this chapter will be the description of the best-known generic methods. Finally, practical aspects about the implementation and the applications of such methods on real datasets will be provided.

2. Basic Concepts

2.1. Haplotypes and Linkage Disequilibrium

To describe the related concepts of haplotype and linkage disequilibrium, it is helpful to consider the simplistic example illustrated in Fig. 1. Suppose that, all the individuals in a population carry chromosomes that have either a wild allele $a1$ or a mutant allele $a2$ for a SNP A (Fig. 1a). Now consider that the three following events occur successively:

1. A mutation occurs in the neighbourhood of locus A, creating a new SNP B with an ancestral allele $b1$ and a new allele $b2$ (Fig. 1b). If the mutation occurs on a chromosome that already carries the allele $a1$ for SNP A, there are three different combinations of alleles in the population: $a1b1$, $a2b1$, and $a1b2$. Note that allele $b2$ is always associated on the same chromosome with allele $a1$, which implies that allele $b2$ is always inherited with allele $a1$ to the next generation in absence of recombination. The allele combination $a1b2$ is a haplotype and the non-random association between alleles of A and B is called linkage disequilibrium (LD).

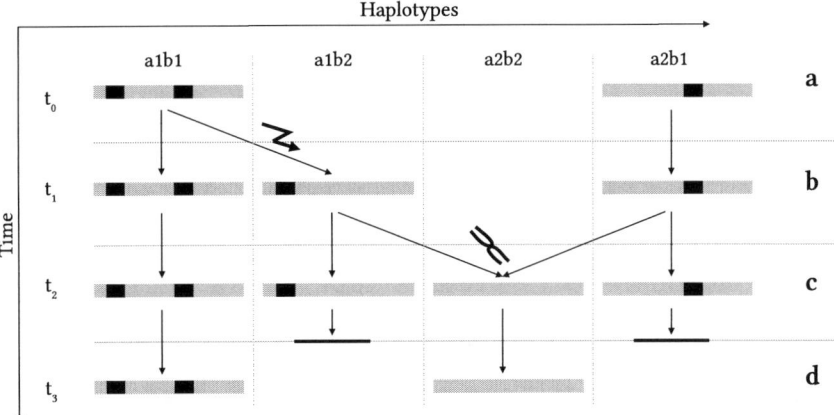

Fig. 1. Toy example of the haplotype diversification process over two SNPs A and B with respective alleles *a1/a2* and *b1/b2*. (**a**) There are two haplotypes in the population *a1b1* and *a2b1*. (**b**) A mutation occurs on *a1b1* creating *a1b2*. (**c**) Haplotypes *a1b2* and *a2b1* are recombined creating thus haplotype *a2b2*. (**d**) Haplotypes *a2b1* and *a1b2* disappear from the population leaving only haplotypes *a1b1* and *a2b2*.

2. A recombination occurs between *A* and *B*, where haplotypes *a2b1* and *a1b2* are combined to form the fourth possible haplotype *a2b2* (Fig. 1c). The recombination breaks the association between alleles *a1* and *b2* and thus decays the LD between locus *A* and *B*. Obviously, the more distant loci *A* and *B* are, the more frequent the recombinations between them may occur, and therefore the more rapidly the LD will decay.

3. Haplotypes *a1b2* and *a2b1* disappear in the population. This occurs generally through random sampling (genetic drift) or variation in fitness (i.e., difference in survival and/or reproduction between haplotypes) leading to the selection of one haplotype (Fig. 1d).

Thus, in a population, haplotypes are created by mutations, affected by recombinations, and sampled by population history. Repeating such events for several SNPs over a large number of generations may lead to complex LD patterns that can be either fully captured by describing exhaustively the haplotype diversity or statistically summarized by measuring pairwise LD levels (10). Ten years ago, it was shown that the vast majority of the human genome is structured by blocks of high LD separated by recombination hotspots around which recombinations occur more frequently (11–13). For instance, it is estimated that ~60% of recombination events were involving only ~6% of the genomic regions (14). This block organization can be interpreted in two ways:

1. From the LD point of view, LD between markers present in the same block is stronger than the LD between markers in different blocks.

2. From the haplotype point of view, the haplotype diversity in a block is reduced and mainly stems from mutations (a block of

m SNPs implies about $m+1$ haplotypes), whereas between blocks, the diversity is important and comes from the combination of haplotypes from each block (schematically two blocks with ~m haplotypes implies ~m^2 haplotypes in total).

This block structure is of crucial interest. Indeed, it provides a reading frame that simplifies the analysis of haplotype diversity since groups of alleles often inherited together can be easily identified.

2.2. Haplotype Reconstruction

In diploid organisms, a genotype is formed by a set of two haplotypes. If an individual carries alleles *a1/a2* and *b1/b2* for two heterozygous SNP *A* and *B* respectively, the problem is to determine if a*1* is associated with *b1* or *b2* on the same chromosome, that is to reconstruct the phase between SNP *A* and *B*. For two heterozygous SNP there are two possible phases, i.e., two candidate pairs of haplotypes. In a general way, reconstructing the haplotypes starting from a genotype consists in finding the real pair of haplotypes among all the possible ones (i.e., 2^{s-1} candidate pairs of haplotypes for *s* heterozygous loci). Since haplotypes correspond to combinations of alleles that are inherited through generations, when genotypes of family members are known, some phases between heterozygous SNPs can be deterministically inferred by applying Mendel inheritance rules. In practice, small family units are used, such as trios (two parents and a child) or duos (one parent and a child). The parental origin of the child alleles is given by making a correspondence with the alleles of the parents. As a consequence, the majority of heterozygous markers can be phased (excepted when all the family members are heterozygous), which greatly reduces the number of candidate haplotypes. For instance, about 75% and 85% of the heterozygous SNPs can be phased for respectively duo and trio genotypes in human populations (3).

In samples of unrelated individuals, direct relatedness between individuals does not exist and it prevents any application of the deterministic phasing procedure described above. However, if the individuals come from the same population, they share a common background that implies some ancestral relatedness on which more sophisticated inference processes can be based.

3. Methods

A haplotype reconstruction method is based on the two following components (15):

1. The genetic **model** regroups realistic assumptions about the pattern of haplotypes that is expected in a population.

2. The computational **algorithm** determines which one, among all the candidate haplotype reconstructions of the sample, is the most consistent with the genetic model.

Many genetic models for haplotypes have been proposed so far, with a degree of complexity that depends on the assumptions made about the haplotype diversification process (cf. Subheading 1.1). Given that individuals share the same background, they share common *recent* ancestors and thus some common haplotype patterns inherited from these ancestors. All genetic models are built upon this "recent ancestor" assumption and differ on the assumptions made about recombinations and mutations that have occurred since the ancestral state. Basically, there are three kinds of genetic models:

1. The **simple** models assume that no mutation and no recombination occurred. Therefore identical haplotypes are shared between individuals. These models are well suited for very short stretches of the genomes since mutations and recombinations occur more likely as the size of the considered region increases.

2. The **mutation** models assume that mutations occurred but no recombination. Therefore, almost identical haplotypes are shared between individuals that differ just in few points that correspond to past mutation events. These models are well suited for stretches of the genome of moderate size that match just one haplotype block.

3. The **mutation and recombination** models assume that mutations as well as recombinations have occurred. Therefore haplotypes are locally clustered into blocks. In a block, there are few common haplotype segments that may differ just by a few mutations. And across blocks, haplotypes are mosaics of these common haplotypes segments. These models are well suited for any region of the genome and remain the most reliable developed so far.

For these models, three kinds of computational algorithms have been developed:

1. The **combinatorial** algorithms consider in turn each possible haplotype reconstruction of the genotypes and then choose the most realistic one according to a score function.

2. The **statistical** algorithms consider the haplotype reconstruction as a set of unknown parameters whose values have to be estimated given the observed genotypes.

3. The **Bayesian** algorithms consider the haplotype reconstruction as a set of discrete random variables and estimate their joint distribution given the observed genotype data and prior assumptions on haplotype distribution.

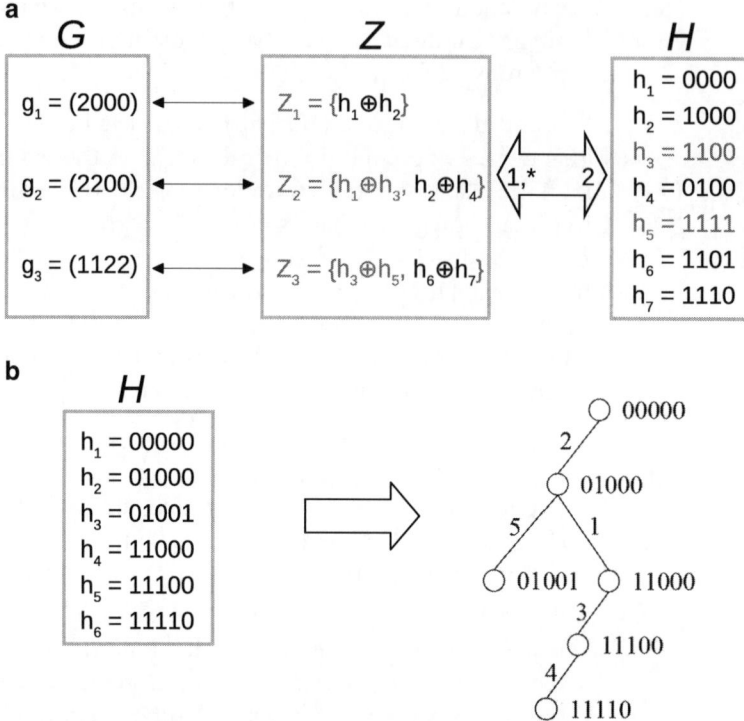

Fig. 2. (a) Example of a dataset with the associated haplotype reconstruction space. *G* is a set of three genotypes defined for four SNPs. *Z* is the set of candidate pairs of haplotypes. *H* is the set of haplotypes. In *grey*, all the possible haplotypes and pairs of haplotypes. In *black*, a single reconstruction. (b) Illustration of a perfect phylogeny of haplotypes. The set of haplotypes *H* fits to a perfect phylogeny.

Over the years, numerous model-algorithm declinations have been developed and implemented (cf. Subheading 4.1), which prevent any attempt to give an exhaustive description of all the methods. We will thus focus on the description of four methods that are the best known and representative of the field.

3.1. Notations

As illustrated in Fig. 2a, a haplotype h is a vector of s values taken in $\{0, 1\}$ that codes for the combination of s alleles. A genotype g is a vector of s values taken in $\{0, 1, 2\}$ that codes the s pairs of alleles (0 for pair 0/0, 1 for pair 1/1 and 2 for pairs 0/1 or 1/0). The sample of n genotypes is denoted $G = \{g_1, ..., g_n\}$ and the associated candidate pairs of haplotypes are denoted $Z = \{Z_1, ..., Z_n\}$. And finally, let H be the unknown haplotype reconstruction of G. Note that hereafter H will take several forms depending on the method considered.

3.2. The Simplest Statistical Model

Over a region of a genome, haplotypes are shared between individuals of a population with a given frequency of occurrence in the sample. To model that, the haplotype reconstruction H of G is considered as a vector of m frequencies $F = \{f_1, ..., f_m\}$ for the m candidate haplotypes of G. Since the true frequency values are unknown, we estimate the most likely values given the data we have, i.e., the genotypes G. To achieve that, we define first a likelihood function $L(F \mid G)$ which is the probability to obtain our sample of genotypes G given some frequencies values F; $Pr(G \mid F)$. And then, we find values of F that maximize this likelihood function. Such likelihood function is first based on a probabilistic model that relates genotype probabilities to haplotype frequencies as follows:

$$Pr(g_i \mid z_i = (h_x, h_y), F) = f_x f_y \qquad (1)$$

The underlying idea is that a candidate pair of haplotypes (h_x, h_y) for a genotype g_i is formed by sampling independently two haplotypes h_x and h_y, involving that the population is at the Hardy-Weinberg equilibrium (16). Since the true pair of haplotypes of a genotype is unknown, the probability to observe a genotype g_i is therefore obtained by summing over all its candidate pairs Z_i:

$$Pr(g_i \mid F) = \sum_{\forall z_i \in Z_i} Pr(g_i \mid z_i, F) \qquad (2)$$

And the likelihood of the whole sample G is finally obtained by using the following multinomial form:

$$L(F \mid G) = Pr(G \mid F) \propto \prod_{i=0}^{n} Pr(g_i \mid F) \qquad (3)$$

Since this likelihood corresponds to a product of many probabilities (comprised between 0 and 1), the log-likelihood is rather used in practice. It is obtained by substituting the product in equation (Eq. 3) by a sum of logarithms. To estimate values of F that maximize this likelihood function, analytical optimization is theoretically possible, but it is infeasible in practice even for very small datasets. Numerical optimization techniques are therefore used and more particularly the Expectation–Maximization (EM) algorithm (17–19). Schematically, the EM algorithm starts from arbitrary parameter values $H^{(0)}$ and goes from iteration t to $t+1$ by the following two steps:

1. E-step: computation of the probabilities of the candidate pairs of haplotypes from current values $F^{(t)}$ by using equation (Eq. 1).

2. M-step: first normalize the probabilities to have a sum of 1 for each genotype g_i. And then, compute new values $F^{(t+1)}$ by weighted gene counting:

$$f_w^{(t+1)} = \frac{1}{n} \sum_{i=1}^{n} \sum_{\forall z_i \in Z_i} \left[\frac{(\delta_{wx} + \delta_{wy})}{2} \times Pr(g_i \mid z_i = (h_x, h_y), F^{(t)}) \right] \quad (4)$$

where $(\delta_{wx} + \delta_{wy})$ indicates the number of times (0, 1 or 2) haplotype h_w appears in pair (h_x, h_y).

These two steps are repeated until convergence when the difference between two successive likelihood values becomes very low: $L(F^{(t+1)} \mid G) - L(F^{(t)} \mid G) < \varepsilon$. Since it could reach a local maxima of the likelihood curve, several random starting points are usually tested and haplotype frequencies values that give the maximum likelihood value are chosen as final optimal estimation. The EM algorithm is quite easy to implement and remains one of the most used, especially for two-SNPs haplotypes frequencies estimations. Optimal values for F may then be used directly in subsequent analyses (5, 6, 20), or, if the focus is more on reconstructing genotypes, the most likely pair of haplotypes can be picked up for each genotype by using equation (Eq. 1).

In practice, the sum in equation (Eq. 2) has to be done on an exponential number of candidate haplotypes which makes naïve implementation of the EM algorithm unsuitable for a relatively moderate number of SNPs. However, several smart strategies have been developed to correct this:

1. The "divide and conquer" strategy called Partition Ligation (PL) (21). It first divides the dataset into small groups of adjacent SNPs in which EM can be easily applied (partition) and then combines inferred "sub-haplotypes" across the whole dataset (ligation).

2. The progressive SNP inclusion strategy, called sometimes as Iterative EM (22). It includes SNPs one at a time rather than all at once in order to progressively estimate most likely haplotypes by successive fast EM algorithm runs.

The underlying idea of these strategies is to concentrate the probability computations just on the few candidate pairs of haplotypes that capture the majority of the cumulative probability mass of the distribution.

3.3. A Phylogeny-Based Combinatorial Method

In absence of recombinations, a new haplotype appears in the population after the occurrence of a mutation on an already existing haplotype (cf. Subheading 1.1). This implies that for s SNPs the occurrence chronology of the corresponding s mutations can be represented by a perfect phylogenetic tree (23, 24) (Fig. 2b) where:

1. A node is a haplotype and the root node is the most ancestral haplotype.

2. A branch between two nodes is a mutation that differentiates the two associated haplotypes.

3. A mutation appears only once in the tree.

The phylogeny-based combinatorial method considers in turn each possible haplotype reconstruction of G and choose the one for which the corresponding $2n$ haplotypes fit a perfect phylogeny. A naïve implementation cannot handle real datasets since the number of possible reconstructions is usually huge even for small datasets. To efficiently solve this problem, consider first the simple situation of a two-SNP sample; if no recombination occurs, there are only three haplotypes (cf. Subheading 1.1). Thus for s SNPs, the haplotype diversity can be explained just by mutations (i.e., by a perfect phylogeny) if it exists in an ordered sequence of $s-1$ pairs of SNPs where each pair is solved by at most three haplotypes (23). However, the haplotype diversity in a real dataset rarely fits a perfect phylogeny except for the common haplotypes found in a block. The perfect phylogeny assumption was thus relaxed in several manners in order to be applied more easily on real datasets, leading to a variety of imperfect phylogeny assumptions (25). From these, the most convenient approach aims to find a haplotype reconstruction of G that has the less imperfect phylogeny, i.e., that minimizes the number of pairs of SNPs that rely on 4-haplotypes.

3.4. Hidden Markov Models of Haplotypes

None of the two previously described methods takes explicitly into account recombinations that necessarily occur when dealing with large genome stretches spanning several haplotypes blocks (cf. Subheading 1.1). Haplotypes in such datasets are "imperfect mosaics" of other haplotypes of the sample whose boundaries depend on the variable recombination rate background (Fig. 3a). The term "mosaic" means that haplotypes result from combinations of small common segments and the term "imperfect" means that the recombinant segments may be altered by mutations (Fig. 3b) (26, 27).

Let $H = \{h_1, \ldots, h_K\}$ denote a set of K known haplotypes and let z denote a particular mosaic of these haplotypes, i.e., a sequence of recombination or no recombination events. Note that there are K^s possible mosaics of K haplotypes of s SNPs. To derive a probabilistic model able to estimate how likely a haplotype h stems from a mosaic z of haplotype segments of H, two probabilistic quantities are needed (Fig. 3c):

1. The probability matrix T of all possible recombinations between haplotypes of H. Thus the probability $Pr(z \mid H, T)$ of mosaic z is given by the product of the probabilities of the recombination events contained in z. These probabilities are called transition probabilities.

2. The probability matrix E of all possible mutations on the haplotypes of H. Thus the probability $Pr(h \mid z, H, E)$ that mosaic z mutates into haplotype h is given by the product of the probabilities of the mutations that distinguish mosaic z of haplotype h.

Since the mosaic z is unknown in practice, the conditional probability of haplotype h given H is obtained by summing over all possible mosaics:

$$Pr(h|H,T,E) = \sum_{\forall z \in Z} Pr(h|z,H,E)\, Pr(z|H,T) \qquad (5)$$

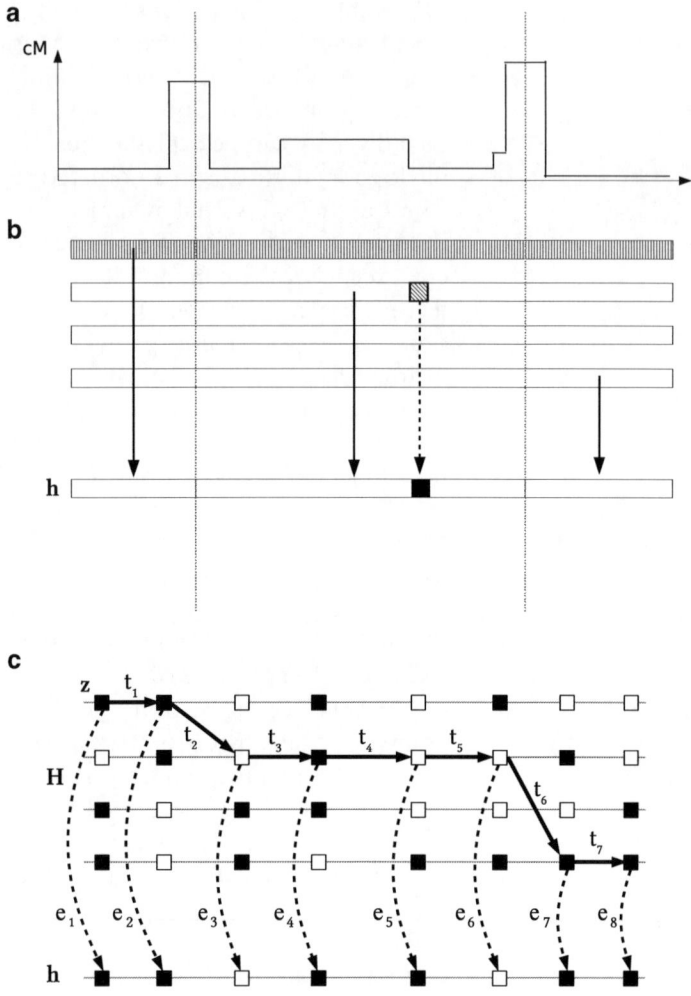

Fig. 3. Schematic illustration of a hidden Markov model of haplotypes. (**a**) The recombination rate background which determines the transition probabilities in the HMM. (**b**) An example of how a haplotype h can be built as an "imperfect mosaic" of known haplotypes. (**c**) A graphical representation of the imperfect mosaic process by a hidden Markov model of haplotypes. The hidden states of the model are given by the alleles of the known haplotypes. The probability of a mosaic z of haplotype segments of H is given by the product of corresponding transition probabilities $Pr(z \mid H, T) = t_1 t_2 t_3 t_4 t_5 t_6 t_7$ where t_1, t_3, t_4, t_5, and t_7 are the probabilities that no recombination occurs and t_2 and t_6 are the probabilities that concerned haplotypes recombine between the two SNPs. Then, the probability that the haplotype h stems from the mosaic z of haplotype segments is given by the product of corresponding emission probabilities $Pr(h \mid z, H, E) = e_1 e_2 e_3 e_4 e_5 e_6 e_7 e_8$ where e_1, e_2, e_3, e_4, e_6, e_7, and e_8 are the probabilities that no mutation occurred and e_5 is the probability that a mutation occurred at this SNP.

Such models are usually called hidden Markov models (HMM) since the computation of $Pr(z \mid H, T)$ relies on a Markov process and since z does not match perfectly to h. Nowadays, HMMs are extensively used to model the block structure of haplotypes in statistical genetics. Naïve computation of equation (Eq. 5) requires to sum over K^s possible mosaics which remains infeasible even for few SNPs and few haplotypes. Fortunately, an efficient algorithm for HMMs, called forward–backward, allows computing recursively the sum with a computational effort that increases only linearly with Ks (28).

3.5. Bayesian Method for HMMs

In the Bayesian context, the haplotype reconstruction H of G is considered as a set of n discrete random variables $\{H_1, ..., H_n\}$ whose respective possible outcomes are $\{Z_1, ..., Z_n\}$ and estimate their joint distribution $Pr(H \mid G)$ given the observed genotype data G. In practice, to estimate $Pr(H \mid G)$, the most common method relies on a Gibbs sampler (GS) (15, 26, 29). The GS is a Monte Carlo Markov chain $H^{(0)}, ..., H^{(t)}, ...$ in which each state $H^{(t)}$ is a reconstruction of G in $2n$ haplotypes. The GS starts from a random reconstruction $H^{(0)}$ and goes from $H^{(t)}$ to $H^{(t+1)}$ through the three following steps:

1. Picking up randomly a genotype g_i in G.
2. Computing the distribution $Pr(H_i \mid H_{-i}^{(t)})$ where $H_{-i}^{(t)}$ is the set of $2n-2$ haplotypes of $H^{(t)}$ that are assigned to the genotypes $\{g_1, ..., g_{i-1}, g_{i+1}, ..., g_n\}$ (All genotypes excepted g_i).
3. Constructing $H^{(t+1)}$ by sampling from $Pr(H_i \mid H_{-i}^{(t)})$ a new pair of haplotypes for g_i.

These three steps are iterated a large number of times, then the first states of the Markov chain are discarded (called burn-in iterations) and an estimation of the target distribution $Pr(H \mid G)$ is obtained by averaging over the remaining states of the Markov chain. Once $Pr(H \mid G)$ is estimated, it is straightforward to obtain a single point estimate of the haplotype reconstruction of G by picking up the n most likely pairs of haplotypes.

The computation of $Pr(H_i \mid H_{-i}^{(t)})$ distribution is to compute the conditional probabilities of the candidate pairs of haplotypes of g_i given the others $2n-2$ haplotypes found in the rest of the sample. Similarly to equation (Eq. 1), a pair of haplotypes is formed by sampling independently the two corresponding haplotypes:

$$Pr(H_i = (h, h') \mid H_{-i}^{(t)}) = Pr(h \mid H_{-i}^{(t)}) \times Pr(h' \mid H_{-i}^{(t)}) \qquad (6)$$

The probabilities $Pr(h \mid H_{-i})$ are computed through an HMM of haplotypes (cf. Subheading 3.4) by summing over all the possible mosaics of $2n-2$ haplotypes as in equation (Eq. 5). This HMM is parametrized by the following transition and emission distributions.

The **transition** probability that haplotype h_x is recombined with haplotype h_y between SNP j and $j+1$ is defined as follows:

$$t_{h_x \to h_y}(j) = \begin{cases} e^{-\rho_j d_j / K} + (1 - e^{-\rho_j d_j / K}) / K & \text{if } h_x = h_y \\ (1 - e^{-\rho_j d_j / K}) / K & \text{if } h_x \neq h_y \end{cases} \tag{7}$$

where d_j and ρ_j are respectively the distance in base pair and the recombination rate between SNPs j and $j+1$. The idea here is that a recombination occurs with probability $1 - e^{-\rho_j d_j / K}$, and when such an event occurs, it is equally likely with any of the target haplotype h_y. Therefore the case of no recombination observed between h_x and h_y (the upper term of Eq. 7) regroups the cases of no recombination ($e^{-\rho_j d_j / K}$) and recombination with another copy of exactly the same haplotype ($(1 - e^{-\rho_j d_j / K}) / K$). Note that a recombination occurs more likely when the product of ρ_j and d_j is large. This captures the decay of LD over long distance or for a high recombination rate. The $s-1$ recombination rates ρ_j can either be inferred from the assigned haplotypes in a parallel MCMC scheme (26, 27) or they can be computed from an already available genetic map (30, 31).

The **emission** probability that the jth allele of haplotype h_x mutates into the jth allele of haplotype h is defined as follows:

$$e_{h_x \to h}(j) = \begin{cases} \dfrac{K}{2(K+\theta)} + \dfrac{\theta}{2(K+\theta)} & \begin{array}{l} \text{if the } j \text{th alleles of } h_x \\ \text{and } h \text{ are identicals} \end{array} \\ \dfrac{\theta}{2(K+\theta)} & \begin{array}{l} \text{if the } j \text{th alleles of } h_x \\ \text{and } h \text{ are differents} \end{array} \end{cases} \tag{8}$$

Where θ is a known scaled mutation rate. This emission probability arises from Ewens sampling formulas (26), where a new allele is sampled with probability $\theta/(K+\theta)$ and an old one with probability $K/(K+\theta)$. Thus, the case where h_x is equal to h (the upper term of Eq. 8) regroups the cases of no mutation ($K/(K+\theta)$) and mutation to the same allele type ($\theta/2 (K+\theta)$) since SNPs are bi-allelic.

The computational difficulties in implementing such a GS algorithm rely first on the exponential numbers of outcomes (pairs of haplotypes) on which the distributions must be estimated, and second on the number of haplotypes on which the HMM has to be built. As for the EM algorithm described above (cf. Subheading 3.2), partition–ligation (15) or progressive SNP inclusion (32) strategies may circumvent this exponential complexity. However, these strategies are gradually replaced by alternative methods that sample directly pairs of haplotypes from a coupled HMM of haplotypes by recovering pairs of mosaics via the Viterbi algorithm (33, 34) or via more sophisticated sampling schemes (35). Moreover, further improvements were recently introduced to reduce the computational burden required by a large number of individuals: now HMMs are usually built just from a limited number of randomly chosen haplotypes in H_{-i} (36) or from haplotypes in H_{-i} closely related to the genotype of interest (31).

3.6. Maximum Likelihood Method for HMMs

In the above Bayesian framework, the HMM transition and emission probabilities are expressed as population genetics quantities. In the frequentist framework, they are considered as unknown parameters $H = \{T, E\}$ for which maximum likelihood estimates can be found by maximizing the probability to obtain the observed data. For this purpose, we assume that haplotype diversity in the dataset results from mutation and recombination events on a limited number K of ancestral haplotypes on which an HMM is built (35, 37). Such an ancestral haplotype is no longer defined as a classical vector of alleles 0/1 but rather as a vector whose values are comprised between 0 and 1 and code for the alleles in a continuous manner. For instance a value of 0.3 indicates that the ancestral haplotype emits allele 1 with probability 0.3 and allele 0 with probability 0.7. Such allele coding allows fuzzy definitions of the ancestral haplotypes and, therefore, allows imperfect matches with observed alleles in order to mimic mutations. The likelihood function has the multinomial form as in equation (Eq. 3), and $Pr(g \mid K, T, E)$ is obtained by summing over all possible mosaics derived of ancestral haplotypes instead of all possible pairs of haplotypes. This has the nice property of being quadratic with K instead of being exponential. The maximum likelihood estimation of the parameters T and E is achieved via an EM algorithm with several starting points. The values that yield the maximum likelihood value are then kept as a final estimation, although some authors prefer to average over the starting points arguing that it yields even better estimates (35). Concerning the unknown number K of ancestral haplotypes, the simplest solution consists in choosing an a priori value (for instance 20) (37), but cross validation techniques have also been described (35). The idea is to artificially introduce missing data in the dataset, then run several times the EM algorithm with increasing values of K (8, 12, 16, etc...), and to select the K value which recovers best the missing data.

4. Implementations and Practical Applications

4.1. Implementations

Historically, the first haplotype inference software was developed in 1990 (38). It was based on a combinatorial algorithm to find a parsimonious reconstruction of the sample (i.e., which relies on few distinct haplotypes). Almost 5 years later, with the growing amount of genetic data for candidate gene studies, several authors proposed simultaneously a statistical alternative based on a simple genetic model able to handle readily real datasets. These methods took abusively the name of the statistical algorithm upon which they were based: the Expectation–Maximization (EM) algorithm (17–19). In the early 2000, several Bayesian alternatives for the EM algorithm were developed still upon simple genetic models (29). From 2003, more realistic and sophisticated genetic models taking into account explicitly mutations and/or recombinations were published. Among them, the more accurate is implemented in the Phase v2.1 software (27). Despite its prohibitive computation times, it was extensively used in genetic studies and popularized the use of HMMs of haplotypes for the purpose of haplotype reconstruction.

We now review the main software implementing haplotype reconstruction:

1. HAPLOVIEW   (http://www.broadinstitute.org/scientific-community/science/programs/medical-and-population-genetics/haploview/haploview)

    An implementation of the simple EM algorithm (cf. Subheading 2.2) with a friendly graphical interface to visualize block structure and LD plots (39).

2. HAP (http://research.calit2.net/hap/)

    A combinatorial algorithm based on imperfect phylogeny (cf. Subheading 2.3). Probably the most adapted combinatorial method for real datasets. It uses a sliding window strategy to handle large datasets and to recover haplotype block structure (25).

3. PHASE   v2.1   (http://stephenslab.uchicago.edu/software.html)

    The gold standard software. A Bayesian HMM based method (cf. Subheadings 2.4 and 2.5) that is the most accurate method developed so far (27). Despite the use of a partition–ligation strategy to handle large datasets, it remains very slow in practice (40).

4. ISHAPE (http://www.griv.org/ishape/)

    A modified implementation of PHASE v2.1 which limits the number of candidate haplotypes by pre-treatment with the EM algorithm in order to accelerate computations (22).

5. SHAPEIT (http://www.griv.org/shapeit/)

   A very rapid implementation of PHASE v2.1 algorithm which uses the progressive SNP inclusion technique (cf. Subheadings 2.2 and 2.5) while maintaining the same accuracy of PHASE v2.1 (32).

6. MACH (http://www.sph.umich.edu/csg/abecasis/MACH/index.html)

   A Bayesian HMM method well adapted for large datasets (cf. Subheading 2.4) (36). Two improvements reduce the computational burden of Phase v2.1. First, the number of previously sampled haplotypes on which the HMM is built is limited by a constant (200 by default). Second, the pair of haplotypes for a genotype is sampled directly from the HMM instead of considering each candidate pairs of haplotypes (cf. Subheading 2.4).

7. BEAGLE (http://faculty.washington.edu/browning/beagle/beagle.html)

   The fastest HMM-based algorithm developed so far. Previously sampled haplotypes are clustered into a graph on which it is very fast to sample new haplotypes for a genotype (41–43).

8. IMPUTE (https://mathgen.stats.ox.ac.uk/impute/impute_v2.html)

   Another HMM-based algorithm, which builds the HMM upon a subset of previously sampled haplotypes that are close in terms of mutations to the current haplotypes assigned for the genotype we are updating (30, 31).

9. FASTPHASE (http://stephenslab.uchicago.edu/software.html)

   A HMM of haplotypes software for which maximum likelihood estimates of transition and emission probabilities are found via an Expectation–Maximization algorithm (cf. Subheading 2.5) (35). To estimate the number K of ancestral haplotypes, a cross validation technique is used; missing data are artificially introduced in the dataset, then the HMM is fitted to the data several times with several values for K, and the value of K that best predicts the missing data is kept as the true optimal value. Moreover, the pair of haplotypes for a genotype is directly sampled from the HMM to accelerate computations.

4.2. Practical Aspects of Haplotype Reconstruction

Given a sample of genotypes for which the haplotypes have to be reconstructed, several questions arise.

1. Are in silico alternatives sufficiently reliable for downstream analyses? That question was first addressed 10 years ago for the EM algorithm with the simple frequency model. Datasets with small number of SNPs for which haplotypes were determined experimentally were used to estimate if the haplotypes and the

corresponding frequencies obtained by EM were reliable (44, 45). These studies showed that reconstructions were very accurate for common haplotypes with frequency above 1%, but remained more uncertain for rare haplotypes (46). These results are consistent since the simple model does not make any assumption about mutation and recombination, and thus it is unable to promote any haplotype pattern for rare haplotypes.

2. How to use the inferred haplotypes in downstream analyses? The direct approach is to base the analyses on the best set of $2n$ haplotypes. Although this simple approach is often used in practice, it is more recommended to propagate the uncertainty of haplotype assignments (i.e., probabilities of pairs of haplotypes) into the downstream analyses when it is possible (6, 20).

3. Which method should be used? Recently, an extensive comparison study was performed on unrelated and trio datasets generated from HapMap (cf. Subheading 3.3) to compare several representative software of haplotype inference (40). The methods that yielded the most accurate haplotypes were based on HMM which remains indeed the most adapted model to capture complex LD pattern. Among the HMM methods, PHASE v2.1 (cf. Subheading 3.1) seems to give the best results with around 99.9% and 95% of phases correctly reconstructed for respectively trio and unrelated datasets with moderate SNP density. PHASE v2.1 is very slow in practice, and several alternatives have been proposed with either algorithmic improvements (SHAPE-IT and ISHAPE) or modifications of the model (MACH and IMPUTE), to accelerate the computations.

4. What are the parameters that decrease accuracy? The LD pattern in a dataset is determinant for the quality of the reconstructed haplotypes, since LD is in some ways the information upon which haplotype inference is based. As LD gets weaker, it becomes harder to recover accurately the haplotypes. As a consequence, the factors decreasing the LD level reduce the accuracy of haplotype inference:

    (a) A low density of SNPs decreases haplotypes accuracy since LD levels are weak between distant SNPs. Thus it is strongly recommended to base inference on all the SNPs available in a region and then extract the haplotypes on the subset of SNPs studied. It was already shown that this strategy substantially improves the accuracy (47).

    (b) The LD level varies between Human populations (3), hence haplotypes may be harder to recover in some populations. For instance in African populations LD is generally lower compared to Asian or European populations.

(c) A particular LD pattern may be very difficult to capture. For example, haplotypes in a specific region containing several recombination hotspots are usually more difficult to infer than in a region that contains just a single haplotype block.

4.3. Haplotype Maps in Human Genetics

Owing to the growing interest of haplotypes in genetic epidemiology or in population genetics, an international consortium has built in three successive phases a dense catalogue, called HapMap (http://hapmap.ncbi.nlm.nih.gov/), of Human genetic variations:

1. For the phase 1 (1), more than one million SNPs were accurately and completely genotyped among 270 individuals from three populations: 30 African trios, 90 unrelated Asians and 30 European trios.

2. For the phase 2 (2), the density of the catalogue was increased to 3.1 millions SNPs among the same 270 individuals.

3. For the phase 3 (3), the depth of the catalogue was increased to 1,301 individuals from 11 distinct populations.

These data have allowed the determination of the common haplotypes among Human populations (48), the characterization of numerous recombination hotspots (49), the definition of the block structure of LD that results in low haplotype diversity and the identification of some loci submitted to natural selection during human evolution (50). All these findings are used in genetic studies in the following way. First, they allow identifying some minimal subsets of SNP able to capture most of the common haplotype diversity of the Human genome (Fig. 4a) upon which modern genotyping chips are designed. This greatly reduces the genotyping costs while maintaining good coverage of the Human genetic diversity. Once the samples are genotyped with such chips, untyped SNPs can be recovered by making some correspondence between typed SNPs and haplotypes from the HapMap panels (Fig. 4b–e). This process is commonly called SNP imputation and is used in practice to improve power of genetic studies and to make possible meta-analyses between genetic studies based on distinct genotyping chips (30, 51).

To complete and extend the HapMap data with rare SNPs, insertions, deletions, copy number variations, and other polymorphisms, a new project called 1,000 genomes (http://www.1000genomes.org) is currently sequencing completely the genomes of around 2,500 individuals from 16 different populations (52).

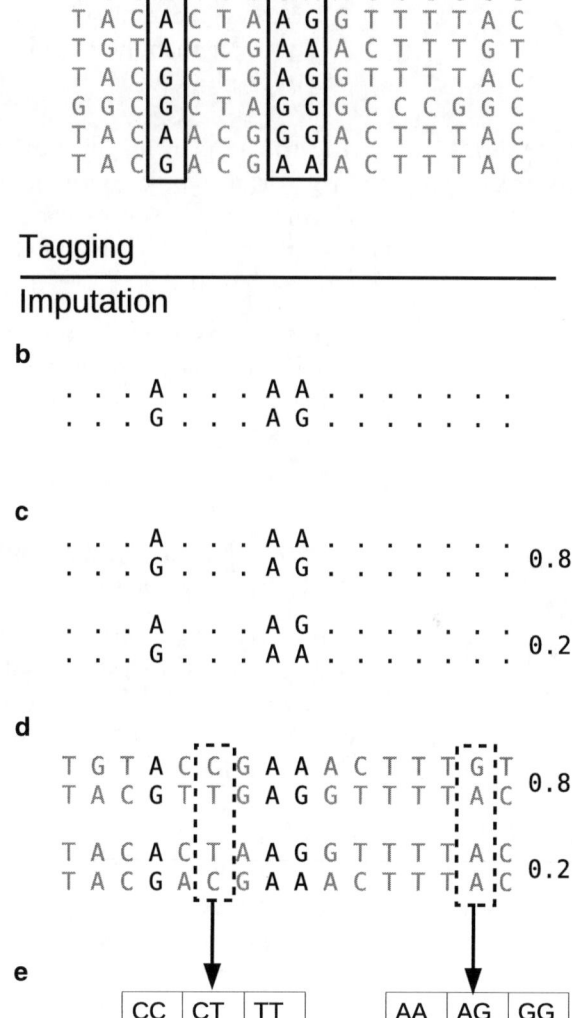

Fig. 4. A schematic illustration of the SNP tagging and imputation processes. (**a**) The seven haplotypes defined on 16 SNPs can be mapped just with a minimal set of three SNPs. These SNPs are called tagging SNPs and are sufficient to capture all the haplotype diversity of the region. (**b**) Suppose that an individual is genotyped with a chip that was built on these three tagging SNPs. This individual is heterozygous for the first and the third SNP and homozygous for the second one. (**c**) The genotype is phased and the two pairs of haplotypes are obtained with respective probabilities 0.8 and 0.2, so there is an uncertainty in the phasing. (**d**) Now given a reference panel (for example HapMap) for which the haplotypes are defined on all the available SNPs of the region, it is straightforward to complete the untyped SNP of the pairs of haplotypes. (**e**) Given the probabilities of the pairs of haplotypes, it is also straightforward to compute the probabilities of the untyped genotypes.

References

1. The HapMap consortium (2003) The international HapMap project. Nature 426:789–796

2. The HapMap consortium (2005) A haplotype map of the human genome. Nature 437: 1299–1320

3. The HapMap consortium (2007) A second generation human haplotype map of over 3.1 million SNPs. Nature 449:851–861

4. The Wellcome Trust Case-Control Consortium (2007) Genome-wide association study of 14,000 cases of seven common diseases and 3,000 shared controls. Nature 447:661–678

5. Zhang S, Pakstis AJ, Kidd KK, Zhao H (2001) Comparisons of two methods for haplotype reconstruction and haplotype frequency estimation from population data. Am J Hum Genet 69:906–914

6. Schaid DJ (2004) Evaluating associations of haplotypes with traits. Genet Epidemiol 27: 348–364

7. Xu J (2006) Extracting haplotypes from diploid organisms. Curr Issues Mol Biol 8:113–122

8. Niu T (2004) Algorithms for inferring haplotypes. Genet Epidemiol 27:334–347

9. Salem RM, Wessel J, Schork NJ (2005) A comprehensive literature review of haplotyping software and methods for use with unrelated individuals. Hum Genomics 2:39–66

10. Pritchard JK, Przeworski M (2001) Linkage disequilibrium in humans: models and data. Am J Hum Genet 69:1–14

11. Daly MJ, Rioux JD, Schaffner SF et al (2001) High-resolution haplotype structure in the human genome. Nat Genet 29:229–232

12. Patil N, DA BernoAJ H et al (2001) Blocks of limited haplotype diversity revealed by high-resolution scanning of human chromosome 21. Science 294:719–1723

13. Gabriel SB, Schaffner SF, Nguyen H et al (2002) The structure of haplotype blocks in the human genome. Science 296:2225–2229

14. Kong A, Gudbjartsson DF, Sainz J et al (2002) A high-resolution recombination map of the human genome. Nat Genet 31:241–247

15. Stephens M, Donnelly P (2003) A comparison of bayesian methods for haplotype reconstruction from population genotype data. Am J Hum Genet 73:1162–1169

16. Mayo O (2008) A century of Hardy-Weinberg equilibrium. Twin Res Hum Genet 11:249–256

17. Excoffier L, Slatkin M (1995) Maximum-likelihood estimation of molecular haplotype frequencies in a diploid population. Mol Biol Evol 12:921–927

18. Long JC, Williams RC, Urbanek M (1995) An E-M algorithm and testing strategy for multiple-locus haplotypes. Am J Hum Genet 56:799–810

19. Hawley ME, Kidd KK (1995) HAPLO: a program using the EM algorithm to estimate the frequencies of multi-site haplotypes. J Hered 86:409–411

20. Zaykin DV, Westfall PH, Young SS et al (2002) Testing association of statistically inferred haplotypes with discrete and continuous traits in samples of unrelated individuals. Hum Hered 53:79–91

21. Qin ZS, Niu T, Liu JS (2002) Partition-ligation-expectation-maximization algorithm for haplotype inference with single-nucleotide polymorphisms. Am J Hum Genet 71: 1242–1247

22. Delaneau O, Coulonges C, Boelle P et al (2007) ISHAPE: new rapid and accurate software for haplotyping. BMC Bioinformatics 8:205

23. Bafna V, Gusfield D, Lancia G, Yooseph S (2003) Haplotyping as perfect phylogeny: a direct approach. J Comput Biol 10:323–340

24. Eskin E, Halperin E, Karp RM (2003) Efficient reconstruction of haplotype structure via perfect phylogeny. J Bioinform Comput Biol 1:1–20

25. Halperin E, Eskin E (2004) Haplotype reconstruction from genotype data using Imperfect Phylogeny. Bioinformatics 20:1842–1849

26. Li N, Stephens M (2003) Modeling linkage disequilibrium and identifying recombination hotspots using single-nucleotide polymorphism data. Genetics 165:2213–2233

27. Stephens M, Scheet P (2005) Accounting for decay of linkage disequilibrium in haplotype inference and missing-data imputation. Am J Hum Genet 76:449–462

28. Rabiner LR (1989) A tutorial on hidden Markov model and selected applications in speech recongnition. Proc IEEE 77:257–285

29. Stephens M, Smith NJ, Donnelly P (2001) A new statistical method for haplotype reconstruction from population data. Am J Hum Genet 68:978–989

30. Marchini J, Howie B, Myers S et al (2007) A new multipoint method for genome-wide association studies by imputation of genotypes. Nat Genet 39:906–913

31. Howie BN, Donnelly P, Marchini J (2009) A flexible and accurate genotype imputation method for the next generation of genome-wide association studies. PLoS Genet 5:e1000529

32. Delaneau O, Coulonges C, Zagury J (2008) Shape-IT: new rapid and accurate algorithm for haplotype inference. BMC Bioinformatics 9:540

33. Kimmel G, Shamir R (2005) The incomplete perfect phylogeny haplotype problem. J Bioinform Comput Biol 3:359–384

34. Sun S, Greenwood CMT, Neal RM (2007) Haplotype inference using a Bayesian Hidden Markov model. Genet Epidemiol 31:937–948

35. Scheet P, Stephens M (2006) A fast and flexible statistical model for large-scale population genotype data: applications to inferring missing genotypes and haplotypic phase. Am J Hum Genet 78:629–644

36. Li Y, Abecasis GR (2006) Mach 1.0: rapid haplotype reconstruction and missing genotype inference. Am J Hum Genet 79:2290

37. Kimmel G, Shamir R (2005) A block-free hidden Markov model for genotypes and its application to disease association. J Comput Biol 12:1243–1260

38. Clark AG (1990) Inference of haplotypes from PCR-amplified samples of diploid populations. Mol Biol Evol 7:111–122

39. Barrett JC, Fry B, Maller J, Daly MJ (2005) Haploview: analysis and visualization of LD and haplotype maps. Bioinformatics 21: 263–265

40. Marchini J, Cutler D, Patterson N et al (2006) A comparison of phasing algorithms for trios and unrelated individuals. Am J Hum Genet 78:437–450

41. Browning SR, Browning BL (2007) Rapid and accurate haplotype phasing and missing-data inference for whole-genome association studies by use of localized haplotype clustering. Am J Hum Genet 81:1084–1097

42. Browning SR (2008) Missing data imputation and haplotype phase inference for genome-wide association studies. Hum Genet 124: 439–450

43. Browning BL, Yu Z (2009) Simultaneous genotype calling and haplotype phasing improves genotype accuracy and reduces false-positive associations for genome-wide association studies. Am J Hum Genet 85:847–861

44. Tishkoff SA, Pakstis AJ, Ruano G, Kidd KK (2000) The accuracy of statistical methods for estimation of haplotype frequencies: an example from the CD4 locus. Am J Hum Genet 67:518–522

45. Fallin D, Schork NJ (2000) Accuracy of haplotype frequency estimation for biallelic loci, via the expectation-maximization algorithm for unphased diploid genotype data. Am J Hum Genet 67:947–959

46. Adkins RM (2004) Comparison of the accuracy of methods of computational haplotype inference using a large empirical dataset. BMC Genet 5:22

47. Coulonges C, Delaneau O, Girard M et al (2006) Computation of haplotypes on SNPs subsets: advantage of the "global method". BMC Genet 7:50

48. Hinds DA, Stuve LL, Nilsen GB et al (2005) Whole-genome patterns of common DNA variation in three human populations. Science 307:1072–1079

49. Myers S, Bottolo L, Freeman C et al (2005) A fine-scale map of recombination rates and hotspots across the human genome. Science 310:321–324

50. Sabeti PC, Varilly P, Fry B et al (2007) Genome-wide detection and characterization of positive selection in human populations. Nature 449:913–918

51. Marchini J, Howie B (2010) Genotype imputation for genome-wide association studies. Nat Rev Genet 11:499–511

52. The 1000 Genomes Project Consortium (2010) A map of human genome variation from population-scale sequencing. Nature 467: 1061–1073

Chapter 12

Allele Identification in Assembled Genomic Sequence Datasets

Katrina M. Dlugosch and Aurélie Bonin

Abstract

Allelic variation within species provides fundamental insights into the evolution and ecology of organisms, and information about this variation is becoming increasingly available in sequence datasets of multiple and/or outbred individuals. Unfortunately, identifying true allelic variants poses a number of challenges, given the presence of both sequencing errors and alleles from other closely related loci. We outline the key considerations involved in this process, including assessing the accuracy of allele resolution in sequence assembly, clustering of alleles within and among individuals, and identifying clusters that are most likely to correspond to true allelic variants of a single locus. Our focus is particularly on the case where alleles must be identified without a fully resolved reference genome, and where sequence depth information cannot be used to infer the putative number of loci sharing a sequence, such as in transcriptome or post-assembly datasets. Throughout, we provide information about publicly available tools to aid allele identification in such cases.

Key words: Allelic variation, Paralogs, Gene duplication, Maximum likelihood clustering, Single-linkage clustering, AllelePipe, Granularity, Transcriptome data, Next-generation sequencing

1. Introduction

Surveys of intra-specific molecular genetic variation now form the core of evolutionary studies of individual organisms. The frequencies of mutations segregating within and across populations can reveal the history of migration, gene flow, demography, recombination, and natural selection in a species, as well as the genetic basis of its phenotypes (1–4). Since the early allozyme studies in the late 1960s, it has been clear that such genetic diversity is pervasive in living organisms (5); yet, only now are we getting a clear picture of the extent and nature of this diversity through a few model species. For example, the latest data from the 1000-genome project identified

François Pompanon and Aurélie Bonin (eds.), *Data Production and Analysis in Population Genomics: Methods and Protocols*, Methods in Molecular Biology, vol. 888, DOI 10.1007/978-1-61779-870-2_12, © Springer Science+Business Media New York 2012

15 million single nucleotide polymorphisms (SNPs), one million short insertion–deletion (indel) mutations, and 20,000 structural variants in the human genome, most of which were previously unknown (6). Similarly, a recent whole-genome resequencing effort revealed >800,000 unique SNPs and ~80,000 unique 1- to 3-bp indels in two divergent strains of the plant model *Arabidopsis thaliana*, relative to its reference genome (7).

At the gene or haplotype level, these individual polymorphisms combine to generate distinct allelic forms. New alleles are continually created with each new mutation, and these rise and fall in frequency in response to both drift and selection. In some cases, selection appears to have favored the retention and proliferation of large numbers of segregating alleles via negative frequency-dependent selection (e.g., those involved in pathogen recognition and plant self-incompatibility systems (8, 9)). For example, no less than 241 different alleles have been identified so far for the gene determining the ABO blood group in humans (10). Importantly, variation in the nature of drift and selection experienced by individual loci has led to striking variation in the number of mutations that separate a given pair of alleles: observations of intra-specific allelic divergence within model eukaryotes range over two orders of magnitude *within* species, from <0.1% to >10%, for synonymous site divergence in coding regions (11–14). Variation in the level of divergence among alleles poses a challenge for our analyses of genomic sequence surveys, which are increasingly available. If we hope to identify alleles that are segregating at the same locus, how similar should we expect their sequences to be? How do we distinguish these from sequences belonging to other loci?

Ideally, sequences representing alleles from different loci would at least show a consistently higher level of divergence from one another than exists among alleles of the same locus. Unfortunately, insights from whole genome sequencing indicate that this criterion will be violated frequently as a result of ongoing gene duplication (15–19). For inbred or haploid genotypes of most eukaryotes studied to date, analyses of synonymous site divergence among genes reveal a characteristic frequency distribution sensu (18), wherein the genome includes many highly similar paralogous loci and fewer and fewer paralogs at higher levels of divergence (Fig. 1). These patterns are consistent with a high rate of both gene duplication and loss (17), and indeed duplication rates have been estimated to meet or exceed rates of SNP mutations per generation for many loci (20). This frequent formation of paralogs means that the genome is populated with loci that are separated by levels of divergence (among one another) that span the divergence among their own alleles (e.g., (21)).

There are some options for working around the problem of alleles that cannot be disentangled among paralogous loci. Prior to widespread genomic studies of nonmodel and natural populations,

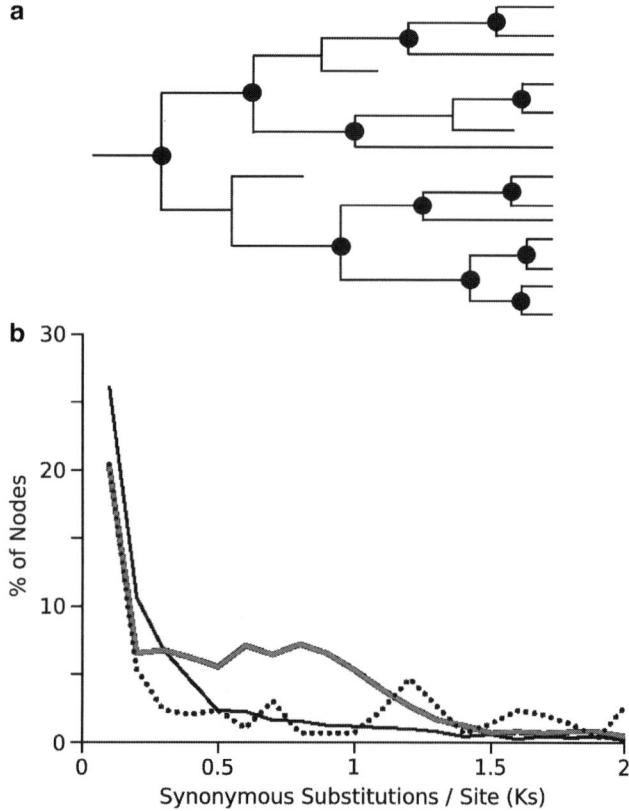

Fig. 1. (**a**) Gene tree showing gene duplication and loss in the history of a single genome. *Dots* indicate nodes that can be reconstructed from extant sequences. (**b**) Frequency distribution of synonymous site divergence at gene duplication nodes for human (*solid black line*, NCBI build 36.43), insect *Drosophila melanogaster* (*dotted line*, Berkeley *Drosophila* Genome Project build 4.3.4) and plant *Arabidopsis thaliana* (*gray line*, TAIR build 9). All species show a characteristic distribution with many recent duplication events; the older peaks in *A. thaliana* reveal ancient genome duplication (42). The most recent duplication events (those below 0.1 Ks divergence) are similar in divergence as are alleles, and these dominate at 20–30% of duplications in these genomes.

allelic variation was simply avoided through the use of inbred or haploid tissue. Once a full genome reference library was in hand, subsequent resequencing efforts identified (and continue to identify) alleles by mapping sequences onto these reference genomes (reviewed in (22, 23)). If a sequence cannot be uniquely mapped to one location on the reference, as expected due to problems with closely related paralogs, it is typically discarded from any further analysis of diversity. Note that variation in paralog number among individuals and incomplete assembly of the reference genome will both introduce error into this mapping process.

For nonmodel organisms without a reference genome, the situation becomes considerably more complicated because sequences must be clustered somehow into groups of putative alleles.

Existing methods for grouping similar sequences within and among species typically rely on the use of arbitrary sequence similarity thresholds or distributions for allowable allelic divergence (e.g., refs. 24–27 and references therein; 28). For surveys of genomic DNA, read depth information (the number of reads that align to a particular position) can provide a valuable indicator of potential problems with paralogs, where sudden increases in read depth, relative to the average, signal that multiple loci might have clustered together as one. This approach is presented in the chapter dealing with RAD tag assembly and analysis (see the chapter by Hohenlohe et al., this volume). For surveys of expressed transcripts (i.e., EST/transcriptome/cDNA sequencing, and RNA-seq), read depth information is unrelated to the frequency of occurrence in the genome and cannot be used to identify problematic clusters.

In this chapter, we examine the steps involved in identifying and clustering alleles in genomic data for which read depth is not informative (transcriptome or already assembled datasets), and a reference genome is not available. This situation is becoming increasingly common in transcriptome surveys of nonmodel organisms and in comparative genomic analyses using published assemblies. We leverage the information afforded by genomic data for *multiple individuals* within a species, when available. We also describe our own publicly available software AllelePipe, a pipeline to aid in moving data through analyses of allelic variation (http://EvoPipes.net/AllelePipe.html).

2. General Procedure to Identify Allelic Clusters

2.1. Sequence Assembly

Sequence assembly is a nontrivial task, making published assemblies a valuable resource for further analyses. To identify alleles in either previously assembled data or new assemblies, it is critical to consider the parameter decisions that have affected the reconstruction of alleles and paralogs during assembly. Fundamentally, sequence assembly is a process of merging reads that are highly similar. Typically, neither de novo nor reference-based assemblies require exact matches before merging reads. A certain amount of sequence divergence is allowed in successful matches because all sequencing methods are prone to error. A high depth of coverage (many reads aligned together at the same position) allows subsequent bioinformatic error estimation and correction via majority-rule, maximum likelihood, or Bayesian methods (e.g., (29–33)) The sequence divergence (or similarity) cut-offs for merging reads are set by the user, and their stringency will necessarily impact the degree to which highly similar alleles/paralogs are seen as error and are merged into the same contigs. A high depth of coverage will also help to avoid problems with allele/paralog merging because most assembly programs detect strong support for multiple versions of a sequence, and separate these into different contigs—although this solution can be problematic if errors occur multiple times at high coverage positions (29). In some cases, authors of the assembly may intentionally adjust settings to try and collapse allelic variation in a sample, in order to obtain a single consensus genomic sequence from outbred individuals or multiple strains. It is also important to note that recently diverged alleles are by their nature separated by a very low density of mutations, which are less likely to be detected by short-read sequences and assembly programs.

There are a variety of ways to assess the dataset quality once the assembly has been completed. The number and length of contigs, the percentage of reads assembling, and the recovery of gene families known from other organisms are all common metrics of the completeness of a genomic or transcriptomic survey (e.g., (34)). None of these metrics explicitly examines the resolution of alleles and paralogs unless these are already identified by mapping to a reference genome. We advocate two approaches for datasets without a reference. First, known highly conserved single copy loci should be represented by roughly the expected number of alleles per individual for the ploidy and heterozygosity of a given species. A number of single copy gene datasets are available for different groups of organisms, including a recently developed list of highly conserved orthologs across all of eukaryotes (35) available at http://compgenomics.ucdavis.edu/compositae_reference.php. These can be searched against an assembly with discontiguous MegaBLAST or tBLASTx ((36, 37); available at http://blast.ncbi. nlm.nih.gov/Blast.cgi), and the number of matches examined closely.

Second, most genomic and transcriptomic datasets should yield the expected distribution of divergence events for paralogs and alleles in gene families (Fig. 1), where there is clear peak at low divergence. These distributions can be created using the DupPipe (38) at Evopipes.net (http://evopipes.net/dup_pipe.html). Over-assembled data will produce truncated curves (Fig. 2, dotted line), where close paralogs and alleles have been merged together and these recent divergence events are not seen. Under-assembled data, where many near-identical copies have failed to assemble will produce extreme front peaks in the distribution (Fig. 2, dashed line). For example, a pattern of under-assembly is common in output from the assembly software MIRA (39), which is otherwise outstanding for its resolution of highly similar copies, but has a tendency to produce many duplicates in areas of high coverage. The pipeline iAssembler (e.g., (40); available at http://bioinfo.bti.cornell.edu/tool/iAssembler/) has been created to combat this problem by iteratively assembling datasets with both MIRA and CAP3 (41), the latter being the standard assembly tool of the Sanger

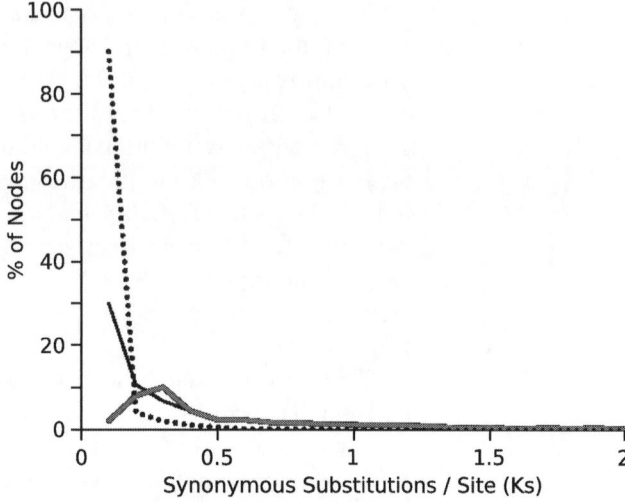

Fig. 2. Example frequency distributions of synonymous site divergence since gene dupli-cation events within a genome for properly assembled (*solid black line*), over-assembled (*gray line*), and under-assembled sequences (*dotted line*). Over-assembled sequences will lack recent duplication events while under-assembled data will be strongly dominated by apparent close duplicates.

sequencing era. Note that additional peaks may appear in these distributions due to past genome duplication events (e.g., (38, 42)), but these will not result in the over- and under-assembly patterns described here.

2.2. Sequence Clustering

The first step in identifying allelic variation across a genomic dataset is to group similar contigs within and/or among individuals into clusters that might represent individual loci. Deciding what is meant by "similar" is a fundamental challenge. As noted above, alleles can vary widely in their level of divergence. Wang and colleagues (27) found that minimum similarity thresholds on the order of 90% might be appropriate for clustering transcripts of the same gene, in transcriptome data from *A. thaliana*. The most detailed information about the typical variation among alleles in a genome under study can come from the data themselves, by assessing the divergence of putative alleles identified between conspecific *individuals* in the dataset. Highly similar sequence matches among conspecifics will either be identical (same allele) or divergent because they represent different alleles segregating in the species, with error introduced by close paralogs. The median similarity of reciprocal best matches between contigs from two individuals should give an idea of the typical similarity expected between any two sequences that are part of the same locus (Fig. 3), and can be used as a starting minimum similarity for clustering contigs in a dataset. A list of reciprocal best hits can be created using the RBH pipeline at EvoPipes.net (http://evopipes.net/rbhpipe.html).

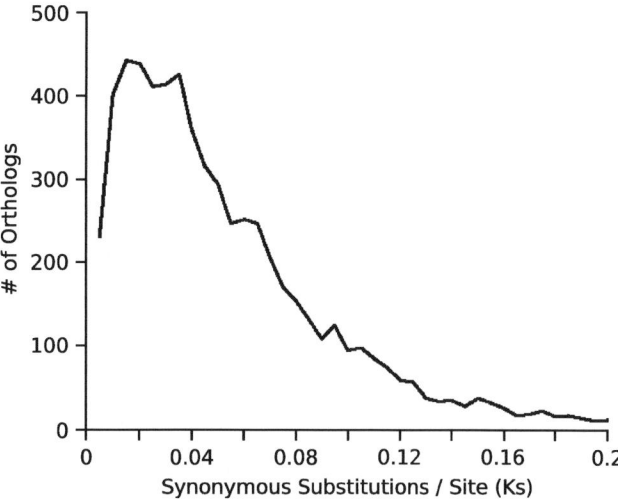

Fig. 3. Frequency distribution of synonymous site divergence of putative alleles of the same locus between individuals, based upon reciprocal best matches between transcriptomes of two *Centaurea solstitialis* plants (see Note 1).

Sequence similarity across the dataset can be found by searching for alignments of all sequences against all sequences (across one or many individuals). Many programs are available for this purpose, such as MegaBLAST (37) in the downloadable form of the BLAST package (available at http://blast.ncbi.nlm.nih.gov/Blast.cgi), and more recently developed high throughput programs, such as SSAHA2 and SMALT, from the Sanger Institute ((43); both available at http://www.sanger.ac.uk/resources/software/). Below, we describe a pipeline that we have developed to execute searching and subsequent clustering steps. In general, two key criteria are important when conducting these types of searches and filtering for desired matches. First, alignments should be continuous and include the entire region of overlap. Small indels may be acceptable, but large gaps are not expected among alleles within a species, with rare exception of intron variation in genomic DNA (44). Complete alignment throughout overlaps is a criterion also used by assembly software, but not by commonly used local alignment searches— such as BLAST—so alignments may have to be verified by custom scripts when these tools are used. Second, alignment lengths need to be sufficiently long to quantify the degree of similarity. For example, 95% "neutral" site similarity implies that an *average* of only five SNPs will separate alleles across 100 bp of alignment in noncoding regions. For coding regions, the number of codons and synonymous sites are a fraction of the total sequence length, and minimum overlaps on the order of ~300 bp (100 codons) may be prudent for properly assessing sequence similarity.

Once similarity between pairs of sequences is known, these must be aggregated into clusters. The simplest form of aggregation

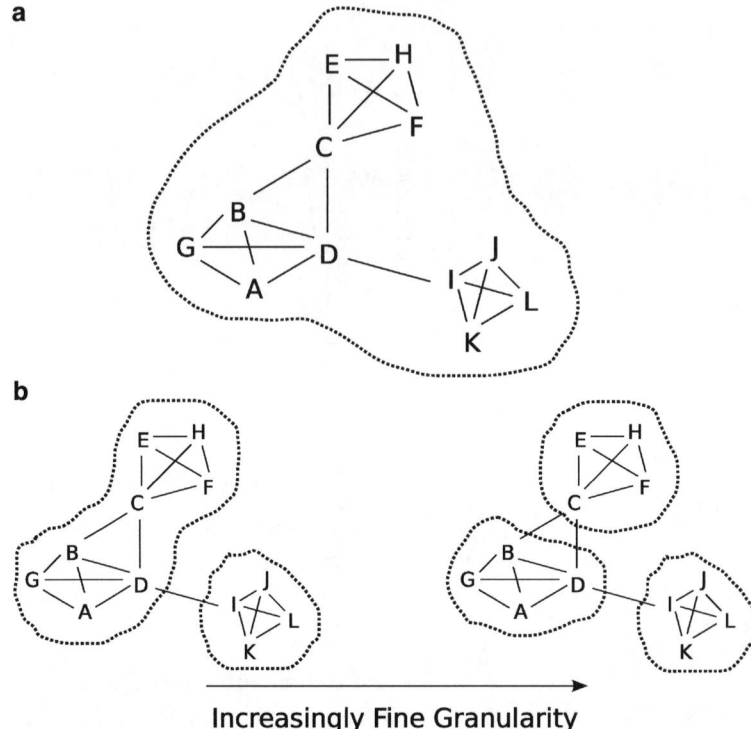

Fig. 4. Diagram of clusters (*dashed outlines*) for alleles (*letters*) connected by sequence similarity (*lines*) using either (**a**) single linkage clustering, or (**b**) graphical clustering with increasingly fine granularities.

is "single-linkage" clustering, where any sequences that share sufficient similarity are placed in the same cluster (Fig. 4a). This is highly inclusive, and can lead to large clusters where increasingly divergent sequences are added because they share sufficient similarity with at least one existing member of a cluster. To avoid the potential problem of over-aggregation of sequences by single-linkage clustering, a maximum likelihood approach to grouping clusters of highly similar sequences has been developed as the Markov Cluster Algorithm provided in the software MCL (http://micans.org/mcl/), and implemented for protein families as OrthoMCL (45). This approach considers all sequences in a network (graph) analysis, where sections of the network with many strong links (here sequence similarities) are identified as clusters. It is possible to control the scale of the clustering—whether groups are large or more fine-scaled—using the "Inflation" (-I) parameter to control granularity (Fig. 4b). Altering this parameter results in inevitable trade-offs between the appropriate separation of clusters that truly represent distinct loci and the maintenance of clusters of divergent alleles of the same locus (and associated Type I and Type II clustering errors, sensu (27)). These trade-offs can be observed by examining known single-copy loci (e.g., (35)), and identifying the numbers of clusters that represent what should be a single locus (see Note 1). This is the single most important parameter for MCL clustering, and should be tuned for each dataset when used.

We point out that there are several sources of errors that can affect the accuracy of clustering, beyond the inevitable aggregation of paralogous loci into the same cluster: (a) large families of genes with a nearly continuous range of divergence among members will form clusters of sequences that cannot easily be separately and that will not form a single multiple alignment (hereafter "superclusters"); (b) highly divergent alleles (e.g., alleles maintained by frequency-dependent selection (8, 9)) are unlikely to cluster together under most clustering scenarios; (c) alternate splicing products in transcriptome data will generate separate clusters that are not true loci (e.g., (46)); and (d) sequencing chimeras, which are polymerase errors, will either produce erroneous loci or cause merging of clusters (47).

2.3. Haplotype (Allele) Calling

With clusters (putative loci) in hand, SNPs and indels can be identified among sequences in the clusters, allowing individual alleles to be defined. The most efficient current approach for identifying sequence variants is to create a consensus sequence for the cluster, and then to map the member sequences against this reference. This reference-guided approach necessarily means that we have created a consensus genomic library for the dataset, which can then conveniently be used to identify alleles in additional individuals as new data are collected. Sequence mapping to a reference is possible with a wide variety of current programs (e.g., (30, 31, 33, 39, 43, 48)). As in the case of sequence similarity parameters, mapping parameters must take into account thresholds for minimum alignment length and maximum sequence divergence, where the latter should be less stringent than for initial clustering, particularly if single-linkage clustering has been used.

At this stage, several types of potential error can be removed when defining alleles. Even the best assemblies can fail to merge some identical contigs, and these redundant sequences can be collapsed into single alleles. Unique SNPs or indels that are only observed once in a large set of individuals (possible sequencing errors) can be ignored if desired. Variants that may be associated with errors unique to a given sequencing platform may also be ignored; for example, erroneous indels may be common near mononucleotide repeats in 454 sequences (49), making these markers less reliable even if observed across multiple individuals. Quality and/or coverage information at polymorphic positions may also be used to help validate SNP and indel validity.

2.4. Single Locus Cluster Sorting

Even an optimal clustering strategy (i.e., best possible minimum sequence similarity requirements and clustering granularity settings) will still produce some allele clusters that circumscribe multiple loci.

Here, information from multiple individuals can offer critical insights. At the minimum, clusters can be filtered for those with no more than the expected number of alleles per individual (two for diploids). This strategy of course becomes more accurate with larger numbers of individuals and higher levels of heterozygosity, such that multilocus clusters will often reveal more than the expected number of alleles. With enough individuals, patterns of association between multiple SNPs in each allele offer the opportunity to infer mutually exclusive sets of recombining alleles (50). Currently, we are not aware of any software that automates this analysis, and it is a ripe area for future development. For the special case where many individuals are from the same population and data coverage is expected to be good for each individual, departure from HWE could be used to infer multilocus clusters (see the chapter by Hohenlohe et al., this volume). Note that particular care must be taken here because many common biological processes can cause departures from HWE, including selection (51), which is often the target of genomic sequencing projects.

2.5. The AllelePipe Software

We provide a bioinformatic pipeline called AllelePipe (Dlugosch et al. in prep; avail at http://EvoPipes.net/AllelePipe.html) to aid in clustering putative alleles across one or more individuals (see also Note 1). Briefly, our pipeline takes in assembled sequence contigs and passes them through the following steps:

1. Similarity is assessed among all sequences using SSAHA2 (43) according to user-defined minimum similarity and alignment length thresholds (Defaults: 95% similarity over 100 bp).

2. Alignment throughout the region of overlap is verified.

3. Sequences are clustered by either single-linkage clustering or MCL as desired, with the option of restarting the clustering with alternative methods/granularities.

4. Multiple alignments are created for sequences within each cluster and their consensus sequence generated, using CAP3 (41). A single consensus genomic reference fasta file is generated for the whole dataset which can be used again in other analyses.

5. Optionally, putatively chimeric clusters are removed, assuming that these are clusters where only one sequence bridges an internal region of the multiple alignment. This step is only appropriate for datasets with many individuals and good coverage of loci, where many sequences should be aligning across the length of each locus

6. SNPs (and optionally indels) are identified using SSAHAsnp (43) against the reference sequence for the same or different sets of individuals, as desired (the program can be restarted from this step for additional analyses).

7. Clusters are sorted as being single or multilocus, based upon user settings for the maximum number of alleles allowed per individual.

3. Summary

The availability of genomic polymorphism data within and among individuals of the same species is one of the most exciting outcomes of recent advances in the ease and affordability of genome-scale sequencing. Tapping that treasure trove of information is another matter, and will continue to challenge bioinformatic analyses until a time when all individuals under study are sequenced completely to their physically accurate full chromosomes. In the meantime, our best strategy is to use all available information to filter clusters of sequences for those most likely to represent true alleles of individual loci. We have proposed a few simple steps along this path, and there is substantial opportunity for further advancement, particularly using inference of recombination events to cluster segregating alleles.

4. Note

1. Single-linkage clustering, where sequences that share sufficient similarity with one member of a cluster are placed in this cluster (Fig. 4a), is the simplest and most inclusive way of clustering closely related sequences. Nevertheless, this approach can lead to problems for some clusters, when increasingly dissimilar sequences are aggregated inappropriately into one group. The current best alternative to single-linkage clustering sequences—without a reference genome for guidance—is maximum likelihood clustering, as implemented in the software MCL (http://micans.org/mcl/). In MCL, clusters are created among the sequences with the strongest links (by any measure of similarity set by the user) and the "Inflation" (-I) parameter controls whether groups are separated at a large or fine scale, known as "granularity" (Fig. 4b). This is the most important parameter for MCL clustering, and should be tuned for each dataset when used.

   We demonstrate this using a pair of transcriptome assemblies for two individuals of the thistle *Centaurea solstitialis*. One library includes 23,267 unigenes (19.3 total Mbp) generated by Sanger sequencing, from a publicly available completed assembly of an individual from North America (38). The second was

Table 1
The number of clusters inferred to be single-locus (no more than two variants per individual), multilocus, monomorphic (no allelic variation), and super clusters (those which will not align), using single linkage clustering and several granularities of MCL clustering of two *Centaurea solstitialis* individuals

	Single linkage clustering	MCL clustering granularities (option -I)			
		1.4	2	4	6
Single-locus	4,208	4,205	4,258	4,423	4,455
Multilocus	2,161	2,171	2,167	2,115	2,099
Monomorphic	288	289	295	319	332
Super cluster	23	21	14	7	5

obtained from publicly available 454 Life Sciences sequence data of an individual from South America (doi: 10.5061/dryad. cm7td/4). We cleaned these data with SnoWhite (http:// evopipes.net), and assembled them using MIRA (39) and CAP3 (41) to yield 43,503 unigenes (32.3 Mbp). Putative alleles between these two individuals were most often separated by 2–4% *synonymous* divergence (total divergence will be lower typically) (Fig. 3).

Sequences were clustered within and between individuals using the AllelePipe software, with a minimum similarity of 95% (i.e., maximum 5% total divergence) and minimum alignment length of 300 bp. We created clusters with single linkage clustering and with MCL granularities (parameter -I) of 1.4, 2, 4, and 6 (Table 1). Most clusters were inferred to be single-locus under all clustering strategies, but a subset of loci were more variable. As granularities became finer, there were fewer cases of inferred multilocus clusters and clusters that would not previously form a single alignment (superclusters), but also more cases of monomorphic (invariant) clusters, which might indicate splitting apart of alleles of the same locus. The potential problem of locus splitting is demonstrated by alignment of cluster consensus sequences with known eukaryotic single copy loci ("Ultra-Conserved Orthologs" or UCOs) (35). A total of 310 of the 357 UCOs matched consensus sequences of clusters in our dataset (based on tblastx comparisons, with maximum e-value of 0.1 and minimum 30 protein residues). The majority of these loci matched a single cluster under single-linkage

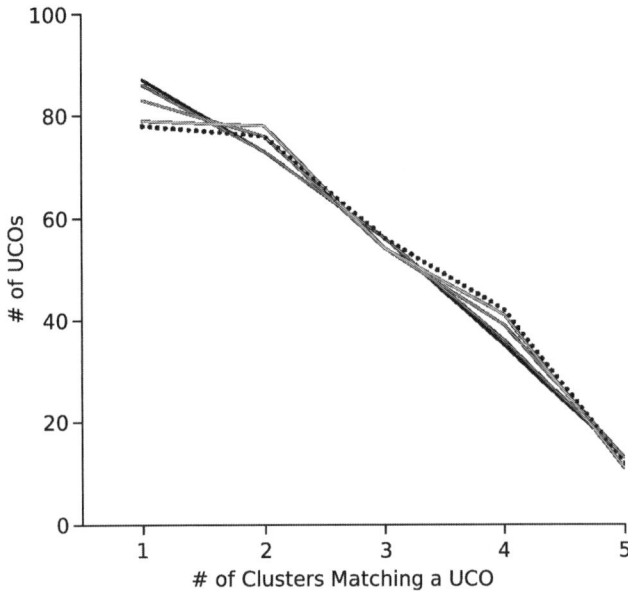

Fig. 5. Histograms of the number of clusters matching conserved single copy eukaryotic loci (UCOs) for single linkage clustering (*solid black line*) and MCL clustering with increasingly fine granularities (granularities 1.4, 2, and 4 shown with increasingly *light gray lines*; finest granularity of 6 shown with *dotted line*), of sequences from two *Centaurea solstitialis* individuals.

clustering, but increasingly fine MCL clustering granularities began to diminish the number of single matches and increase the number of UCOs matching 2–4 clusters each (Fig. 5). Matches to 2–4 clusters are consistent with the undesirable partitioning of single-locus clusters into individual alleles. We conclude that, while single linkage clustering results in some losses of useful data due to super clusters and multilocus clusters, this aggressive clustering strategy does retain most valid clusters of alleles while avoiding problems of locus-splitting. Likelihood-based clustering may be most appropriate for further partitioning of clusters that appear to be multilocus.

Acknowledgments

We thank MS Barker, LH Rieseberg, I Mayrose, and SP Otto for insightful discussions on this topic. We also thank Z Lai and LH Rieseberg for making available multi-individual genomic datasets that prompted our interests in this area.

References

1. Avise JC (2004) Molecular markers, natural history, and evolution. Sinauer Associates, Sunderland
2. Lynch M, Walsh B (1998) Genetics and analysis of quantitative traits. Sinauer Associates, Sunderland
3. Wakeley J (2008) Coalescent theory: an introduction. Roberts & Company, Greenwood Village
4. McCarthy MI, Abecasis GR, Cardon LR et al (2008) Genome-wide association studies for complex traits: consensus, uncertainty and challenges. Nat Rev Genet 9:356–369
5. Zuckerkandl E, Pauling L (1965) Evolutionary divergence and convergence in proteins. In: Bryson V, Vogel H (eds) Evolving genes and proteins. Academic, New York
6. Altshuler DL, Durbin RM, Abecasis GR et al (2010) A map of human genome variation from population-scale sequencing. Nature 467:1061–1073
7. Ossowski S, Schneeberger K, Clark RM et al (2008) Sequencing of natural strains of *Arabidopsis* thaliana with short reads. Genome Res 18:2024–2033
8. Charlesworth D, Vekemans X, Castric V, Glemin S (2005) Plant self-incompatibility systems: a molecular evolutionary perspective. New Phytol 168:61–69
9. Hulbert SH, Webb CA, Smith SM, Sun Q (2001) Resistance gene complexes: evolution and utilization. Annu Rev Phytopathol 39:285–312
10. Patnaik SK, Blumenfeld OO (2011) Patterns of human genetic variation inferred from comparative analysis of allelic mutations in blood group antigen genes. Hum Mutat 32:263–271
11. Bergelson J, Kreitman M, Stahl EA, Tian D (2001) Evolutionary dynamics of plant R-genes. Science 292:2281–2285
12. Lawlor DA, Ward FE, Ennis PD et al (1988) HLA-A and B polymorphisms predate the divergence of humans and chimpanzees. Nature 335:268–271
13. Li WH, Sadler LA (1991) Low nucleotide diversity in man. Genetics 129:513–523
14. Moriyama EN, Powell JR (1996) Intraspecific nuclear DNA variation in *Drosophila*. Mol Biol Evol 13:261–277
15. Demuth JP, De Bie T, Stajich JE et al (2006) The evolution of mammalian gene families. PLoS One 1:e85
16. Hahn MW, De Bie T, Stajich JE et al (2005) Estimating the tempo and mode of gene family evolution from comparative genomic data. Genome Res 15:1153–1160
17. Hahn MW, Han MV, Han S-G (2007) Gene family evolution across 12 *Drosophila* genomes. PLoS Genet 3:e197
18. Lynch M, Conery JS (2000) The evolutionary fate and consequences of duplicate genes. Science 290:1151–1155
19. Sebat J, Lakshmi B, Troge J et al (2004) Large-scale copy number polymorphism in the human genome. Science 305:525–528
20. Lynch M (2007) The origins of genome architecture. Sinauer Associates, Sunderland
21. Fredman D, White SJ, Potter S et al (2004) Complex SNP-related sequence variation in segmental genome duplications. Nat Genet 36:861–866
22. Bentley DR (2006) Whole-genome re-sequencing. Curr Opin Genet Dev 16:545–552
23. Charlesworth B (2010) Molecular population genomics: a short history. Genet Res 92:397–411
24. Li L, Stoeckert CJ, Roos DS (2003) OrthoMCL: identification of ortholog groups for eukaryotic genomes. Genome Res 13:2178–2189
25. Nagaraj SH, Gasser RB, Ranganathan S (2007) A hitchhiker's guide to expressed sequence tag (EST) analysis. Brief Bioinform 8:6–21
26. Tang J, Vosman B, Voorrips RE et al (2006) QualitySNP: a pipeline for detecting single nucleotide polymorphisms and insertions/deletions in EST data from diploid and polyploid species. BMC Bioinformatics 7:438
27. Wang J-PZ, Lindsay BG, Leebens-Mack J et al (2004) EST clustering error evaluation and correction. Bioinformatics 20:2973–2984
28. Hazelhurst S, Hide W, Lipták Z et al (2008) An overview of the wcd EST clustering tool. Bioinformatics 24:1542–1546
29. Lynch M (2009) Estimation of allele frequencies from high-coverage genome-sequencing projects. Genetics 182:295–301
30. Malhis N, Jones SJM (2010) High quality SNP calling using Illumina data at shallow coverage. Bioinformatics 26:1029–1035
31. Li H, Ruan J, Durbin R (2008) Mapping short DNA sequencing reads and calling variants using mapping quality scores. Genome Res 18:1851–1858
32. Zerbino DR, Birney E (2008) Velvet: algorithms for de novo short read assembly using de Bruijn graphs. Genome Res 18:821–829

33. Li H, Durbin R (2009) Fast and accurate short read alignment with Burrows–Wheeler transform. Bioinformatics 25:1754–1760

34. Gibbons JG, Janson EM, Hittinger CT et al (2009) Benchmarking next-generation transcriptome sequencing for functional and evolutionary genomics. Mol Biol Evol 26:2731–2744

35. Kozik A, Matvienko M, Michelmore RW (2010) Effects of filtering, trimming, sampling and k-mer value on de novo assembly of Illumina GA reads. In: Plant and Animal Genomes XVIII Conference, San Diego

36. Altschul SF, Gish W, Miller W et al (1990) Basic local alignment search tool. J Mol Biol 215: 403–410

37. Zhang Z, Schwartz S, Wagner L, Miller W (2000) A greedy algorithm for aligning DNA sequences. J Comput Biol 7:203–214

38. Barker MS, Kane NC, Matvienko M et al (2008) Multiple paleopolyploidizations during the evolution of the Compositae reveal parallel patterns of duplicate gene retention after millions of years. Mol Biol Evol 25: 2445–2455

39. Chevreux B, Pfisterer T, Suhai S (2000) Automatic assembly and editing of genomic sequences. In: Suhai S (ed) Genomics and proteomics: functional and computational aspects. Kluwer Academic/Plenum Publishers, New York

40. Guo S, Zheng Y, Joung JG et al (2010) Transcriptome sequencing and comparative analysis of cucumber flowers with different sex types. BMC Genomics 11:384

41. Huang X, Madan A (1999) CAP3: a DNA sequence assembly program. Genome Res 9:868–877

42. Barker MS, Vogel H, Schranz ME (2009) Paleopolyploidy in the brassicales: analyses of the cleome transcriptome elucidate the history of genome duplications in *Arabidopsis* and other brassicales. Genome Biol Evol 1:391–399

43. Ning Z, Cox AJ, Mullikin JC (2001) SSAHA: a fast search method for large DNA databases. Genome Res 11:1725–1729

44. Omilian AR, Scofield DG, Lynch M (2008) Intron presence-absence polymorphisms in Daphnia. Mol Biol Evol 25:2129–2139

45. Enright AJ, Van Dongen S, Ouzounis CA (2002) An efficient algorithm for large-scale detection of protein families. Nucleic Acids Res 30:1575–1584

46. Gupta S, Zink D, Korn B et al (2004) Genome wide identification and classification of alternative splicing based on EST data. Bioinformatics 20:2579–2585

47. Bragg LM, Stone G (2009) k-link EST clustering: evaluating error introduced by chimeric sequences under different degrees of linkage. Bioinformatics 25:2302–2308

48. Li R, Yu C, Li Y et al (2009) SOAP2: an improved ultrafast tool for short read alignment. Bioinformatics 25:1966–1967

49. Margulies M, Egholm M, Altman WE et al (2005) Genome sequencing in microfabricated high-density picolitre reactors. Nature 437: 376–380

50. Griffin PC, Robin C, Hoffmann AA (2011) A next-generation sequencing method for overcoming the multiple gene copy problem in polyploid phylogenetics, applied to *Poa* grasses. BMC Biol 9:19

51. Hartl DL, Clark AG (2006) Principles of population genetics, 4th edn. Sinauer Associates, Sunderland

52. Lai Z, Kane N, Kozik A et al (2012) Genomics of compositae weeds: EST libraries, microarrays, and evidence of introgression. American Journal of Botany 99:209–218

Chapter 13

Multiple Testing in Large-Scale Genetic Studies

Matthieu Bouaziz, Marine Jeanmougin, and Mickaël Guedj

Abstract

Recent advances in Molecular Biology and improvements in microarray and sequencing technologies have led biologists toward high-throughput genomic studies. These studies aim at finding associations between genetic markers and a phenotype and involve conducting many statistical tests on these markers. Such a wide investigation of the genome not only renders genomic studies quite attractive but also lead to a major shortcoming. That is, among the markers detected as associated with the phenotype, a nonnegligible proportion is not in reality (false-positives) and also true associations can be missed (false-negatives). A main cause of these spurious associations is due to the multiple-testing problem, inherent to conducting numerous statistical tests. Several approaches exist to work around this issue. These multiple-testing adjustments aim at defining new statistical confidence measures that are controlled to guarantee that the outcomes of the tests are pertinent. The most natural correction was introduced by Bonferroni and aims at controlling the family-wise error-rate (FWER) that is the probability of having at least one false-positive. Another approach is based on the false-discovery-rate (FDR) and considers the proportion of significant results that are expected to be false-positives. Finally, the local-FDR focuses on the actual probability for a marker of being associated or not with the phenotype. These strategies are widely used but one has to be careful about when and how to apply them. We propose in this chapter a discussion on the multiple-testing issue and on the main approaches to take it into account. We aim at providing a theoretical and intuitive definition of these concepts along with practical advises to guide researchers in choosing the more appropriate multiple-testing procedure corresponding to the purposes of their studies.

Key words: Multiple testing, Genetic, Association, Biostatistics, GWAS, Bonferroni, FWER, FDR

1. Introduction

During the last decade, advances in Molecular Biology and substantial improvements in microarray and sequencing technologies have led biologists toward high-throughput genomic studies. In particular, the simultaneous genotyping of hundreds of thousands of genetic markers such as single nucleotide polymorphisms (SNPs) on chips has become a mainstay of biological and

François Pompanon and Aurélie Bonin (eds.), *Data Production and Analysis in Population Genomics: Methods and Protocols*, Methods in Molecular Biology, vol. 888, DOI 10.1007/978-1-61779-870-2_13, © Springer Science+Business Media New York 2012

biomedical research. The analysis of these large-scale data involves performing statistical tests at a given significance level to select "significant" markers of interest. These candidates will need to be verified in a follow-up study. Because validation can often only be performed at low throughput, selecting the most relevant set of markers from the primary analysis is critical.

A fundamental difficulty in the interpretation of large-scale genomic studies is given by the simple but tricky statistical issue known as multiple testing which states that among the large number of markers investigated, a nonnegligible proportion will be spuriously selected as significant. For instance, there are still serious problems in practice due to the fact that few associations between markers and phenotypes are consistently and convincingly replicated (1). Such non-replication can have many causes (2), and it is widely considered that failure to fully account for the consequences of multiple testing is one of them. In other words, the availability of high-throughput molecular technologies represents a tremendous opportunity for finding the causative genes among those tested, but multiple-testing problems may produce excessive false-positives or overlook true-positive signals. Hence, there is a need for statistical methodologies accounting for multiple-testing issues.

Many statistical approaches, generally referred to as multiple-testing procedures, have been developed to deal with multiple testing and the inherent problem of false-positives. They consist in reassessing probabilities obtained from statistical tests by considering more interpretable and suited statistical confidence measures such as the FWER, the FDR, or the local false-discovery-rate (local-FDR).

Several recent reviews dealt with the multiple-testing problem in the context of large-scale molecular studies (3, 4, 5, 6, 7, 8). Our goal in this chapter is to provide an intuitive understanding of these confidence measures, an idea on how they are computed and some guidelines about how to select an appropriate measure for a given experiment. First we give some statistical basics about hypothesis testing and multiple testing. Then we describe the main confidence measures (FWER, FDR, and local-FDR) by giving for each a formal and an intuitive definition, the way to apply them, and a discussion of their respective advantages and drawbacks. We also provide information about the p-values distribution which is an important point seldom considered in practice.

Several software solutions dedicated to genome-wide association studies such as PLINK[1] consider the multiple-testing problem with simple and common procedures. In this chapter, applications

---

[1]http://pngu.mgh.harvard.edu/~purcell/plink.

are proposed with the statistical software R, by providing simple R commands. For an introduction to R, one should refer to the R website[2] and to available tutorials.[3]

2. Statistical Background

2.1. Hypothesis Testing

As introduced by Chen et al. (8), a statistical hypothesis test between two groups is a formal scientific method to examine the plausibility of a statement regarding the comparison of a measurement between the two groups. The statement about the comparison is typically formulated as a "null hypothesis" (H_0) that there is no difference in the values of the measurement between the groups. An "alternative hypothesis" (H_1) represents the conclusion we are interested in, such as there is a difference between the measurements conducted within the groups. In our context of population-based genetic studies, we will assume that the null hypothesis is that a given genetic variation has no effect on a trait of interest (i.e., it is not associated with the trait), and the alternative is that there is an association. More concretely, we can consider a null hypothesis H_0 that all the possible genotypes at one locus (aa, aA, or AA) lead to the same risk of having the disease and as alternative H_1 that certain genotypes are more at risk.

Usually, a statistic (S) is defined to assess the level of difference between the groups to be compared. A way to consider how meaningful is S is to assess the probability that a particular value of this statistic (S^{obs}) would occur by chance under the null hypothesis (7).

The null hypothesis (H_0) is used to derive the null distribution of the test statistic (S). This distribution serves as a reference to describe the variability of S due to chance under H_0. The hypothesis-testing procedure compares the observed statistic (S^{obs}) to the null distribution and computes a statistical confidence measure called p-value to summarize the result. The p-value is defined as the probability that a test statistic at least as large as the observed one would occur in data drawn according to H_0:

$$p\text{-value} = \mathbb{P}_{H_0}(S \geq S^{obs}).$$

A small p-value indicates that the test statistic lies in the extremities of the null distribution which suggests that the null hypothesis does not accurately describe the associated observation. In practice, determining whether an observed test statistic is significant

---

[2]http://cran.r-project.org.
[3]http://www.statmethods.net.

Table 1
Outcomes of a statistical test performed at the level α

	H_0 is not rejected	H_0 is rejected
H_0 is true	true-negative $(1-\alpha)$	false-positive / type-I error (α)
H_0 is false	false-negative / type-II error (β)	true-positive / power $(1-\beta)$

requires comparing its corresponding p-value to a confidence threshold (α) also known as "level of significance". When the p-value is less than α the null hypothesis is rejected with sufficient confidence. Conversely, if the p-value is above the threshold, the result is considered as nonsignificant and the null hypothesis is not rejected. This does not necessarily mean that the null hypothesis is true, it only suggests that there is not sufficient evidence against H_0.

Many studies use a threshold $\alpha = 5\%$, historically suggested by Fisher (9) who argued that one should reject a null hypothesis when there is only 1 in 20 chance that it is true. It is, however, possible to select other thresholds as there are no particular statistical reasons for using $\alpha = 5\%$. The choice of the significance threshold actually depends on the costs associated with false-positives and false-negatives, and these costs may differ from one experiment to another (7).

When a statistical test is performed, depending on whether the null hypothesis is true or false and whether the statistical test rejects or does not reject the null hypothesis, one of four outcomes will occur: (1) the test rejects a true null hypothesis (false-positive or type-I error), (2) the test does not reject a true null hypothesis (true-negative), (3) the test rejects a false null hypothesis (true-positive), and (4) the test does not reject a false null hypothesis (false-negative or type-II error). The true state and the decision to accept or reject a null hypothesis are summarized in Table 1.

Defined that way, the p-value of a test appears simply as the probability of a false-positive and the corresponding threshold α corresponds to the false-positive (or type-I error) rate.

Another value is presented in Table 1 and corresponds to the type-II error rate β, that is the probability of false-negative. We usually consider $1-\beta$, the statistical power of the test instead. Without presenting in depth this notion, the statistical power corresponds to the probability of rejecting the null hypothesis when the alternative hypothesis is true, i.e., in the case of genetic studies, the probability of detecting an association when there is actually one. The power of a statistical test is used to determine what is its ability to detect true-positives. As statistical power and significance level are dependent, a trade-off has to be found to properly manage false-positive and false-negative rates. This is a

Table 2
Outcomes of *n* statistical tests performed at the level α

	H_0 is not rejected	H_0 is rejected	Total
H_0 is true	true-negatives (tn)	false-positives (fp)	$n_0 = tn + fp$
H_0 is false	false-negatives (fn)	true-positives (tp)	$n_1 = fn + tp$
Total	$n_U = tn + fn$	$n_R = fp + tp$	n

way of selecting the significance level relevant to the data studied and also to compare different approaches. Given a common significance threshold α, the best test at the level α is the one maximizing the power $1 - \beta$.

2.2. Multiple Testing

When only one test is conducted, the probability of having a false-positive is controlled at the level of significance α. When more than one test is performed, it can no longer be interpreted as the probability of false-positive of the overall tests but rather as the expected proportion of tests providing false-positive results.

The generic situation is the following: when *n* statistical tests are performed, depending on whether each hypothesis tested is true or false and whether the statistical tests reject or does not reject the null hypotheses, each of the *n* results will fall in one of four outcomes defined in Table 1. It leads to the equivalent Table 2 corresponding to multiple tests, indicating the actual number of false-positives and false-negatives (*fp* and *fn*) instead of their respective rates of occurrence (α and β).

Let us consider *n* tests performed in a single experiment. According to the **Hypothesis-testing** section, with just one test ($n=1$) performed at the usual 5% significance level, there is a 5% chance of incorrectly rejecting H_0. However, with $n=20$ tests in which all the null hypotheses are true, the expected number of such false rejections is $20 \times 0.05 = 1$. Now, with $n=100,000$ tests, the expected number of false-positives is 5,000 which is much more substantial.

In the case of genetic studies, this number of false-positives is higher than the expected number of true discoveries and unfortunately, it is not possible to know which null hypothesis is correctly or incorrectly rejected. As a result, the control of the proportion of false-positives in the context of multiple testing is a crucial issue and the false-positive rate does not appear adapted anymore to define a confidence threshold to apply to the *p*-values. We therefore need multiple-testing correction procedures aiming at controlling certain more relevant statistical confidence measures such as the FWER, the FDR, and the local-FDR introduced in the following sections.

3. Controlling the Family-Wise Error Rate

3.1. Definition

The first alternative confidence measure proposed to handle the multiple-testing problem is referred to as the FWER criterion. It is defined as the probability of falsely rejecting at least one null hypothesis over the collection of hypotheses (or "family") that is being considered for joint testing:

$$\text{FWER} = \mathbb{P}_{H_0}(fp \geq 1 \text{ at the level } \alpha).$$

Controlling the FWER at a given level is to control the probability of having at least one false-positive, which is very different from the false-positive rate. In practice, as the number (n) of tests increases, the false-positive rate remains fixed at the level α whereas the FWER generally tends toward 1.

Several procedures exist to control the FWER; here we introduce the Bonferroni procedure that can be considered as the reference method.

3.2. The Bonferroni Procedure

Certainly the simplest and most widely used method to deal with multiple testing is to control the FWER by applying the Bonferroni adjustment (10, 11) which accounts for the number of tests. It is based on the following simple relation between the p-value of a test i (p_i), the number of tests performed (n), and the FWER at the level p_i (FWER$_i$):

$$
\begin{aligned}
\text{FWER}_i &= \mathbb{P}_{H_0}(fp \geq 1 \text{ at the level } p_i) \\
&= \mathbb{P}_{H_0}(\{\text{ test } 1 \text{ is } fp \text{ at the level } p_i\} \text{ or} \ldots \\
&\quad \text{or } \{\text{test } n \text{ is } fp \text{ at the level } p_i\}) \\
&\leq \sum_{k=1}^{n} \mathbb{P}_{H_0}(\{\text{test } k \text{ is } fp \text{ at the level } p_i\}) \\
&\leq np_i,
\end{aligned}
$$

because for each test i, $\mathbb{P}_{H_0}(\text{test } i \text{ is } fp \text{ at the level } p_i) = p_i$. The new confidence values $p_i^{\text{Bonf}} = np_i$ correspond to the p-values adjusted by the Bonferroni procedure, and represents a majoration of the FWER. Controlling the FWER at a 5% level requires to apply a 5% threshold to the adjusted p-values corresponding to the product of each p-value with the number of tests. One can also prefer to apply a threshold of 5%/n to the unadjusted p-values. For instance, to ensure that the FWER is not greater than 5% when performing 100 tests, each result can be considered as significant only if the p-value is less than $0.05/100 = 0.0005$. It makes no difference in terms of results.

3.3. In Practice

To control the FWER at 5%, one has to multiply the *p*-values resulting from the application of the test on each genetic marker by the number (*n*) of markers tested and to apply a 5% level to these adjusted *p*-values to determine the significant ones. Several softwares such as PLINK implement this simple procedure. Here we introduce how to apply it in R to a set of *p*-values (p) to obtain a set of adjusted *p*-values (pBonf):

```
> n = length(p)

> pBonf = n*p
```

In practice, this can provide probabilities greater than 1 so formally we need to force such probabilities to 1:

```
> pBonf[which(pBonf > 1)] = 1
```

This step is optional if the aim is only to select a set of markers reaching the significance level. Finally we can identify the set of markers significant at the 5% level after using the Bonferroni procedure by applying the R command:

```
> level = 0.05

> which(pBonf <= level)
```

3.4. Notes on the FWER

The major advantage of the Bonferroni procedure is that it is simple and straightforward to calculate and can easily be used in any multiple-testing application (4). However, some authors argue that one major disadvantage of the Bonferonni procedure is that it over adjusts the *p*-values, resulting in a control of the FWER slightly more stringent than expected. An alternative was developed in the case of independent tests by Sidak (12) and is based on:

$$\text{FWER}_i = \mathbb{P}_{H_0}(fp \geq 1 \text{ at the level } p_i)$$
$$= 1 - \mathbb{P}_{H_0}(fp = 0 \text{ at the level } p_i)$$
$$= 1 - (1 - p_i)^n$$

This adjustment results in a more precise control of the FWER. However, this approach assumes that all the tests performed are independent and may therefore not be suitable for every situation while the Bonferonni procedure does not make any assumption about the relation between the tests. Moreover, in the case of a large number of markers tested, such as in genome-wide association studies, and small *p*-values in which we are interested in, the $1 - (1 - p_i)^n$ proposed by Sidak can be reasonably approximated by the np_i proposed by Bonferroni. For those interested in obtaining exact confidence measures in the case the tests can be considered as independent, they can apply the simple following R command instead of the one used in the previous section:

```
> pSidak = 1-(1-p)^length(p).
```

As a matter of fact, tests are often dependent when testing genetic markers that are statistically associated by linkage disequilibrium (LD) over the genome. In such situations, a practical and common alternative is to approximate the exact FWER using a permutation procedure (3, 4). Here, the genotype data are retained but the phenotype labels are randomized over the individuals to generate a data set that has the observed LD structure but satisfies the null hypothesis of no association with the phenotype. By analyzing many of such data sets, the exact FWER can be approximated. The method is conceptually simple but can be computationally intensive, particularly as it is specific to a particular data set and the whole procedure has to be repeated if the data set is somehow altered (3). Moreover, it requires complex programming skills and statistical knowledge. However, the availability of relatively inexpensive and fast computers and the use of built-in permutation utilities available for several genetics programs (such as PLINK) obviate these problems to some extent.

Actually the main disadvantage with the use of the FWER as a confidence measure is its unreliability with certain data sizes. This procedure works well in settings involving a few tests (e.g., 10–20, usual for candidate gene studies) and even when the number of tests is somewhat larger (e.g., a few hundreds as in genome-wide microsatellite scans) (4). Yet the control of the FWER is not ideal when the number of tests is very large, as the level of significance becomes too stringent and as a result true associations may be overlooked and a consequent loss of test power occurs (5). For instance, in the context of genome-wide association studies, if 1 million genetic markers are tested, the p-value threshold for each marker must be set to 5×10^{-8} to control the FWER at 5% which is very low.

Even though such levels certainly would provide a safeguard against false-positives, they would also lead to an unacceptable number of false-negatives, particularly for complex traits where the loci effects are expected to be moderate at best. Consequently, less stringent confidence measure-based methods are designed to find a proper balance between false-positives and false-negatives in large-scale association studies. This constitutes one of the burning methodological issues in contemporary Genetic Epidemiology and Statistical Genetics (4).

4. Analyzing the Distribution of *p*-Values to Understand the Basis of More Advanced Multiple-Testing Corrections

4.1. Mixture Distribution of p-Values

Analyzing the distribution of *p*-values provides a way to determine how many tests are declared under the null hypothesis (H_0) or under the alternative (H_1) and therefore to assess the multiple-testing problem. A very interesting result is that under H_0, the null distribution of *p*-values $(\mathcal{D}_0)$ corresponds to a standard Uniform distribution on the interval $[0, 1]$ (Fig. 1-a).

On the other hand, the alternative distribution of *p*-values $(\mathcal{D}_1)$ under H_1 corresponds to a distribution that tends to accumulate toward 0 (Fig. 1-b). In practice, one deals with *p*-values drawn under H_0 and H_1 which corresponds to a mixture distribution $\mathcal{D}$ of $\mathcal{D}_0$ and $\mathcal{D}_1$ (Fig. 1-c) (13):

$$f = \pi_0 f_0 + \pi_1 f_1,$$

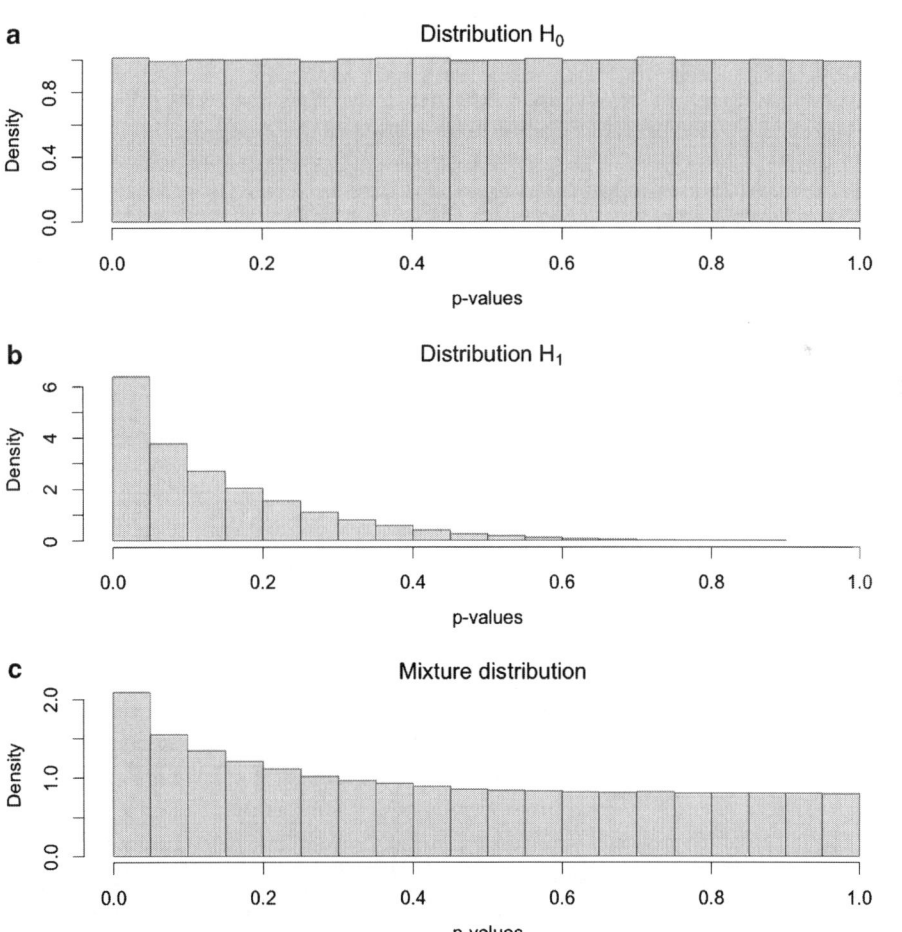

Fig. 1. Distribution of *p*-values under H_0 **(a)**, H_1 **(b)** and the resulting mixture distribution **(c)**.

where f, f_0, and f_1 are, respectively, the probability density functions defining the distributions $\mathcal{D}$, $\mathcal{D}_0$, and $\mathcal{D}_1$; π_0 and π_1 are the proportion of p-values generated under H_0 and H_1, respectively, with $\pi_0 + \pi_1 = 1$.

4.2. Finding Evidence of the Existence of True-Positives

A usual question addressed at the outset of the analysis of large-scale genetic data is whether there is evidence that any of the markers tested and declared as significant at a given confidence level α are true-positives. Investigating the distribution of p-values can help finding an answer.

4.2.1. Plotting the Observed Distribution of p-Values

Plotting the distribution of p-values is a first intuitive approach to assess approximatively the evidence of true-positives. Indeed, a simple histogram of the p-values can indicate whether the distribution is made of a mixture of p-values drawn under H_0 and H_1 (Fig. 1-c) or is composed of p-values drawn under H_0 only (Fig. 1-a). In the case the distribution is a mixture then some of the tests detected as significant may be true findings. The R command to obtain an histogram of a set of p-values (p) is :

> hist(p)

Figure 2-a represents the histograms of p-values obtained for three increasing proportions of H_1.

An alternative and widely used approach is to compute a Quantile–Quantile plot (Q–Q plot) of the p-values comparing their distribution to the standard uniform distribution expected under H_0. If the two distributions are similar, the points in the Q–Q plot will approximately lie on the line $y=x$; if the observed quantiles are markedly more dispersed than the Uniform quantiles, this suggests that some of the significant results may be true-positives. In practice, QQ-plots can be generated by applying the R function qqplot on a set of p-values (p):

> qqplot(x = p, y = runif(100000))

> abline(a = 0, b = 1, lty = "dashed", col = "orange", lwd = 2)

Figure 2-b represents Q–Q plots of p-values obtained for the three increasing proportions of H_1.

4.2.2. Estimating the Proportion of Markers Under H_0 and H_1

As we discussed previously, plotting the distribution of p-values can provide a qualitative and approximative criterion for assessing the proportion of genetic markers associated with the disease. Also, the actual proportions of p-values drawn under H_0 and H_1 (π_0 and $\pi_1 = 1 - \pi_0$, respectively), which are too rarely considered, can provide such information. They are also important to assess the FDR, a confidence measure that will be presented in the next section. Furthermore, a reliable estimate of π_0, the number of truly null tests, is of great relevance for calculating the sample size when

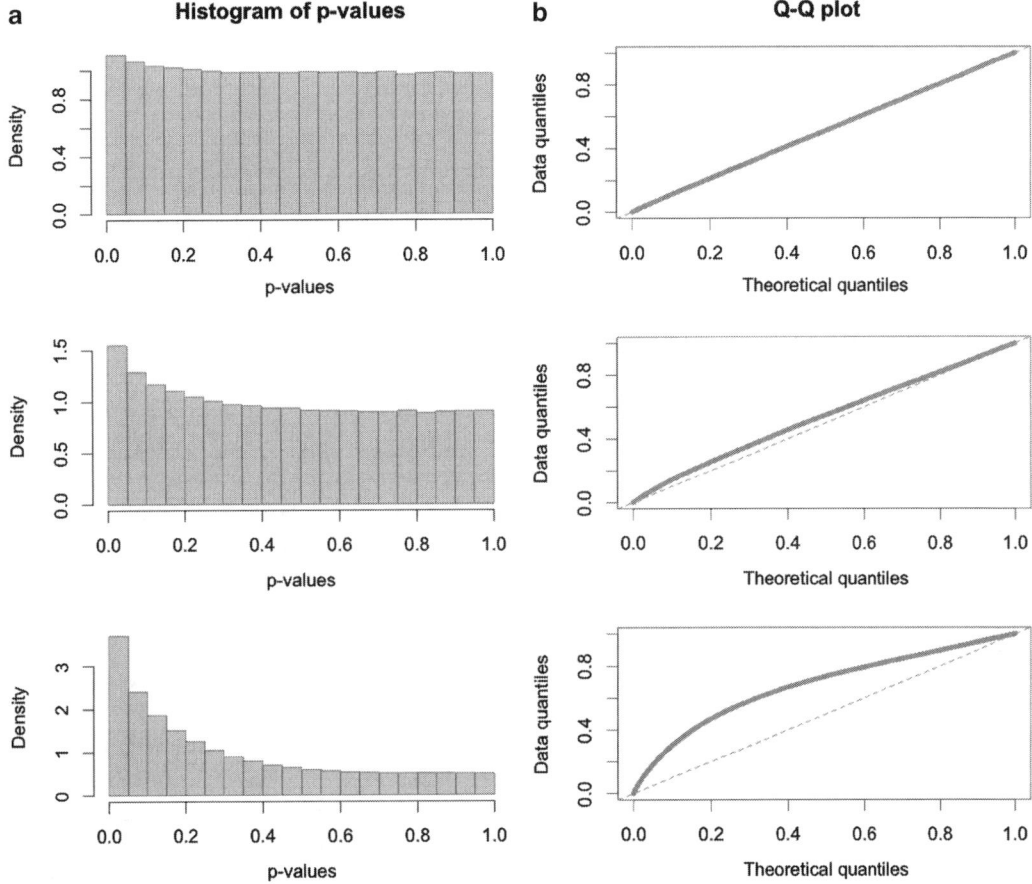

Fig. 2. Histogram of p-values (**a**) and Q–Q plot (**b**) for three increasing proportion of markers under H_1 (π_1 = 2%, 10%, and 50%).

designing the study (14, 15). A variety of methods have been proposed for estimating π_0 based generally on statistical techniques such as mixture model estimation (16, 17, 18), nonparametric methods (19, 20, 21, 22), and Bayesian approaches (23).

The most simple and adopted method is inspired from the approach proposed by Storey and Tibshirani (24). The idea is to estimate π_0 from a part of the distribution between a given value λ and 1 where the distribution $\mathcal{D}_1$ drawn under H_1 is negligible, and then normalized between 0 and 1:

$$\hat{\pi}_0(\lambda) = \frac{\text{number of } p\text{-values} > \lambda}{n \times (1 - \lambda)},$$

where n is the number of genetic markers tested. Some authors showed that a value of λ of 0.5 is a good compromise between bias and variability to assess π_0. With p a given set of p-values, π_0 can be assessed in R by applying:

```
> lambda = 0.5
> pi0 = mean(p > lambda, na.rm = TRUE) /
(1 - lambda)
```

An application of this procedure to the three distributions represented in Fig. 2 estimates π_0 of 98%, 90%, and 50%, and hence π_1 of 2%, 10%, and 50%, respectively.

4.2.3. The Poisson Test

To assess whether any alternative hypothesis is true, it is common to perform a Poisson test. Indeed, when it is assumed that the tests are independent, a Poisson distribution can be used as a model for the number of significant results at a given level α. If one assumes that all the n markers are under the null hypothesis then it is possible to determine, via the level α, the expected number of markers that should come out as significantly associated n_R^{exp}. Given the observed number of markers associated n_R^{obs}, the Poisson distribution allows us to compute a p-value indicating if n_R^{obs} is significantly greater than n_R^{exp}, in which case it is likely that some of the markers associated are true-positives. For example, if 500 independent tests are performed, each at the level $\alpha = 0.05$, we expect $n_R^{exp} = 500 \times 0.05 = 25$ significant tests to occur when all null hypotheses are true. If we observe $n_R^{obs} = 38$ significant markers, the p-value of such an observation is given by the Poisson distribution with mean 25, and is equal to 5.7×10^{-3}. So if we observe 38 significant markers out of 500 markers tested at a 5% level, it is very likely that some of them truly are associated with the disease. In practice and with the R software, let p be a set of p-values and pPois the p-value resulting from the Poisson distribution (given by the R function ppois):

```
> level = 0.05
> pPois = 1-ppois(q = length(which(p <= level)),
lambda = length(p)*level)
```

An application of this procedure to the three distributions represented in Fig. 2 based on 2,000 tests results in p-values of 0.12, 3×10^{-5}, and 2×10^{-16}. A drawback of this approach is that it assumes that the tests are independent. In the case they are not, this method overstates the evidence that some of the alternative hypotheses are true when the test statistics are positively correlated, which commonly occurs in practice in genome-wide association studies due to LD. In such a situation, an equivalent test taking dependencies into account can be implemented based on permutations as was explained for the FWER, but remains computationally expensive.

4.3. Notes on the Mixture Distribution of p-Values

Understanding and investigating the mixture distribution of p-values drawn under H_0 and H_1 is seldom considered in the analysis of large-scale genetic data. However, it can provide a substantial

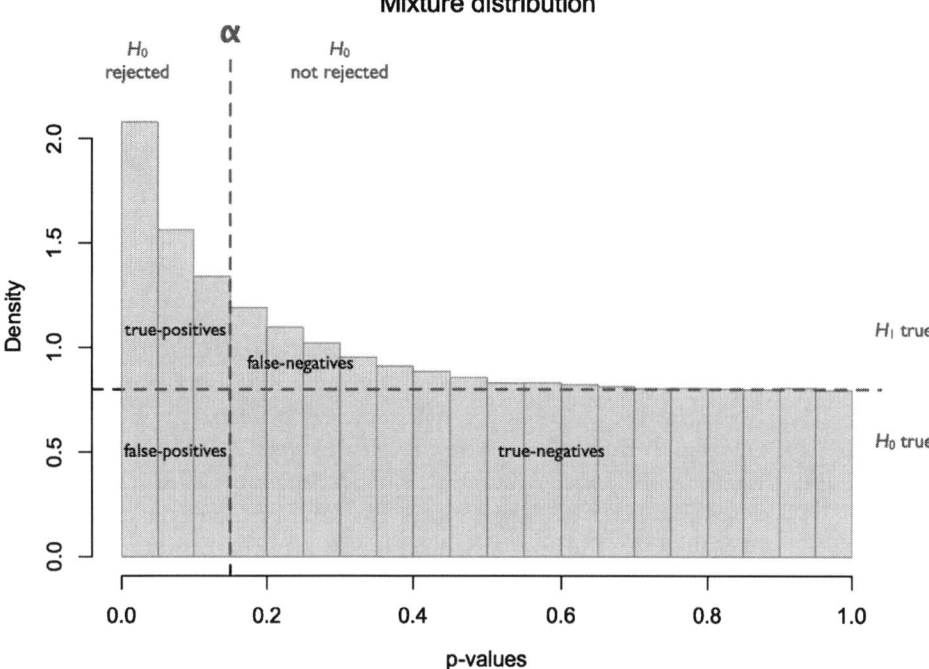

Fig. 3. Mixture distribution of *p*-values with corresponding proportions of true-positives, true-negatives, false-positives, and false-negatives at the level α.

amount of information. Plotting the distribution of *p*-values, assessing the proportion of markers under H_0 and H_1, and conducting a test based on the Poisson distribution are useful confidence indicators to confirm whether there is actually something to find in the data. Figure 3 provides a graphical representation of a mixture distribution of *p*-values along with the potential outcome of the corresponding statistical tests. Moreover, the mixture model assumption and the estimation of the proportion of markers under H_0 and H_1 are very important to understand and apply more advanced multiple-testing correction approaches (see the following sections).

All these approaches make the theoretical, and often observed in practice, assumption that the distribution under H_0 is a standard uniform distribution. Any strong deviation from this expected distribution can alert about the appropriateness of the statistical test chosen for the analysis and should be examined by a statistician. It can be the case for instance when testing markers on small sample sizes for which the χ^2 test is biased, and the Fisher exact test results lead to a discrete distribution of *p*-values that might not fit the expected continuous Uniform distribution under H_0.

5. Controlling the False-Discovery Rate

5.1. Definition

Controlling at 5%, the FWER is widely used in statistics. However, preventing against any single false-positive in large-scale genetic studies leads often to too stringent p-value corrections and therefore many missed findings. For these reasons, there is a certain reluctance to use FWER-based multiple-testing corrections such as the one proposed by Bonferonni (5). In practice, most researchers would reasonably accept a higher risk of having a false-positive in return for a greater statistical power (3). To overcome the limitations of the FWER, Benjamini and Hochberg introduced a more intuitive statistical concept as an alternative: the FDR (25). FDR-based multiple-testing corrections use the following idea: instead of considering that one wants to be sure at 95% that none of the tests declared as significant is a false-positive (i.e., considering the FWER), the FDR method focuses on the expected proportion of truly null hypotheses that are falsely rejected.

Considering the n tests presented in Table 2, the total number of rejections (n_R), and the total number of nonrejections ($n_U = n - n_R$), at the level α are observable. However, the values of tp, fp, tn, and fn are unknown. With these notations, Benjamini and Hochberg defined the FDR, that is the expected proportion of tests falsely declared significant among all tests declared significant, as:

$$\mathrm{FDR} = \mathbb{E}\left(\frac{fp}{fp + tp}\right)$$

$$= \mathbb{E}\left(\frac{fp}{n_R}\right),$$

with logically $FDR = 0$ if $n_R = 0$. Intuitively, if 100 tests are predicted to be significant, i.e., $n_R = 100$, and if the FDR is controlled at 5% then $fp = 5$ of these tests should be false-positives.

5.2. The Benjamini–Hochberg Procedure

The definition of the FDR proposed by Benjamini and Hochberg (25) is the most often used as it comes with a simple procedure to apply it: for each test i, the FDR corresponding to the level p_i (FDR_i) is controlled by the expected number of test statistics declared as significant under H_0 ($n \times p_i \times \pi_0$) over the observed number of test statistics declared as significant (n_{Ri}):

$$\mathrm{FDR}_i \leq \frac{n \times p_i \times \pi_0}{n_{Ri}}$$

To obtain a correct estimation of the FDR, it is therefore necessary to have a precise estimation of the proportion π_0. This implies having some knowledge about the alternative hypothesis which is not always the case. The idea proposed by Benjamini and Hochberg to avoid the explicit calculation of π_0 is to use the fact that $\pi_0 \leq 1$ and therefore:

$$\text{FDR}_i \leq \frac{n \times p_i}{n_{Ri}}.$$

5.3. In Practice

The Benjamini–Hochberg procedure is easy to apply as, like for the FWER corrections, it corresponds to applying a threshold to a set of adjusted p-values $p_i^{\text{BH}} = \frac{n \times p_i}{n_{Ri}}$. Basically, the adjusted p-values correspond to the p-values sorted in ascending order and divided by their own rank. Here, we introduce how to apply it in R to a set of p-values (p) to obtain a set of adjusted p-values (pBH):

```
> n = length(p)
> pBH = p*n/rank(p)
```

In practice, this can result in probabilities upper than 1 so formally we need to force such probabilities to 1:

```
> pBH[which(pBH > 1)] = 1
```

5.4. Notes on the FDR

The main advantage of FDR estimation is that it allows identifying a set of "candidate positives", of which a weak and controlled proportion are likely to be false-positives. The false-positives within the candidate set can then be identified in a follow-up study. Note that the FDR and the false-positive rate are often mistakenly equated, but their difference is actually very important (24). Given a level of confidence for declaring the test statistics significant, the false-positive rate is the rate that test statistics obtained under H_0 are called significant by chance, whereas the FDR is the rate that significant test statistics are actually null. For instance, a false-positive rate of 5% means that on average 5% of the test statistics drawn under H_0 in the study will be declared as significant; an FDR of 5% means that among all test statistics declared as significant, 5% of these are actually drawn from H_0 on average.

One possible problem with the procedure of Benjamini and Hochberg is that when considering a sorted list of p-values in ascending order, it is possible for the FDR associated with the p-value at rank m to be higher than the FDR associated to the one at rank $m + 1$ (7). This "nonmonotonicity" (i.e., the FDR does not consistently increase) can make the resulting FDR estimates difficult to interpret. A simple alternative available in the R function p. adjust handles this case:

```
> i = n:1
> o = order(p, decreasing = TRUE)
> ro = order(o)
> pBH = pmin(1, cummin(n/i * p[o]))[ro]
```

Moreover, a main assumption in the Benjamini–Hochberg approach is that the tests are independent. As already discussed before, this may not be the case when analyzing large-scale genetic data. To account for such situations, Benjamini and Yekutieli (26) developed a quite similar procedure where the FDR is controlled by

$$\text{FDR}_i \leq \frac{n \times p_i}{n_{Ri} \times c(n)},$$

where $c(n)$ is a function of the number of tests depending on the correlation between the tests. If the tests are positively correlated, $c(n) = 1$, there is no difference with the approach presented before, otherwise if the tests are negatively correlated, $c(n) = \sum_{k=1}^{n} \frac{1}{k}$.

Although the Benjamini–Hochberg procedure is simple and sufficient for many studies, one can argue that an upper bound of 1 for π_0 leads to a loss of precision in the estimation of the FDR. Such estimations are actually probably conservative with respect to the proportion of test statistics drawn under H_0 and H_1; that is, if the classical method estimates that the FDR associated with a collection of p-values is 5%, then on average the true FDR is lower than 5%.

Consequently, a variety of more sophisticated methods introducing the estimation of π_0 have been developed for achieving more accurate FDR estimations. Depending on the data, applying such methods may make a big difference or almost no difference at all (7). To this end, Storey introduced the q-value defined to be the FDR analogue of the p-value (24). Storey's procedure can be considered as analogue to the procedure of Benjamini and Hochberg except that it incorporates an estimate of the null proportion π_0. This new procedure is implemented in the R package qvalue.

Finally, as mentioned in the previous section, FDR-based corrections strongly depend on the assumption that the p-values are uniformly distributed if there is no association, which may not always be the case in practice. Methods providing more exact estimations of the FDR exist and rely on more advance statistical and algorithmic notions (27).

6. Controlling the Local False-Discovery-Rate

6.1. Definition

During the last decade, the FDR criterion introduced by Benjamini and Hochberg has received the greatest focus, due to its lower conservativeness compared to the FWER. The FDR is defined as the mean proportion of false-positives among the list of rejected hypotheses. It is therefore a global criterion that cannot be used to assess the reliability of a specific genetic marker. More recently, a strong interest has been devoted to the local version of the FDR, called "local-FDR" (28) and denoted hereafter fdr. The idea is to quantify the probability for a given null hypothesis to be true according to the specific p-value of each genetic marker tested. For a given p-value p_i associated to a test i:

$$
\begin{aligned}
\text{fdr}_i &= \mathbb{P}(\text{ test statistic } i \text{ is under } H_0 \text{ knowing } p_i) \\
&= \frac{\mathbb{P}(H_0)\mathbb{P}_{H_0}(p_i)}{\mathbb{P}(p_i)} \\
&= \frac{\pi_0 f_0(p_i)}{f(p_i)} \\
&= \frac{\pi_0 f_0(p_i)}{\pi_0 f_0(p_i) + \pi_1 f_1(p_i)},
\end{aligned}
$$

with π_0 and π_1 the proportion of p-values generated under H_0 and H_1, respectively, and f_0, f_1, and f the density functions corresponding to the distributions $\mathcal{D}_0$, $\mathcal{D}_1$, and $\mathcal{D}$ as described in Section 4. In general, the local-FDR is more difficult to estimate than the FWER and the FDR due to the difficulty in estimating density functions. Many approaches that are fully parametric (29, 23, 17, 16), semi-parametric (30), Bayesian (31, 32), or empirical Bayes (28) have been proposed, and most of them rely on the mixture model assumption described in Subheading 4.

6.2. The Kerfdr Procedure

The practical problem with local-FDR estimation is that if the distribution under H_0 ($\mathcal{D}_0$) can reasonably be considered as Uniform, the distribution under H_1 ($\mathcal{D}_1$) needs systematically to be estimated so does π_0.

The semi-parametric approach developed by Robin et al. (30) uses the distribution $\mathcal{D}_0$ to provide a flexible kernel-based estimation of the alternative distribution. This approach is implemented in the R package kerfdr[4] (33).

---

[4]http://stat.genopole.cnrs.fr/sg/software/kerfdr.

6.3. In Practice

kerfdr can be installed via the R utility for installing add-on packages:

```
>install.packages("kerfdr")
```

Once the package is installed, it can be called by applying the R command:

```
> library(kerfdr)
```

Then kerfdr can be applied to a given set of p-values (p):

```
> results = kerfdr(p)
```

The R object returned by kerfdr contains a lot of information. In particular, one can retrieve an estimation of the proportion of p-values drawn under H_0 (or H_1), the kerfdr estimation of the local-FDR, and a summary of the number of significant p-values after Bonferroni, BH, and kerfdr corrections at different significance level with "results$pi0" (or "results$pi1"), "results$localfdr," and "results$summary," respectively.

6.4. Notes on the Local-FDR

The main advantage of the local-FDR over the more classical FDR measure is that it assesses for each genetic marker its own measure of significance. In this sense, it appears more intuitive and precise than the FDR. However, its estimation requires to determine with precision π_0 and the distribution under H_1 ($\mathcal{D}_1$) which requires more advanced statistical skills and the use of fully developed packages such as kerfdr. This *semi-parametric* approach appears flexible since it does not demand to formulate restrictive assumptions on $\mathcal{D}_1$ and shows good performances compared to other approaches (34). Moreover, the package proposes several practical extensions such as the possibility to use prior information in the estimation procedure (*semi-supervised*) and the ability to handle truncated distributions such as those generated by Monte-Carlo estimation of p-values. Simulations showed that such information can significantly improve the quality of the estimates (33). Finally, like for the estimation of the FWER and the FDR, kerfdr and most of the local-FDR estimation procedures assume that the test of n genetic markers are independent.

7. Conclusions

Large-scale genetic association studies offer a systematic strategy to assess the influence of common genetic variants on phenotypes (35). The main property that makes these experiments so attractive is their massive scales. Paradoxically, this can also result in spurious discoveries, which must be guarded against to prevent that time and resources are spent on leads that would prove irrelevant (7). To this end, multiple-testing corrections have been developed, based on alternative measures of significance adapted to genome-wide studies. Nowadays, FWER, FDR, and local-FDR constitute key statistical concepts that are widely applied in the study of high-dimensional data.

The FWER controls the probability of having one or more false-positives and turns out to be too stringent in most of the situations. As a more intuitive alternative, the FDR considers the proportion of significant results that are expected to be false-positives. More recently through the local-FDR, authors have proposed to consider the actual probability for a result of being under H_0 or H_1. Considering the fact that several confidence measures can be applied to account for multiple testing, a question that naturally arises is which method one should use and also if an FWER correction—due to its stringency—is even appropriate. Noble provided a practical solution to the problem (7): based on the same way that a significance threshold is chosen, choosing which multiple-testing correction method to use depends on the cost associated with false-positives and false-negatives. For example, if one's follow-up analyses will focus on only few experiments, then an FWER-based correction is appropriate. Alternatively, if one plans on performing a collection of follow-up experiments and tolerate having some of them that fail, then the FDR correction may be more appropriate. Finally, if one is interested in following up on a single gene, local-FDR may be precisely the more suited method.

Correlations between genetic markers, due to LD for instance, are still a difficult issue when considering multiple testing. Such local dependencies between the tests lead to a smaller number of independent tests than SNPs examined. As we have seen, some approaches allow taking dependent tests into account such as permutation-based methods (5, 26, 27), but require far more advanced programming skills and statistical knowledge than the simple Bonferroni and Benjamini–Hochberg procedures.

However, theoretical and simulation studies suggest that multiple-testing corrections assuming the independence of the tests perform quite well in cases of weak positive correlations, which is common in many genetic situations (5, 26). The issue of dependency in the analysis of large-scale molecular data is likely to be a live field of research in the near future. As statistical techniques are still developing to account for the complexities involved in controlling false-positives in exploratory researches, independent replications and validations still remain necessary steps in the discovery process (6).

References

1. Ioannidis JP, Ntzani EE, Trikalinos TA, Contopoulos-Ioannidis DG (2001) Replication validity of genetic association studies. Nat Genet 29:306–309.

2. Page GP, George V, Go RC, Page PZ, Allison DB (2003) "Are we there yet?": Deciding when one has demonstrated specific genetic causation in complex diseases and quantitative traits. Am J Hum Genet 73:711–719.

3. Balding DJ (2006) A tutorial on statistical methods for population association studies. Nat Rev Genet 7:781–791.

4. Rice TK, Schork NJ, Rao DC (2008) Methods for handling multiple testing. Adv Genet 60:293–308.

5. Moskvina V, Schmidt KM (2008) On multiple-testing correction in genome-wide association studies. Genet Epidemiol 32:567–573.

6. van den Oord EJCG (2008) Controlling false discoveries in genetic studies. Am J Med Genet B Neuropsychiatr Genet 147B:637–644.

7. Noble WS (2009) How does multiple testing correction work? Nat Biotechnol 27:1135–1137.

8. Chen JJ, Roberson PK, Schell MJ (2010) The false discovery rate: a key concept in large-scale genetic studies. Cancer Control 17:58–62.

9. Fisher RA (1925) Statistical methods for research workers, 11th edn.(rev.). Oliver & Boyd, Edinburgh.

10. Bonferroni C (1935) Studi in Onore del Professore Salvatore Ortu Carboni, chapter Il calcolo delle assicurazioni su gruppi di teste. pp. 13–60.

11. Bonferroni C (1936) Teoria statistica delle classi e calcolo delle probabilita. Publicazioni del R Instituto Superiore de Scienze Economiche e Commerciali de Firenze 8:3–62.

12. Sidak Z (1967) Rectangular confidence region for themeans of multivariate normal distributions. J Am Stat Assoc 62:626–633.

13. McLachlan G, Peel D (2000) Finite mixture models. Wiley, New York

14. Jung SH (2005) Sample size for fdr-control in microarray data analysis. Bioinformatics 21:3097–3104.

15. Wang SJ, Chen JJ (2004) Sample size for identifying differentially expressed genes in microarray experiments. J Comput Biol 11:714–726.

16. Pounds S, Morris SW (2003) Estimating the occurrence of false positives and false negatives in microarray studies by approximating and partitioning the empirical distribution of p-values. Bioinformatics 19:1236–1242.

17. McLachlan G, Bean R, Ben-Tovim Jones L (2006) A simple implementation of a normal mixture approach to differential gene expression in multiclass microarrays. Bioinformatics 22:1608–1615.

18. Markitsis A, Lai Y (2010) A censored beta mixture model for the estimation of the proportion of non-differentially expressed genes. Bioinformatics 26:640–646.

19. Mosig MO, Lipkin E, Khutoreskaya G, Tchourzyna E, Soller M, et al. (2001) A whole genome scan for quantitative trait loci affecting milk protein percentage in israeli-holstein cattle, by means of selective milk dna pooling in a daughter design, using an adjusted false discovery rate criterion. Genetics 157:1683–1698.

20. Scheid S, Spang R (2004) A stochastic downhill search algorithm for estimating the local false discovery rate. IEEE/ACM Trans Comput Biol Bioinform 1:98–108.

21. Langaas M, Lindqvist BH, Ferkingstad E (2005) Estimating the proportion of true null hypotheses, with application to dna microarray data. J R Stat Soc Ser B 67:555–572.

22. Lai Y (2007) A moment-based method for estimating the proportion of true null hypotheses and its application to microarray gene expression data. Biostatistics 8:744–755.

23. Liao JG, Lin Y, Selvanayagam ZE, Weichung JS (2004) A mixture model for estimating the local false discovery rate in dna microarray analysis. Bioinformatics 20:2694–2701.

24. Storey JD, Tibshirani R (2003) Statistical significance for genomewide studies. Proc Natl Acad Sci U S A 100:9440–9445.

25. Benjamini Y, Hochberg Y (1995) Controlling the false discovery rate: a practical and powerfull approach to multiple testing. JRSSB 57:289–300.

26. Benjamini Y, Yekutieli D (2001) The control of the false discovery rate in multiple testing under dependency. Ann Stat 29:1165–1188.

27. Wojcik J, Forner K (2008) Exactfdr: exact computation of false discovery rate estimate in case-control association studies. Bioinformatics 24:2407–2408.

28. Efron B, Tibshirani R (2002) Empirical bayes methods and false discovery rates for microarrays. Genet Epidemiol 23:70–86.

29. Allison DB, Gadbury G, Heo M, Fernandez J, Lee CK, et al. (2002) Mixture model approach for the analysis of microarray gene expression data. Comput Statist Data Anal 39:1–20.

30. Robin S, Bar-Hen A, Daudin JJ, Pierre L (2007) A semi-parametric approach for mixture models: Application to local false discovery rate estimation. Comput Statist Data Anal 51: 5483–5493.

31. Broet P, Lewin A, Richardson S, Dalmasso C, Magdelenat H (2004) A mixture model-based strategy for selecting sets of genes in multiclass response microarray experiments. Bioinformatics 20:2562–2571.

32. Newton MA, Noueiry A, Sarkar D, Ahlquist P (2004) Detecting differential gene expression with a semiparametric hierarchical mixture method. Biostatistics 5:155–176.

33. Guedj M, Robin S, Celisse A, Nuel G (2009) Kerfdr: a semi-parametric kernel-based approach to local false discovery rate estimation. BMC Bioinformatics 10:84.

34. Strimmer K (2008) A unified approach to false discovery rate estimation. BMC Bioinformatics 9:303.

35. Risch N, Merikangas K (1996) The future of genetic studies of complex human diseases. Science 273:1516–1517.

Chapter 14

Population Genomic Analysis of Model and Nonmodel Organisms Using Sequenced RAD Tags

Paul A. Hohenlohe, Julian Catchen, and William A. Cresko

Abstract

The evolutionary processes of mutation, migration, genetic drift, and natural selection shape patterns of genetic variation among individuals, populations, and species, and they can do so differentially across genomes. The field of population genomics provides a comprehensive genome-scale view of these processes, even beyond traditional model organisms. Until recently, genome-wide studies of genetic variation have been prohibitively expensive. However, next-generation sequencing (NGS) technologies are revolutionizing the field of population genomics, allowing for genetic analysis at scales not previously possible even in organisms for which few genomic resources presently exist. To speed this revolution in evolutionary genetics, we and colleagues developed *Restriction site Associated DNA* (RAD) sequencing, a method that uses Illumina NGS to simultaneously type and score tens to hundreds of thousands of single nucleotide polymorphism (SNP) markers in hundreds of individuals for minimal investment of resources. The core molecular protocol is described elsewhere in this volume, which can be modified to suit a diversity of evolutionary genetic questions. In this chapter, we outline the conceptual framework of population genomics, relate genomic patterns of variation to evolutionary processes, and discuss how RAD sequencing can be used to study population genomics. In addition, we discuss bioinformatic considerations that arise from unique aspects of NGS data as compared to traditional marker based approaches, and we outline some general analytical approaches for RAD-seq and similar data, including a computational pipeline that we developed called Stacks. This software can be used for the analysis of RAD-seq data in organisms with and without a reference genome. Nonetheless, the development of analytical tools remains in its infancy, and further work is needed to fully quantify sampling variance and biases in these data types. As data-gathering technology continues to advance, our ability to understand genomic evolution in natural populations will be limited more by conceptual and analytical weaknesses than by the amount of molecular data.

Key words: Genetic mapping, Population genetics, Genomics, Evolution, Genotyping, Single nucleotide polymorphisms, Next-generation sequencing, RAD-seq

1. Evolutionary Studies at a Genome-Wide Scale

Population genetics has long been a powerful theoretical framework for understanding the action of evolution (1–4). As population

François Pompanon and Aurélie Bonin (eds.), *Data Production and Analysis in Population Genomics: Methods and Protocols*, Methods in Molecular Biology, vol. 888, DOI 10.1007/978-1-61779-870-2_14, © Springer Science+Business Media New York 2012

genetics has matured over the last century new techniques for tracking alleles in populations have emerged. Phenotypic markers, allozymes, microsatellites, and ultimately DNA sequence data have all been used to quantify genetic variation (5). However, despite the robust mathematical theory of classical population genetics, a significant shortcoming has been its focus on allelic variation at a small number of loci. Evolutionary processes, such as natural selection and genetic drift, act in concert with genetic factors such as dominance and epistasis, linkage and recombination, and gene and genome duplication, to produce the structure of genomic variation observed in natural populations (6–8). Population genetics theory has addressed these interactions among loci as well, but most often in the context of two or a small number of loci.

The advance of molecular genetic technologies has allowed a more complete and comprehensive view of the evolution of genetic variation within and among populations. We can now view many classical statistics, such as nucleotide diversity and population divergence, as continuous variables distributed across a genome, an area of inquiry that has been termed *population genomics* (9–13). A critical feature of genomic distributions is their spatial autocorrelation—correlation among measurements at neighboring genomic regions—resulting from linkage disequilibrium (LD) (14, 15). The degree to which this autocorrelation itself changes along the genome reflects selection and recombination as well as other evolutionary forces (16). Because of this autocorrelation, inferring the evolutionary history of any single locus is complicated by the influence of its genomic neighbors (17–19). However, this genomic structure also opens the door to new tests of selection based specifically on statistics describing the local extent of LD (8, 20).

Population genomics allows the simultaneous identification of a genome-wide average and outliers for any given statistic. The genome-wide average provides a baseline view of neutral processes, both demographic (e.g., population size, migration rate) and genetic (e.g., mutation rate, recombination). Outliers from the background indicate the action on specific loci of evolutionary forces like natural selection (21, 22), providing an apparently clean separation between neutral and nonneutral processes (12, 13, 23). However, the genetic effects of selection can mimic those of demographic factors so that effective population size and migration rate are also continuous variables along the genome (24). Demographic processes can also affect the variance as well as the average of genome-wide distributions (25, 26). Therefore, the signatures of all of these neutral and nonneutral processes on both the genome-wide distribution of population genetic statistics and on specific genomic regions should be considered simultaneously when making inferences from genomic data.

Population genomic studies have already been successful in providing surprising new insights, such as uncovering loci responding to selection in natural populations, providing candidate genes for functional studies and emphasizing the ubiquity of selection in natural populations (27). However, several avenues for future progress in strengthening population genomic approaches should be kept in mind. These considerations are reviewed elsewhere ((12, 24, 28, 29)—but see Note 1 for a general discussion). The goal of this chapter is to introduce the use of a specific next-generation sequencing (NGS) application, Restriction site Associated DNA (RAD) sequencing (see Etter and Johnson, this volume), for evolutionary genomic analyses, including the general framework for the analysis of RAD-seq data. We discuss several primary bioinformatic considerations, such as inferring marker loci and genotypes from RAD tag sequences for organisms with and without sequenced genomes, distinguishing SNPs from error, and inferring heterozygosity in the face of sampling variance. We also introduce a flexible software pipeline for these analyses called Stacks. New bioinformatic advances will certainly be made in this area, and we therefore present these analytical approaches less as a set of concrete protocols, and more as a conceptual framework for using RAD-seq in evolutionary genomics studies.

2. NGS and RAD Analyses for Population Genomics

A limitation for the maturation of the field of population genomics has been that, for most organisms, the cost of obtaining true genome-scale data has been prohibitively expensive. A significant advance for population genomics has been the advent over the last 5 years of NGS technologies (30, 31). These new tools open the possibility of gathering genomic information across multiple individuals at a genome-wide scale, both in mapping crosses and in natural populations (32, 33). This breakthrough technology is becoming increasingly important for evolutionary genetics, even in organisms for which few genomic resources presently exist (34, 35). This innovation has already led to studies in QTL mapping (36), population genomics (27), and phylogeography (37, 38) that were not possible even a few years ago.

The ultimate promise of NGS for population genomics is being able to gather perfect genetic information—the entire genome sequences of multiple individuals (33, 39). However, complete genome resequencing is still prohibitively expensive for most evolutionary studies, and for many purposes is actually unnecessary. For example, because linkage blocks are often quite large in quantitative trait locus (QTL) mapping crosses, progeny can be adequately typed with genetic markers of sufficient density (36). Similarly, depending upon the size of LD blocks in natural populations, population genomic analyses can be performed with dense panels of genotypes without complete coverage of the genome (27, 37, 39).

The large numbers of short reads provided by platforms, such as Illumina, are ideal for dense genotyping in population genomics (32, 34). A methodological difficulty is focusing these reads on the same genomic regions in order to maximize the probability that most individuals in a study will be assayed at homology regions. We developed a procedure called *R*estriction site *A*ssociated *D*NA sequencing (RAD-seq) that accomplishes this goal of genome sub-sampling (see refs. 36, 40 and this volume for methodological details). RAD-seq technology allows single nucleotide polymorphisms (SNPs) to be identified and typed for tens or hundreds of thousands of markers spread evenly throughout the genome. Therefore, RAD-seq provides a flexible, inexpensive platform for the simultaneous discovery of tens of thousands of genetic markers in model and nonmodel organisms alike. We confine our discussions to using RAD-seq to infer genotypes of haploid and diploid individuals by using barcoding and multiplexing, but undoubtedly other pooling strategies could be used for many evolutionary genomic studies (see Note 2).

3. RAD Analyses for Evolutionary Genomics: A Wide Range of Possibilities

Because RAD sequencing simultaneously identifies and types SNP markers, it is a powerful tool for many different evolutionary genomic studies. Although the methodological details for generating RAD-seq data have become more robust over recent years, the specifics of how to analyze these data in the most efficient manner are still far from clear. At a most basic level, RAD sequencing is a powerful tool for SNP discovery to develop fixed genotyping platforms that can be used in additional studies. For example, by RAD sequencing just tens of individuals from a few distinct populations of interest, a researcher could identify thousands of SNP markers that could be genotyped in a large sample of individuals using high-throughput qPCR-based assays. This approach has been recently applied to hybridizing species of trout in the Western US (41).

RAD sequencing by itself can be used for a variety of projects without developing additional, specific genotyping platforms. One of the most popular uses of RAD sequencing is for linkage mapping in laboratory crosses (36, 42–44). A panel of backcross or F_2 individuals can be sequenced along with the progenitors of the cross, and these data can be used to create a high density genetic

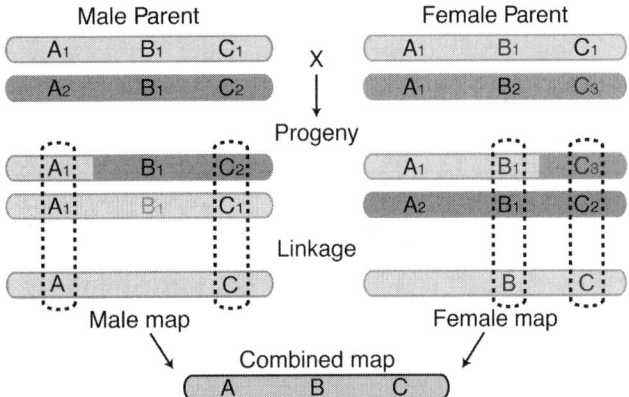

Fig. 1. A RAD tag genetic map can be generated using an F1 family created from heterozygous parents. Markers that are heterozygous in one parent and homozygous in the other (i.e., A and B) can be mapped for each parent as a backcross in what is known as a pseudo-testcross format. Because markers A and C are heterozygous in the male parent, and B and C are heterozygous in the female parent, each can be mapped with respect to one another to create a male and female map. A subset of the markers will be differentially heterozygous in both parents, and the shared allele (C) can link sex-specific maps into a combined map. This pseudo-testcross format requires a large number of genetic markers, such as is provided by the RAD genotyping platform.

linkage map. This linkage map can be used immediately for mapping of induced mutations in the laboratory, or Quantitative Trait Loci (QTL) mapping of evolutionarily significant phenotypes. Significantly, the RAD approach permits the addition of new mapping families from distantly related populations or different species without developing a completely new set of informative markers. Because RAD-seq produces a large number of genetic markers (10,000 to 100,000), enough markers can be scored in an F_1 family produced from a single pair of heterozygous parents to allow the creation of a genetic map in pseudo-testcross format (Fig. 1; see Note 3). This approach has already been used to produce a very high density genetic map in the spotted gar (45). In addition, the large number of RAD markers can be used to facilitate the assembly of physical maps of genomes. With even a low level of sequence coverage (1× or 2×) of an average sized genome, a large number of contigs will contain one or more identifiable RAD tags which can then be used to anchor them to the genetic map.

Another powerful use of RAD genotyping is for high precision population genetic, phylogeographic, and phylogenetic studies. Many closely related populations and species have evolutionary histories that are obscured by incomplete lineage sorting in the context of rapid cladogenesis, significant residual gene flow that continues after lineage splitting, or different evolutionary histories for different genomic regions (e.g., mtDNA vs. nuclear genomic regions under sex-specific migration). Because of methodological considerations, traditional population genetic and phylogeography studies have utilized a small number of loci that have not permitted precise point estimates for many population genetic parameters. With the orders of magnitude increase in marker number provided by RAD sequencing, the resolution of phylogenetic relationships increases significantly. For example, whereas the postglacial phylogeography of the pitcher plant mosquito Wyeomyia smithii was poorly estimated using mtDNA sequence data, the relationship of 15 populations from the Eastern seaboard of North America was well-resolved with only a modest amount of RAD sequence data (37). Similarly, RAD markers could theoretically be used within populations to make high quality inferences of pedigrees among individuals.

A final, significant benefit of RAD sequencing for evolutionary genetics is the ability to perform genome-wide studies in natural populations (27). Because RAD markers are spread throughout genomes at high density, they can be used in genomic analyses of differentiated populations to address questions, such as whether genomic regions appear to be under balancing or directional selection. For example, regions that are under balancing selection, either because of overdominance or frequency dependent selection, would be detected by genomic regions that exhibit significant increases in genetic diversity and heterozygosity over that seen in the rest of the genome. In the case of divergent selection, comparisons between two populations would reveal genomic regions with significant increases in divergence among populations (for example, increased F_{ST}) and decreases in genetic diversity within populations. Importantly, because selective sweeps may be large relative to the RAD marker density, regions of linked markers will provide robust evidence of selection as compared to the identification of single, outlier markers. We have applied this approach to threespine stickleback and identified signatures of both balancing and directional selection (27).

Genome scans across populations can help identify genomic regions that are subject to selection, an important goal for many evolutionary genetic studies. However, many phenotypes are likely subject to differing types of selection leading to changes in numerous genomic regions, making the association of phenotypes with causative genetic changes a difficult task. RAD genome scans can also be performed on phenotypes within populations or hybrid zones to associate phenotypic variation with causative genomic regions. Because of the increased number of generations of recombination, the size of linkage blocks may often be significantly smaller in phenotypically variable populations. Although this may require an increased density of RAD markers, it also may significantly increase the mapping precision (as compared to laboratory crosses) such that just a few candidate genes are within each identified region. These so-called Genome-Wide Association Studies (GWAS) have until recently been confined to humans or a small number of model organisms for which dense genotyping panels have been produced.

RAD sequencing opens the possibility of quickly identifying genomic regions contributing to variation in just about any organism, therefore rapidly increasing the set of candidate genes known to contribute to evolutionarily important phenotypic differences in the wild.

4. Analytical Considerations for Evolutionary Genomics Using RAD Sequencing

Although evolutionary genomic studies using RAD sequencing are feasible for most evolutionary genetics labs with a moderate level of molecular sophistication, several aspects of data analysis should be considered at the very beginning of the project. The first concerns the computational facilities necessary to complete a project, which are significant for NGS. As the number of reads and read lengths continue to increase, computational resources will only become more important (see Note 4). Second, the desired use of the RAD sequences should be considered. Creating a linkage map may require only a couple of thousand markers because of the relatively small numbers of meioses in a cross, whereas a genome scan or GWAS study would require tens or hundreds of thousands of markers. The choice of enzyme for RAD analysis (see Etter and Johnson, this volume), computational demands and necessary software are significantly different for each type of study. Third, a significant consideration is whether the organism of interest has a reference genome against which the reads can be aligned, or whether the reads will need to be assembled de novo into a linkage map. In both cases, the inferences of real nucleotide variants from sequencing error needs to be done in a statistically rigorous framework, which is described in detail in Note 5. Below, we describe the general format for the analysis of RAD genotyping data in evolutionary genomic studies for the two most common types of situations, organisms with and without a reference genome. To ease the analysis of these data, we describe the use of a freely available software platform called Stacks that can be run on a local server, a computer cluster, or in a cloud computing scenario.

Critical to statistical analysis of RAD sequencing data is the recognition that it is a multilevel sampling process (Fig. 2). As a library of DNA fragments is created, selectively amplified, and sequenced as a pool, the number of sequence reads from a particular allele at a specific locus in a particular individual is the result of sampling at each step of the protocol. As a result, we observe sampling variance at all levels: total reads per individual in a pooled lane, number of loci represented in each individual, mean and variance of read counts across loci, and read counts of alternative alleles at polymorphic loci. Added to this is random sequencing error. All of these factors can be dealt with using relatively straightforward statistical models, such as a multinomial sampling model to assign

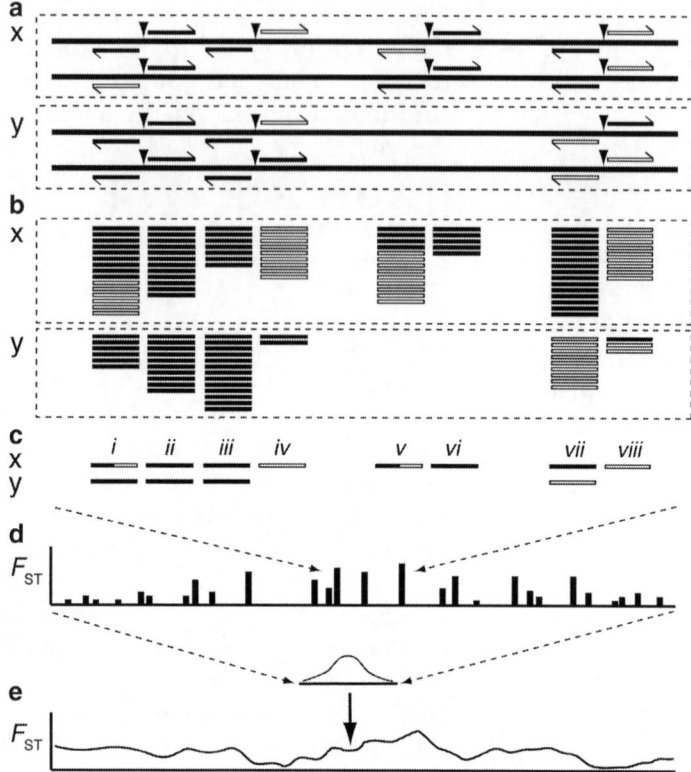

Fig. 2. Schematic diagram of RAD-seq sampling variance, genotyping, and analysis with a reference genome sequence. (**a**) RAD digestion of genomic DNA from two diploid individuals *x* and *y*. Four RAD sites (*arrowheads*) are illustrated here, meaning that eight homology RAD tags may be observed per RAD site. There may be polymorphism within a sample for the presence or absence of RAD sites, and SNPs within RAD tag sequences are indicated by filled versus open bars. (**b**) The count of reads across RAD tags has several sources of sampling variance: across individuals, across loci within individuals, and across alternative alleles at heterozygous loci. (**c**) The maximum likelihood genotyping method accounts for this sampling variance in assigning diploid genotypes along each RAD tag within each individual. Here, the correct genotype is assigned at homozygous (e.g., tag ii in both individuals) or heterozygous (e.g., tags i and v in individual *x*) loci, but note that haplotype phase between heterozygous RAD tags cannot be determined. At other loci (e.g., tags iv and viii in individual *y*), the depth of sequencing happens to be too small to assign a genotype with statistical confidence, so no genotype is assigned. At other loci, polymorphism in the presence/absence of RAD sites can lead to missing data (e.g., tags v and vi in individual *y*) or to the potentially incorrect inference of a homozygote (e.g., tags iii and iv in individual *x*). All of these issues occur whether or not a reference genome sequence is used. (**d**) Population genetic parameters, such as F_{ST}, can be calculated at each individual nucleotide position or SNP. (**e**) A kernel-smoothing sliding window uses a weighting function (e.g., Gaussian) to average across neighboring loci and produce a continuous genomic distribution of population genetic measures. Bootstrap statistical significance can be assigned to each window average by resampling individual loci from across the genome to create a distribution of window average values given the sample size of loci within a given window.

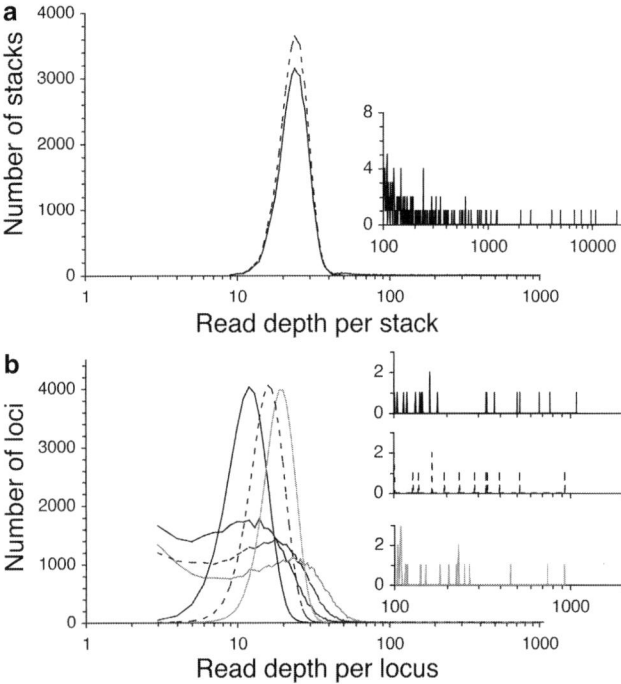

Fig. 3. Sources of increased variance in read depth across loci, illustrated with RAD-seq data from threespine stickleback. These data use the restriction enzyme SbfI, which produces 45,547 predicted RAD tags in the reference genome sequence. (a) Simulation of Poisson sampling across RAD tags shows the effect of duplicate regions across the genome. The *solid line* shows the naive distribution when the genomic locus for each read is known and mean sequencing depth is 25×. The *dotted line* shows the distribution of read depth across stacks if reads were assembled in the absence of a reference genome sequence, allowing up to two nucleotide mismatches over the 60 bp RAD tag sequence. Stacks with excess depth, or lumberjack stacks, incorrectly assemble reads from multiple loci and create a long tail to the distribution (the overall distribution is truncated, with the extremes of the tails for each distribution shown in the *inset*). (b) Observed distribution of read depth across RAD tags for four individuals, after aligning to the reference genome, removing any reads aligned to duplicate loci with up to two nucleotide mismatches (i.e., those loci that grouped together in (a)), and removing any loci with fewer than three reads. Actual read depth (*lower curves*; truncated here, with a maximum depth of 1,972 in one individual) exhibits much greater variance than the Poisson expectation (*upper curves*, calculated from the empirical mean depth and total number of reads for each individual). Similar to panel (a), the extreme tails of the distributions are represented in the *inset*. These four stickleback individuals were run in a single lane of 16 individuals (27).

diploid genotypes at each locus when individuals are separately barcoded (see Note 5). Such a model allows us to assign measures of statistical confidence to inferred genotypes, and carry those measures of confidence through subsequent analyses.

More problematic are potential issues of PCR error early in the protocol, or biases in probability of sequencing among tags, all of which can increase the variance in sequencing depth across tags (Fig. 3). These problems can more closely mimic biological variation and be difficult to detect (e.g., an early PCR error can be amplified and appear in multiple reads, resembling an alternative allele at a locus). One consequence of higher than expected variance in read depth across tags relates to polymorphism in the presence or absence of the RAD site itself. One might expect that read depth would follow a bimodal distribution, such that individuals that are heterozygous for the presence/absence of a RAD site would exhibit a sequencing depth at the associated tags that is roughly half that for homozygous individuals. Such a distribution could allow the identification of genotypes for RAD site presence/absence in the population. However, the variance commonly observed in read depth likely does not allow for this level of discrimination. These issues need further empirical as well as analytical attention.

4.1. Population Genomics in Organisms with a Reference Genome

If the organism of interest has a reference genome, then RAD sequences can be aligned against the genome directly, SNPs called, and summary statistics calculated (for overview, see Fig. 2). Although conceptually simple, a major problem can be introduced by nonhomology repetitive regions, or paralogous regions that are the product of recent gene or genome duplication events. Because the length of reads is still relatively short on most NGS platforms, at least some of the reads will fall into repetitive regions that are similar across several parts of the genome, which will be evidenced by potential assignment with equal probability to these locations (Fig. 3). These reads can be removed from the analysis. Alternatively, paired-end sequencing can be performed to help infer the correct location of each read in the genome. Once homology reads are determined with confidence, SNPs can be called and used for a variety of population genomic studies. One of the most likely, at least in the near term, is for tests of natural selection. A powerful approach is to compare genomic statistics across multiple populations to identify signatures of selection (see Note 6).

A reference genome allows genotype data to be positioned along the genome so that continuous distributions of population genetic statistics can be inferred. Because of the density of SNPs in most RAD sequencing datasets, correlation is expected in population genetic statistics at loci that are physically close together. Rather than a SNP-by-SNP perspective, this allows a population genomics perspective of continuous distributions along the genome. This is achieved using a sliding window approach, with statistical significance of divergent values assessed by comparisons against distributions derived from resampling of values across the genome (see Note 7). Except for studies with very high depth of coverage, we expect some individuals will not be genotyped at some loci simply by chance. As a result, the sample size of individuals may change at each locus, and the contribution of each locus to a window average can be weighted by sample size.

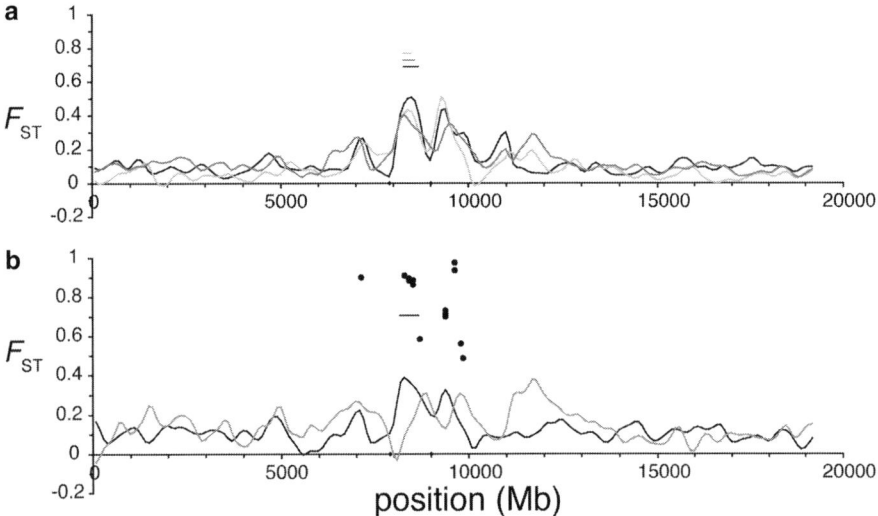

Fig. 4. Population differentiation along Linkage Group VIII, assessed with RAD sequencing, illuminates divergent selection in replicate populations of threespine stickleback. (**a**) Differentiation (F_{ST}) between each of three independently derived freshwater populations and their oceanic ancestor. *Bars* above indicate genomic regions of differentiation significant in each comparison at $p < 10^{-5}$ using bootstrap resampling. Coincident genomic regions of differentiation indicate parallel genetic evolution. (**b**) Overall differentiation between the combined freshwater populations and the oceanic ancestor (*solid line*), and differentiation among the freshwater populations (*gray line*). The bar above indicates significant differentiation in the overall comparison ($p < 10^{-5}$), and *dots* indicate F_{ST} values at highly significant individual SNPs (see ref. 27 for further details).

Window averages can also form the basis of statistical tests for genomic regions that are outliers for a particular statistic. A given genomic region can be tested against a distribution derived from resampling across the whole genome. This approach obviates the need for assuming any sort of null model, for instance of population structure, to derive expectations for population genetic statistics.

Strong inferences about selection can be made when examining replicate populations that evolve in parallel across habitats or putative selective regimes (see Note 8). Replicate oceanic and freshwater populations of threespine stickleback provide an example of parallel genomic evolution between habitat types (27). RAD sequencing revealed evidence for divergent selection between freshwater and oceanic habitats near the middle of the linkage group in each of three independently evolved freshwater populations (Fig. 4a). Further, the response to selection at this location exhibits a pattern consistent with different alleles being selected to high frequency in each of the freshwater populations (Fig. 4b), evidenced by differentiation among freshwater populations as well as between each freshwater population and the common oceanic ancestor. In this case, it appears that divergent selection has acted on different genetic variants in the different freshwater populations (27).

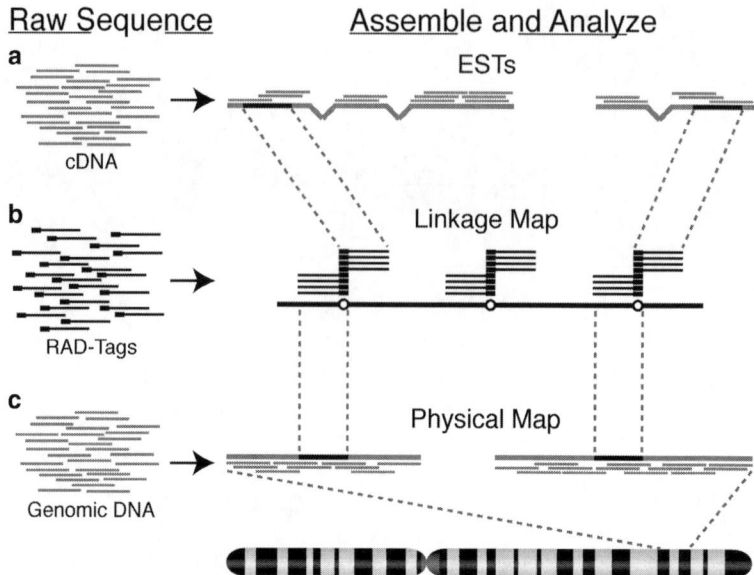

Fig. 5. RAD sequencing in the absence of a reference genome. (**a**) Using cDNA from the organism, assemblies of the Expressed Sequence Tags (ESTs) of the expressed genes can be generated by sequencing cDNA, which can then be linked to a genetic linkage map. (**b**) This genetic map can be created using RAD tags linked in a mapping cross, and the expressed genes can also be compared against other gene databases to identify orthologs, and used to annotate the genomic assembly. (**c**) Sequencing of genomic DNA can be used for the assembly of moderately sized contigs, which can then be ordered and further assembled and tiled with respect to the RAD-Tag-based genetic linkage map. Therefore, RAD sequencing can be used to create a backbone genetic map that can link genomic contigs with ESTs into an informative initial genomic assembly for many organisms.

4.2. De Novo RAD Analyses Using Stacks

When a reference genome does not exist, RAD genotyping can still be performed by assembling reads with respect to one another (see Fig. 5). For genomic regions that are unique, this assembly works as well as aligning against the genome. The aligned stack of reads can be analyzed to determine SNPs and genotypes that can be used in a genetic map or population genomic analysis. All of the sampling issues described above, as well as repetitive regions, are again problematic (Fig. 3a). However, the additional information provided by aligning to multiple genomic regions in the reference genome is absent, making the identification of these repetitive regions even more difficult (see Note 9). In situations where paralogous regions are mistakenly assembled, one of several things can be done. First, the length of the reads can be increased, or fragments can be paired-end sequenced, with the hope of obtaining unique information that can be used to tell paralogous regions apart. These solutions increase the cost of the sequencing to an extent that may not be justified by the increase in information. If data are being collected from individuals in a population or mapping cross, tests of Hardy–Weinberg Equilibrium (HWE) may allow the identification of problematic, incorrectly identified "genotypes" that are really paralogous regions. For example, when two monomorphic, paralogous regions are fixed for alternative nucleotides and then mistakenly assembled as a single locus, a SNP will be inferred but no or few homozygotes will be identified (41). Note that detecting such situations requires individuals to be separately barcoded during sequencing. Lastly, a network-based approach can be employed with the expectation that true SNPs should be at significant frequency and surrounded in sequence space by a constellation of low frequency sequencing errors. All of these solutions are imperfect, and extracting information from de novo RAD tags and other NGS data will be a significant area of research in the near future.

We have developed a software pipeline for the de novo assembly of RAD tags called Stacks (http://creskolab.uoregon.edu/stacks/). The name of the program is derived from the assembly of millions of reads into identical stacks of sequences (Fig. 6). Stacks is flexible enough to be used for different mapping crosses (BC, F_2, RI line, and QTL), as well as for organisms sampled from populations. Stacks begins by taking all of the high quality sequences as judged by sequencer quality scores with a 90% or better probability of being correct within a moving, sliding window average (46, 47) from a project and making the complete set of unique stacks that comprises all of the true alleles of all loci, as well as the constellation of 1 and 2 bp mismatch sequences that happen to have exactly the same sequence because of PCR or sequencing error. For an average sequencing depth, however, each true allele should be close to the mean number of reads, differing only because of sampling variation, whereas the sequencing error stacks will have much lower coverage. Therefore, only sequences that have three or more reads are considered unique stacks and carried through subsequent analyses (note that the number of minimum reads is a configurable parameter). From this set of unique stacks, a k-mer reference dictionary is created against which all sequences can be compared. K-mer sequences have been used extensively in computational biology, as alignment "seeds" in BLAST (48), in the ultra-efficient BLAT (49), as well as in alignment-free sequence comparison (50), and in short-read sequence assemblers, such as AbYSS (51) and Velvet (52) among many other applications. As compared to doing an all-by-all pairwise comparison, this reference dictionary approach is much more computationally feasible. Each sequence is broken into its set of k-mer reads, which are then compared against the reference dictionary, and all matches recorded.

All connections of a calculated minimum number of k-mer matches are retained, therefore creating a connectivity matrix of all reads and stacks. This network has high connectivity nodes of presumptive loci with alleles surrounded by reads with sequencing error, and each of these nodes are loosely connected to other such nodes. These nodes are then evaluated and disassociated from the

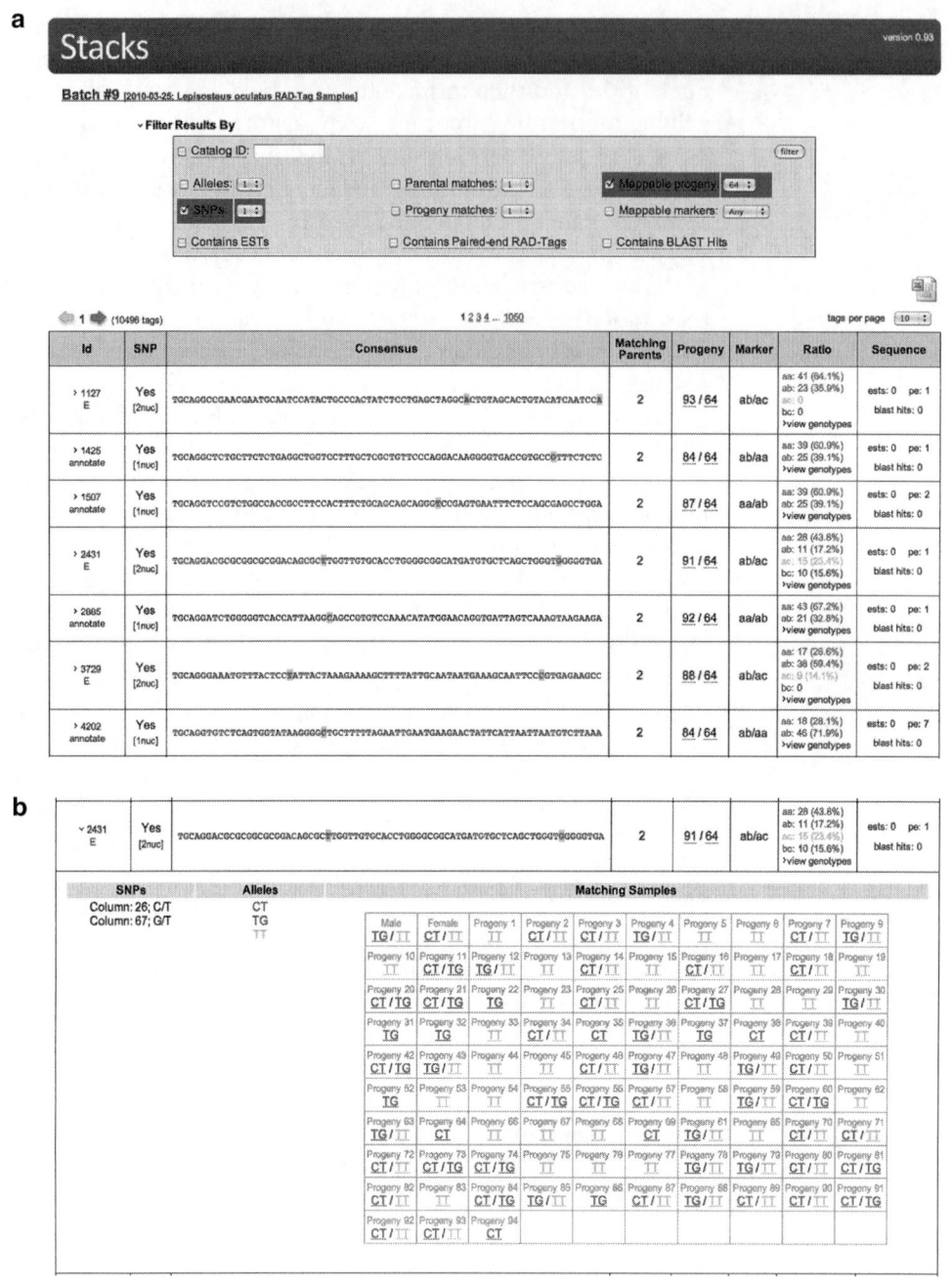

Fig. 6. Screen shot from *Stacks*: an image of the Web-based user interface. (**a**) The main view of the Stacks catalog shows a set of loci along with the consensus sequence for each locus, as well as the number of parents and progeny that have this locus, and whether a SNP was detected within it. (**b**) For a particular locus, Stacks provides a map of which alleles are present in every member of the mapping family or population.

overall network by isolating subgraphs that are more than two or three nucleotide steps away from other such subgraphs (again, a configurable parameter). Each subgraph is then evaluated as a presumptive locus to determine alleles and genotype the individual by examining each nucleotide position in the locus and differentiating SNPs from errors.

Although the Stacks approach works well for most RAD tags, in some situations similar sequences from nonhomology regions will be mistakenly assembled together in a situation prosaically called "paralog" or "homeolog hell" (see Note 9). Unfortunately, for organisms without a reference genome, differentiating repetitive stacks from true, homology loci cannot be determined with absolute confidence at the time of the initial assembly of the reads. However, a goal of Stacks is to utilize additional information to either redistribute problematic tags to the correct subgraphs, or remove from the analysis altogether (see Note 9). The easiest way to handle identified problematic tags that occur because of assembling nonhomology regions is to remove them from the analysis. The RAD approach provides such a large number of markers that removing even 20% of the markers in this manner still provides plenty of markers for most analyses. However, some of these loci can be salvaged by additional techniques. Similar to the strategy for organisms with a reference genome, the read lengths of the libraries can be increased. Likewise, paired-end reads can be used to double the amount of sequence, and because the mate-pair sequences are a few hundred base pairs away from the RAD sequence, they have an increased chance of being outside of repetitive regions that are bedeviling the RAD reads themselves. Moreover the reads from the sheared end can be assembled into one or two contigs, indicating the reads that come from homology versus paralogous regions, respectively. The two separate assemblies can then indicate the real loci for subsequent analyses.

A good example for using RAD genetic mapping comes from the basally branching teleost (ray-finned fish) the spotted gar, *Lepisosteus oculatus* (45). Teleost fish are good model organisms for basic biological research, as well as medical models of human diseases. However, a genome duplication at the base of the teleost radiation can sometimes obscure the relationship between human and teleost orthologs. Gar are thought to have diverged before the genome duplication, and RAD sequence data and the Stacks pipeline were used to construct a meiotic map containing over 8,000 SNPs. Through the use of Illumina sequencing of cDNA, and then analyzing the coding sequence contigs that emerged for RAD sites, the map also includes the locations of nearly a thousand protein-coding markers. These data conclusively show that the gar genome resembles more closely the organization of the human genome than that of the teleost genome. Therefore, with minimal investment of time and money resources, RAD sequencing and analysis using the Stacks pipeline was able to conclusively support the hypothesis that the gar lineage diverged from the rest of the teleost lineage just before the whole genome duplication event that defines the rest of the crown ray-finned fishes.

5. Summary

Ever since the integration of Mendel's laws with evolutionary theory during the Modern Synthesis of the 1930s, biologists have dreamed of a day when perfect genetic knowledge would be available for almost any organism. Nearly a century later, NGS technologies are fulfilling that promise and opening the possibility for genetic analyses that have heretofore been impossible. Perhaps the most critical aspect of these breakthroughs is the unshackling of genetic analyses from traditional model organisms, allowing genomic studies to be performed in organisms for which few genomic resources presently exist. We presented here one application of NGS, RAD sequencing, a focused reduced representation methodology that uses Illumina NGS to simultaneously type and score tens to hundreds of thousands markers in a very cost-efficient manner. The core RAD-seq protocol can be performed in nearly any evolutionary genetic laboratory. Undoubtedly, numerous modifications of the core RAD molecular protocols can be made to suit a variety of additional research problems. Despite the ease of use of RAD and other NGS protocols, a significant challenge facing biologists is developing the appropriate analytical and bioinformatic tools for these data. Although we outline some general analytical approaches for RAD-seq, we fully anticipate that the development of bioinformatic tools for RAD and similar data will be a rich area of research for many years to come.

6. Notes

1. Additional general considerations for population genomic studies

   A few general principles are emerging to maximize statistical power and accuracy in population genomic studies. First, tests based on multiple aspects of genomic structure should be applied to each dataset to increase statistical power (18, 20) and provide more opportunity to separate the roles of demographic and genetic factors from selection (53). Second, the null hypothesis for any test of selection with genome scale data should be derived from the genome-wide distribution rather than a simple *a priori* neutral model. For instance, while neutrality predicts an expectation for Tajima's D of zero, the several assumptions underlying this prediction may be violated more often than not (26, 54, 55). Demographic history and population structure can have a disturbingly large effect on genome-wide expectations and variances and on the rate of false positives (17, 56). To account for this issue, statistical models are required that can accommodate arbitrarily complex, nonequilibrium demographic scenarios, estimate relevant parameters from the genome-wide data, and then identify outlier regions that exhibit signatures of selection. Some progress is being made along this path (57–59).

2. Alternative approaches for inferring population genetic parameters

For most of our work, we use a barcoding and multiplexing strategy that allows us to index every read to a particular individual in a study, placing the inference of population genetic parameters into a more traditional framework. However, some evolutionary genetic applications will dictate alternative experimental designs, including sequencing of pools of individuals to produce point estimates (with error) of allele frequencies (60). Two examples are pools of individuals from natural populations for phylogeography (37), or groups of individuals of different phenotypic classes in a QTL mapping cross (36). These approaches lose information about individual genotypes, but because of the volume of data produced by NGS, techniques or analyses that lose some information remain highly effective. For instance, in the previously mentioned Emerson et al. (37) paper, the fine-scale phylogeographic relationships was estimated among populations of the pitcher plant mosquito *Wyeomyia smithii* that originated postglacially along the eastern seaboard of the USA. Because of the small amount of DNA in each mosquito, six individuals from each population were pooled and genotyped with barcoded adaptors. Rather than directly estimating allele frequencies, which could not be done with high confidence, the analysis focused only on those SNPs for which a statistical model indicated that allele frequency differed significantly between populations. This produced a set of nucleotides that were variable among, and fixed (or nearly so) within, populations. These data were used in subsequent standard phylogenetic analyses. The majority of the potentially informative SNPs were removed from the study, but because such a large number of RAD sites were identified and typed, the remaining 3,741 sites resulted in a well-resolved phylogeny, with high branch support and congruity to previous biogeographic hypotheses for this species. Nonetheless, for many population genomic applications it remains critical to barcode each individual and generate individual-level genotype data; such applications include studies of LD (8), screening for paralogs (41), or accounting for hidden population structure (56).

3. F1 pseudo-testcross

Many organisms produce significant numbers of progeny for genetic mapping, but cannot be efficiently bred through multiple generations because of a long lifespan of the focal organisms. However, enough RAD markers can be scored in an F_1 family produced from a single pair of heterozygous parents to allow the creation of a genetic map in very long-lived organisms, which are not amenable to an F_2 or back-cross map. RAD analyses performed in an F_1 pseudo-testcross format produce two independent genetic maps based only on the recombination occurring during meiosis in each parent, which are then linked together by the smaller proportion of markers shared between the maps (Fig. 1). This approach has already been used to produce a very high-density genetic map in the spotted gar (45).

4. Analyzing RAD-seq and other NGS data
At the time of this writing, a single RAD sequencing run on an Illumina HiSeq2000 can produce nearly 1.4 billion RAD reads, each potentially 150 bp in length that can only be efficiently analyzed on a large-scale computer cluster, such as a shared memory machine with many cores, or a cluster of smaller machines networked together. This is nearly 100 human genomes worth of data, the volume of which has increased by nearly an order of magnitude from just 1 year ago. Even fast desktop computers cannot handle basic analyses (e.g., sorting by barcode) of these many reads in a reasonable amount of time, and are unable to conduct some analyses, such as genomic assembly, at all.

5. Inferring genotypes in the face of sequencing error and sampling variance
NGS introduces sequencing error into many of the reads. Although the sequencing error rate is quite low, on the order of 0.1–1.0% per nucleotide, it can be a significant source of inferential confusion when millions of reads are considered simultaneously. Unfortunately, sequencing error compounds the problems of assembling similar paralogous regions, outlined in the previous section, by increasing the probability of mis-assembly. Error rates can vary across samples, RAD sites, and positions in the reads for each site. In addition, the sampling process of a heterogeneous library inherent in NGS introduces sampling variation in the number of reads observed across RAD sites as well as between alleles at a single site (Fig. 2). These issues could be overcome by greatly increasing total sequencing depth, but of course this approach will also increase the cost of a study. A better approach to differentiating true SNPs from sequencing error is a statistical framework that accounts for the uncertainty in genotyping. Undoubtedly significant progress will be made in this area in the near future; here, we present one approach as an example of a straightforward, flexible statistical model.

The following maximum-likelihood framework is based upon Hohenlohe et al. (27), designed for genotyping diploid individuals sampled from a population. The expectation is that errors should be differentiated from heterozygous SNPs by the frequency of nucleotides, with errors being represented at low frequency while alleles at heterozygous sites in an individual will be present in near equal frequencies in the total number of reads. Modifications to this approach would be required in other cases, such as haploid organisms, recombinant inbred lines, backcross mapping crosses, or single barcodes representing pools of individuals.

For a given site in an individual, let n be the total number of reads at that site, where $n = n_1 + n_2 + n_3 + n_4$, and n_i is the read count for each possible nucleotide at the site (disregarding ambiguous reads). For a diploid individual, there are ten possible genotypes (four homozygous and six (unordered) heterozygous genotypes). A multinomial sampling distribution gives the probability of observing a set of read counts (n_1, n_2, n_3, n_4) given a particular genotype, which translates into the likelihood for that genotype. For example, the likelihoods of a homozygote (genotype 1/1) or a heterozygote (1/2) are, respectively:

$$L(1/1) = P(n_1, n_2, n_3, n_4 \mid 1/1) = \frac{n!}{n_1! n_2! n_3! n_4!} \left(1 - \frac{3\varepsilon}{4}\right)^{n_1} \left(\frac{\varepsilon}{4}\right)^{n_2 + n_3 + n_4} \quad (1a)$$

and

$$L(1/2) = P(n_1, n_2, n_3, n_4 \mid 1/2) = \frac{n!}{n_1! n_2! n_3! n_4!} \left(0.5 - \frac{\varepsilon}{4}\right)^{n_1 + n_2} \left(\frac{\varepsilon}{4}\right)^{n_3 + n_4} \quad (1b)$$

where ε is the sequencing error rate. If we let n_1 be the count of the most observed nucleotide, and n_2 be the count of the second-most observed nucleotide, then the two equations in (Eq. 1) give the likelihood of the two most likely hypotheses out of the ten possible genotypes. From here, one approach is to assign a diploid genotype to each site based on a likelihood ratio test between these two most likely hypotheses with one degree of freedom. For example, if this test is significant at the $\alpha = 0.05$ level, the most likely genotype at the site is assigned; otherwise, the genotype is left unassigned for that individual. In effect this criterion removes data for which there are too few sequence reads to determine a genotype, instead of establishing a fixed threshold for sequencing coverage (27). An alternative approach is to carry the uncertainty in genotyping through all subsequent analyses. This can be done by incorporating the likelihoods of each genotype in a Bayesian framework in subsequent calculation of population genetic measures, such as allele frequency, or by using genotype likelihoods in systematic resampling of the data. In either case, information on LD and genotypes at neighboring loci could also be used to update the posterior probabilities of genotypes at each site.

A central parameter in the model above is the sequencing error rate ε. One option for this parameter is to assume that it is constant across all sites (60), and either estimate it from the data by maximum likelihood or calculate it from sequencing of a control sample in the sequencing run. However, there is empirical evidence that sequencing error varies among sites, and alternatively ε can be estimated independently from the data at each site. Maximum likelihood estimates of ε can be calculated directly at each site by differentiation of equations (Eq. 1), and the likelihood of each genotype hypothesis calculated as above. This technique has been applied successfully to RAD-seq data (27). More sophisticated models could be applied here as well, for instance, assuming a probability distribution from which ε is drawn independently for each site. This probability distribution could be iteratively updated by the data, and it could also be allowed to vary by nucleotide position along each sequence read or even by cluster position on the Illumina flow cell. Further empirical work is needed to assess alternative models of sequencing error.

The analytical method described in this note accounts for sequencing error and the random sampling variance inherent in NGS data. However, it does not account for any systematic biases in, for instance, the frequency of sequence reads for alternative alleles at a heterozygous site. For example, biased representation could occur because PCR amplification occurs more readily on one allele or barcode. Barcoding and calling genotypes separately in each individual alleviates some of this bias. In addition, sampling variation among sites and alleles that occurs early in the process can be propagated and amplified in the RAD-seq protocol. Optimizing the protocol to minimize the number of PCR cycles required is an important component of dealing with this issue. However, to date no analytical theory or tools have been developed to handle these sources of variation, and numerical simulations and empirical studies are needed to quantify them. Most simply, individuals of known sequence could be repeatedly genotyped by RAD-seq, using replicate libraries and barcodes, to estimate and partition the resulting variance in observed read frequencies.

6. Using F_{ST} and similar approaches to infer signatures of selection

The most common approach to multiple-population comparisons is based on F_{ST}, the partitioning of variance in allele frequency among versus within populations (2, 61). Other measures of the partitioning of genetic variance have been proposed (62), and LD can also be partitioned among populations in an analogous manner (63, 64). If differential selection is operating on a locus between two populations, a greater proportion of variance is expected between populations, resulting in higher values of statistics that measure population structure, such as F_{ST} (23). This is a result of positive selection within one or both populations producing shorter within-population coalescence intervals, leading to a greater proportion of coalescent events occurring within each population rather than between them. In contrast, balancing selection within one or both populations maintains polymorphism and pushes coalescence intervals back in time so that the branches leading back from present-day alleles are more likely to experience a migration event, crossing from one population to the other, or even to coalesce when the populations were panmictic. However, F_{ST} is a ratio of allelic diversities, and as such is sensitive to other processes that can affect distribution of nucleotide diversity across the genome, such as recombination and background selection (23, 53).

The statistic F_{ST} is often estimated as a parameter in a specific demographic model, and is thus translated into a neutral estimate of the effective migration rate among many populations, of which the observed populations are a sample (65). However, F_{ST} can also be viewed from a coalescent perspective (66), reflecting the distribution of coalescent events within versus among populations. Separation of the time scales of these two coalescent processes provides the basis for tests of selection using F_{ST} (67). Again, this coalescent view illustrates the analogies between selection and demographic processes that are specific to particular loci. Differential selection at a locus is reflected in relatively fewer coalescent events between populations and more within, which is analogous to a lower migration rate between populations at the selected locus.

7. Sliding window and resampling analysis of population genetic parameters

Sliding window averaging can be used to group neighboring loci and produce continuous distributions of population genetic parameters across the genome (see Fig. 2). Window averages can apply several weighting schemes (27). First, a kernel smoothing approach helps to narrow the focus on genomic regions of interest; data from loci within the window are weighted by some function (such as Gaussian) so that loci in the middle of the window are weighted more heavily than those at the edges. Second, as discussed above, the sampling nature of RAD sequencing means that each locus likely represents a different sample of individuals from the population(s). Sample size can be used to weight loci so that loci with larger sample sizes (and therefore greater confidence in population genetic estimates) have a greater influence on the window average. There are options for choosing the size of windows as well. A single size can be used across the genome, in which case some testing needs to be done to find the window size that best captures the extent of correlations among neighboring loci, without lumping uncorrelated loci (i.e., those approaching linkage equilibrium). Alternatively, the extent of LD can be used to assign variable window sizes along the genome, using statistics, such as extended haplotype homozygosity (EHH) (68). In this case, because RAD sequencing does not explicitly provide haplotype phase information, haplotype phase must be inferred prior to estimating the extent of LD.

Sliding window averages also provide a mechanism for assessing the statistical significance of genomic regions beyond standard tests on individual SNPs. One approach is bootstrap resampling to assign a p-value to each window average. The general protocol is as follows (27). Because of the random distribution of RAD sites, each window will have a different number of loci at different relative positions, and these factors should be taken into account. Data from single SNPs from across the genome are sampled at random with replacement, and assigned to the positions within the focal window. The average is then calculated, weighted as above by a kernel function and by sample size (in this case, the sample size of the resampled loci). This is done multiple times to create a distribution of window averages, and the p-value of the focal window average is its position in this distribution. If loci are resampled from the same dataset (e.g., the same set of populations for F_{ST}), this accomplishes the goal described in Note 1 of assessing significance against the genome-wide distribution rather than against a null model. It is important to note that this method compares each window against a distribution calculated in the absence of LD (because individual SNPs are resampled from across the genome). Some degree of LD is expected in most genomes, so the p-values estimated by this method tend to be extreme (27). Nonetheless, the advantage of this approach over, for instance, comparing each window average to the observed distribution of window averages is that it allows for varying proportions of the genome to exceed a given significance threshold. This proportion reflects the degree to which genomic neighborhoods are responding to evolutionary forces in concert because of LD.

8. The power of multiple independent comparisons for inferring selection

While F_{ST} outliers in a single population comparison can be the result of selection, other factors such as hidden population structure can result in false positives (e.g., ref. 56). However, if replicate evolved populations can be sampled, F_{ST} outliers localized to the same genetic region across separate pairwise comparisons can provide much stronger evidence for the selective significance of that region. In this case, multiple comparisons among populations provide a more complete picture of selection. For instance, genomic regions exhibiting elevated F_{ST} in multiple comparisons across habitat types, but average or even reduced F_{ST} within habitat types, suggest that the same alleles are responding in parallel to selection within each habitat (27, 69). In contrast, genomic regions with elevated F_{ST} both within and among habitats suggest either selection on different alleles or differential selection that is uncorrelated with habitat type (27).

9. Paralog hell for de novo assembly

Stacks of reads may be abnormally deep because the focal RAD site was by chance sampled significantly more than average, or because paralogous regions with similar sequences were erroneously assembled together. For stacks that fall outside the range of the expected number of reads, the data can simply be removed from the analysis. More problematic is a situation where two or a small number of paralogous regions are assembled, and the total number of reads is large, but not significantly so, because of sampling variation across all RAD loci.

Although the Stacks approach works well for most RAD tags, in some situations, similar sequences from nonhomology regions will be mistakenly assembled together. Unfortunately, for organisms without a reference genome, these over-assembled tags cannot be identified with absolute confidence at the time of initial assembly of the reads. However, a goal of Stacks is to utilize additional information to either redistribute problematic tags to the correct subgraphs or remove them from the analysis altogether. A first step is to clearly identify the problematic tags, and several lines of evidence can be used to do so. First, the read depth of each sequence can be explicitly incorporated into the subgraph evaluation models, but also can be used to flag outlier stacks that have many more reads than would be expected simply due to sampling variance. For example, some RAD tags will fall in repetitive sequences of multiple loci that will be assembled into very large lumberjack stacks. These stacks can be tagged and blacklisted from the analyses, or examined in more detail as described below. In addition to causing differences in read depth, mis-assembly of nonhomology regions will cause a locus to appear to have impossible sets of alleles, or ratios of alleles to genotypes in a family or population. For most outbred multicellular populations with average population sizes, each nucleotide should only be segregating two variants, and thus seeing three or four alleles at the exact same nucleotide position provides a clue of the assembly of nonhomology regions. A more likely scenario is two monomorphic paralogous regions that differ from one another only by a single nucleotide, giving the appearance of a single locus with two alleles. However, these "alleles" will always appear to be in a heterozygous state because each individual will always have each allele. Thus, a significant test of divergence from HWE within mapping families or populations can provide strong evidence of the incorrect clustering of nonhomology regions.

Acknowledgments

We thank members of the Cresko and Johnson laboratories, and numerous other University of Oregon researchers, for discussions about the use of NGS data for population genomic studies. This work has been generously supported by research grants from the US National Institutes of Health (1R24GM079486-01A1, F32GM076995 and F32GM095213) and the US National Science Foundation (IOS-0642264, IOS-0843392 and DEB-0919090).

References

1. Fisher RA (1958) The genetical theory of natural selection. Dover, New York
2. Wright S (1931) Evolution in Mendelian populations. Genetics 16:97–159
3. Kimura M (1991) Recent development of the neutral theory viewed from the Wrightian tradition of theoretical population genetics. Proc Natl Acad Sci USA 88:5969–5973
4. Wright S (1978) Evolution and the genetics of populations. University of Chicago Press, Chicago
5. Avise JC (2004) Molecular markers, natural history and evolution, 2nd edn. Sinauer Associates, Sunderland, MA
6. Birney E, Stamatoyannopoulos JA, Dutta A et al (2007) Identification and analysis of functional elements in 1% of the human genome by the ENCODE pilot project. Nature 447:799–816
7. Stranger BE, Nica AC, Forrest MS et al (2007) Population genomics of human gene expression. Nat Genet 39:1217–1224
8. Sabeti PC, Varilly P, Fry B et al (2007) Genome-wide detection and characterization of positive selection in human populations. Nature 449:913–918
9. Beaumont MA, Balding DJ (2004) Identifying adaptive genetic divergence among populations from genome scans. Mol Ecol 13:969–980
10. Liti G, Carter DM, Moses AM et al (2009) Population genomics of domestic and wild yeasts. Nature 458:337–341
11. Rockman MV, Kruglyak L (2009) Recombinational landscape and population genomics of Caenorhabditis elegans. PLoS Genet 5:e1000419
12. Butlin RK (2010) Population genomics and speciation. Genetica 138:409–418
13. Luikart G, England PR, Tallmon D, Jordan S, Taberlet P (2003) The power and promise of population genomics: from genotyping to genome typing. Nat Rev Genet 4:981–994
14. Slatkin M (2008) Linkage disequilibrium – understanding the evolutionary past and

mapping the medical future. Nat Rev Genet 9:477–485

15. Pritchard JK, Pickrell JK, Coop G (2010) The genetics of human adaptation: hard sweeps, soft sweeps, and polygenic adaptation. Curr Biol 20:R208–R215

16. Charlesworth B, Betancourt AJ, Kaiser VB, Gordo I (2009) Genetic recombination and molecular evolution. Cold Spring Harb Symp Quant Biol 74:177–186

17. Boitard S, Schlotterer C, Futschik A (2009) Detecting selective sweeps: a new approach based on hidden markov models. Genetics 181:1567–1578

18. Nielsen R, Williamson S, Kim Y et al (2005) Genomic scans for selective sweeps using SNP data. Genome Res 15:1566–1575

19. Pickrell JK, Coop G, Novembre J et al (2009) Signals of recent positive selection in a world-wide sample of human populations. Genome Res 19:826–837

20. Grossman SR, Shylakhter I, Karlsson EK et al (2010) A composite of multiple signals distinguishes causal variants in regions of positive selection. Science 327:883–886

21. Przeworski M, Coop G, Wall JD (2005) The signature of positive selection on standing genetic variation. Evolution 59:2312–2323

22. Hermisson J, Pennings PS (2005) Soft sweeps: molecular population genetics of adaptation from standing genetic variation. Genetics 169:2335–2352

23. Storz JF (2005) Using genome scans of DNA polymorphism to infer adaptive population divergence. Mol Ecol 14:671–688

24. Hohenlohe PA, Phillips PC, Cresko WA (2010) Using population genomics to detect selection in natural populations: key concepts and methodological considerations. Int J Plant Sci 171(9):1059–1071

25. Teshima KM, Coop G, Przeworski M (2006) How reliable are empirical genomic scans for selective sweeps? Genome Res 16:702–712

26. Wares JP (2010) Natural distributions of mitochondrial sequence diversity support new null hypotheses. Evolution 64:1136–1142

27. Hohenlohe P, Bassham S, Stiffler N, Johnson EA, Cresko WA (2010) Population genomics of parallel adaptation in threespine stickleback using sequenced RAD tags. PLoS Genet 6:e1000862

28. Akey JM (2009) Constructing genomic maps of positive selection in humans: where do we go from here? Genome Res 19:711–722

29. Pool JE, Hellmann I, Jensen JD, Nielsen R (2010) Population genetic inference from genomic sequence variation. Genome Res 20:291–300

30. Mortazavi A, Williams BA, McCue K, Schaeffer L, Wold B (2008) Mapping and quantifying

mammalian transcriptomes by RNA-Seq. Nat Methods 5:621–628

31. Marguerat S, Wilhelm BT, Bahler J (2008) Next-generation sequencing: applications beyond genomes. Biochem Soc Trans 36:1091–1096

32. Mardis ER (2008) Next-generation DNA sequencing methods. Annu Rev Genomics Hum Genet 9:387–402

33. Shendure J, Ji H (2008) Next-generation DNA sequencing. Nat Biotechnol 26:1135–1145

34. Mardis ER (2008) The impact of next-generation sequencing technology on genetics. Trends Genet 24:133–141

35. Van Tassell CP, Smith TP, Matukumalli LK et al (2008) SNP discovery and allele frequency estimation by deep sequencing of reduced representation libraries. Nat Methods 5:247–252

36. Baird NA, Etter PD, Atwood TS et al (2008) Rapid SNP discovery and genetic mapping using sequenced RAD markers. PLoS One 3:e3376

37. Emerson KJ, Merz CR, Catchen JM et al (2010) Resolving post-glacial phylogeography using high throughput sequencing. Proc Natl Acad Sci USA

38. Gompert Z, Lucas LK, Fordyce JA, Forister ML, Nice CC (2010) Secondary contact between *Lycaeides idas and L. melissa* in the Rocky Mountains: extensive admixture and a patchy hybrid zone. Mol Ecol 19:3171–3192

39. Rokas A, Abbot P (2009) Harnessing genomics for evolutionary insights. Trends Ecol Evol 24:192–200

40. Miller MR, Dunham JP, Amores A, Cresko WA, Johnson EA (2007) Rapid and cost-effective polymorphism identification and genotyping using restriction site associated DNA (RAD) markers. Genome Res 17:240–248

41. Hohenlohe PA, Amish JS, Catchen MJ, Allendorf WF, Luikart G (2011) Next-Generation RAD Sequencing Identifies Thousands of SNPs for Assessing Hybridization Between Rainbow and Westslope Cutthroat Trout. Molecular Ecology Resources 11 (Suppl 1):117–122

42. Dettman JR, Anderson JB, Kohn LM (2010) Genome-wide investigation of reproductive isolation in experimental lineages and natural species of Neurospora: identifying candidate regions by microarray-based genotyping and mapping. Evolution 64:694–709

43. Lewis ZA, Shiver AL, Stiffler N et al (2007) High-density detection of restriction-site-associated DNA markers for rapid mapping of mutated loci in Neurospora. Genetics 177:1163–1171

44. Miller MR, Atwood TS, Eames BF et al (2007) RAD marker microarrays enable rapid mapping of zebrafish mutations. Genome Biol 8:R105

45. Amores A, Catchen J, Ferrara A, Fontenot Q, Postlethwait JH (2011) Genome Evolution and Meiotic Maps by Massively Parallel DNA Sequencing: Spotted Gar, an Outgroup for the Teleost Genome Duplication. Genetics 188(4): 799–808

46. Ewing B, Green P (1998) Base-calling of automated sequencer traces using phred. II. Error probabilities. Genome Res 8:186–194

47. Ewing B, Hillier L, Wendl MC, Green P (1998) Base-calling of automated sequencer traces using phred. I. Accuracy assessment. Genome Res 8:175–185

48. Altschul SF, Madden TL, Schaffer AA et al (1997) Gapped BLAST and PSI-BLAST: a new generation of protein database search programs. Nucleic Acids Res 25:3389–3402

49. Kent WJ (2002) BLAT – the BLAST-like alignment tool. Genome Res 12:656–664

50. Vinga S, Almeida J (2003) Alignment-free sequence comparison-a review. Bioinformatics 19:513–523

51. Simpson JT, Wong K, Jackman SD et al (2009) ABySS: a parallel assembler for short read sequence data. Genome Res 19:1117–1123

52. Zerbino DR, Birney E (2008) Velvet: algorithms for de novo short read assembly using de Bruijn graphs. Genome Res 18:821–829

53. Charlesworth B, Nordborg M, Charlesworth D (1997) The effects of local selection, balanced polymorphism and background selection on equilibrium patterns of genetic diversity in subdivided populations. Genet Res 70:155–174

54. Tajima F (1989) Statistical method for testing the neutral mutation hypothesis by DNA polymorphism. Genetics 123:585–595

55. Thornton K (2005) Recombination and the properties of Tajima's D in the context of approximate-likelihood calculation. Genetics 171:2143–2148

56. Excoffier L, Hofer T, Foll M (2009) Detecting loci under selection in a hierarchically structured population. Heredity 103:285–298

57. Bedford T, Cobey S, Beerli P, Pascual M (2010) Global migration dynamics underlie evolution and persistence of human influenza A (H3N2). PLoS Pathog 6:e1000918

58. Beerli P, Palczewski M (2010) Unified framework to evaluate panmixia and migration direction among multiple sampling locations. Genetics 185:313–326

59. Gutenkunst RN, Hernandez RD, Williamson SH, Bustamante CD (2009) Inferring the joint demographic history of multiple populations from multidimensional SNP frequency data. PLoS Genet 5:e1000695

60. Lynch M (2009) Estimation of allele frequencies from high-coverage genome-sequencing projects. Genetics 182:295–301

61. Holsinger KE, Weir BS (2009) Genetics in geographically structure populations: defining, estimating and interpreting FST. Nat Rev Genet 10:639–650

62. Schlotterer C, Kauer M, Dieringer D (2004) Allele excess at neutrally evolving microsatellites and the implications for tests of neutrality. Proc Biol Sci 271:869–874

63. Kelly JK (2006) Geographical variation in selection, from phenotypes to molecules. Am Nat 167:481–495

64. Storz JF, Kelly JK (2008) Effects of spatially varying selection on nucleotide diversity and linkage disequilibrium: insights from deer mouse globin genes. Genetics 180:367–379

65. Weir BS, Cockerham CC (1984) Estimating F-statistics for the analysis of population structure. Evolution 38:1358–1370

66. Slatkin M (1991) Inbreeding coefficients and coalescence times. Genet Res 58:167–175

67. Beaumont MA (2005) Adaptation and speciation: what can Fst tell us? Trends Ecol Evol 20:435–440

68. Sabeti P, Reich DE, Higgins JM et al (2002) Detecting recent positive selection in the human genome from haplotype structure. Nature 419:832–837

69. Kane NC, Rieseberg LH (2007) Selective sweeps reveal candidate genes for adaptation to drought and salt tolerance in common sunflower, *Helianthus annuus*. Genetics 175: 1823–1834

Chapter 15

Analysis and Management of Gene and Allelic Diversity in Subdivided Populations Using the Software Program METAPOP

Andrés Pérez-Figueroa, Silvia T. Rodríguez-Ramilo, and Armando Caballero

Abstract

METAPOP (http://webs.uvigo.es/anpefi/metapop/) is a desktop application that provides an analysis of gene and allelic diversity in subdivided populations from molecular genotype or coancestry data as well as a tool for the management of genetic diversity in conservation programs. A partition of gene and allelic diversity is made within and between subpopulations, in order to assess the contribution of each subpopulation to global diversity for descriptive population genetics or conservation purposes. In the context of management of subdivided populations in *in situ* conservation programs, the software also determines the optimal contributions (i.e., number of offspring) of each individual, the number of migrants, and the particular subpopulations involved in the exchange of individuals in order to maintain the largest level of gene diversity in the whole population with a desired control in the rate of inbreeding. The partition of gene and allelic diversity within and between subpopulations is illustrated with microsatellite and SNP data from human populations.

Key words: Inbreeding, Genetic drift, Migration, Population differentiation

1. Introduction

Most populations of wild animal and plant species are spatially fragmented and this structuring of populations in reduced, and sometimes isolated, groups has an impact on the erosion of genetic variation and on the increase in inbreeding (1, 2). In addition, most species kept in captivity are generally maintained in independent nuclei (zoos, botanic gardens, germplasm centers, etc.) with restricted migration. Methods are thus required to take into account all available information for the management of genetic variation. The METAPOP software (3) provides a general analysis

François Pompanon and Aurélie Bonin (eds.), *Data Production and Analysis in Population Genomics: Methods and Protocols,*
Methods in Molecular Biology, vol. 888, DOI 10.1007/978-1-61779-870-2_15, © Springer Science+Business Media New York 2012

of gene and allelic diversity in subdivided populations (4, 5) as well as a method for the management of genetic diversity in conservation programs (6). In this chapter, we describe the main aspects for the application of the software to genetic marker data.

The analysis of genetic diversity in subdivided populations can be made in terms of gene diversity (expected heterozygosity; (4, 7)) or allelic richness (the number of different alleles segregating in the population; (5)). These parameters give complementary information on the genetic diversity of the population. Whereas the short-term response and inbreeding depression depend directly on gene diversity, long-term selection will be potentially higher in populations with larger allelic diversity. Allelic richness is, in addition, more sensitive to demographic changes (8, 9) and selective sweeps (10) than heterozygosity. The partition of allelic diversity in within and between subpopulation components allows for defining an allele differentiation coefficient between subpopulation, A_{ST}, analogous to Wright's F_{ST} (11). The parameter A_{ST} gives a measure proportional to the number of alleles in which two randomly chosen subpopulations differ. In the same way that a population with a larger number of alleles has a higher adaptive potential than another with a lower number of alleles, A_{ST} quantifies the degree of differential potentiality among subpopulations, stressing the differences among subpopulations regarding rare alleles. This contrasts with F_{ST}, which is generally little dependent on rare alleles.

The partition of gene and allelic diversity in components within- and between subpopulations has several applications in population and conservation genetics. First, the contribution of each subpopulation to the components of genetic diversity has a descriptive interest. This contribution can be obtained by decomposing the total gene and allelic diversity in its components (4). An alternative method is that proposed by Petit et al. (12), where the contribution of each subpopulation to total diversity and its components is estimated by disregarding that subpopulation and recalculating the global average diversity from the remaining pool (see also refs. 4, 13, 14).

A second application of the partition of diversity in within and between-subpopulation components regards the optimization of the contribution of each subpopulation to a synthetic population or gene pool of maximum gene or allelic diversity. This can be useful when germplasm stocks are created, and it is also a way to prioritize subpopulations for conservation (4, 15, 16). These optimal contributions can be applied considering a weighted combination of the within- and between-subpopulation components of gene diversity, where a factor λ can be used to give the desired weight to the within-subpopulation component (see Subheading 2.3.3).

The third application of the partition regards the management of subdivided population in conservation programs. The consensus method for controlling inbreeding and coancestry is to optimize the contribution of each parent (i.e., the number of offspring that each individual leaves to the next generation) by minimizing the global coancestry weighted by the parental contributions (e.g., (14, 17, 18)). The application of this method to the management of subdivided populations was developed by Fernández et al. (6). Basically, the method is aimed at looking for the optimal contributions from each individual to maximize total gene diversity with a given restriction (if desired) in the rate of increase in local inbreeding and the maximum number of migrations allowed among subpopulations. The method also provides the optimum translocations between subpopulations to achieve the highest overall gene diversity. A mate selection strategy is applied (see ref. 19, and references therein), so that the optimization algorithm provides not only the optimal number of offspring contributed by a given parent but also the particular individual to which it should be mated so as to perform minimum coancestry mating (see ref. 20 for an empirical application).

The software METAPOP performs a partition of genetic diversity in a subdivided population focusing on the three above applications. In what follows, we describe the main aspects of the software and provide an example of the first application using extensive microsatellite and SNP data in humans.

2. Program Usage

2.1. Overview and Installation

METAPOP (http://webs.uvigo.es/anpefi/metapop/) is a desktop application that provides an analysis of gene and allelic diversity in a subdivided population from molecular genotype data or just from a coancestry matrix for the whole population. It also allows for population management based on maximization of gene diversity under several methods. The application is written in Java 6 so it can be directly executed in any system (Windows, GNU/Linux, Mac OS) with a compatible Java Virtual Machine. It is distributed in a zip file. To install it, just unzip the file, and to run it, just double-click in the Metapop.jar file (sometimes it is necessary to check that .jar files are associated with the Java interpreter). The basic procedure to perform a population analysis is to load a data file, set the appropriate options and run the analysis. The Graphic User Interface (GUI) consists of a window with three elements: a main menu (as seen in Fig. 1), a panel for settings and controls (as seen in Fig. 2), and a tabbed panel for input/output display (see Note 1). All the actions can be done from the main menu.

2.2. Input Format

Given that METAPOP can use molecular data as well as genealogical data (coancestry matrix), the standard input formats are not suitable so METAPOP has its own file format for input data.

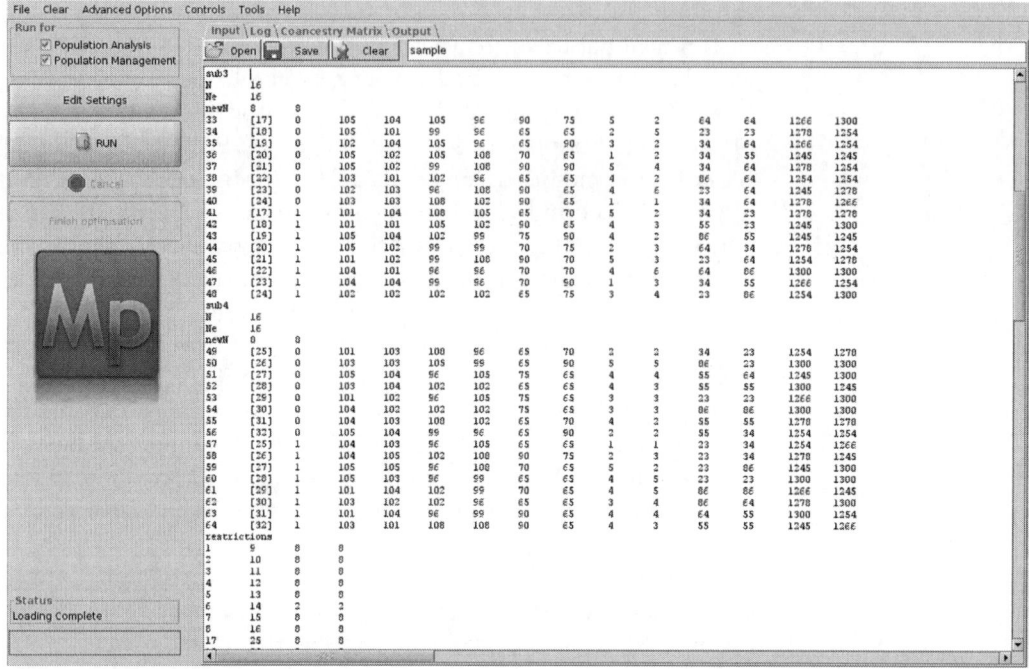

Fig. 1. Screenshot of METAPOP after loading a sample data set.

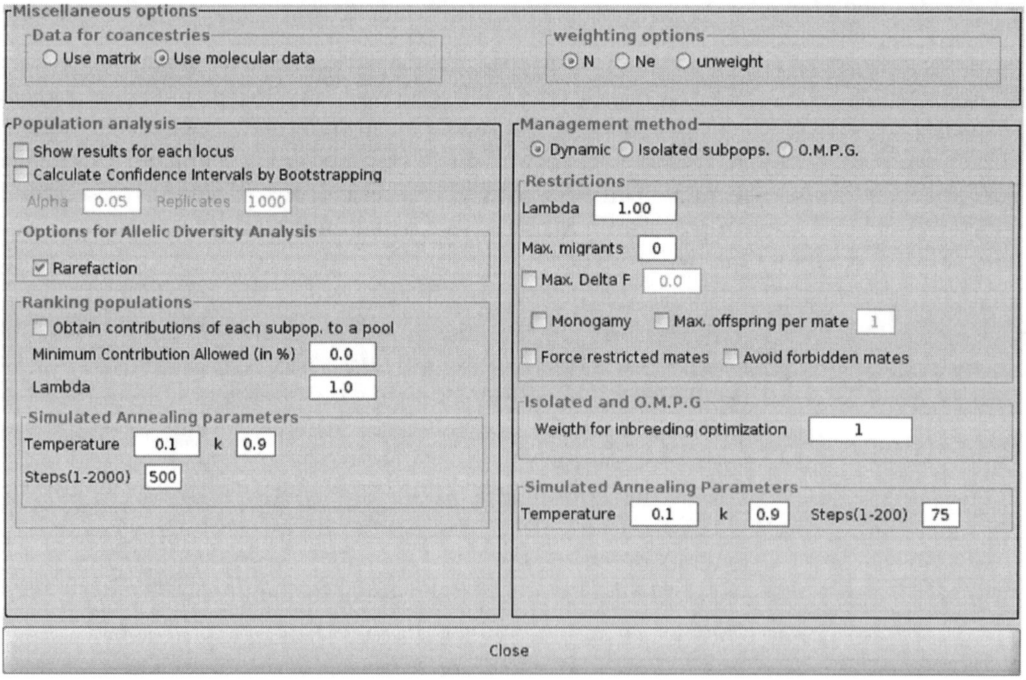

Fig. 2. Screenshot of METAPOP's settings window.

However, file converters are available on METAPOP's Web site, which allow converting Genepop and Structure format files into METAPOP's file format based on molecular data (see Note 2 for further differences between native and transformed files). The content of file is loaded into the tab called "input" where it can be edited, saved or discarded (Fig. 1). The program can load any file but running it will fail if the format is not the one specified here. In the following paragraph, we show the basic format for a small example, with comments between parentheses. Note that the labels of each parameter are case-sensitive. All optional lines can be skipped.

n 2 (the number of subpopulations; in this case two)

nloci 1 (number of loci: in this case just one)

PopA (a user-defined label for the first subpopulation)

N 2 (number of individuals in subpopulation PopA; in this case two)

Ne 2 (effective size for this subpopulation that can be provided for weighting—optional)

newN 3 3 (number of females and then males in the next generation; in this case, the first subpopulation will be expanded to 3 females and 3 males—optional)

1 A1 0 34 38 (information for the first individual called A1. It is a female, as defined by the code 0, heterozygous for alleles 34 and 38; if there were more loci, then an additional pair of numbers per locus should be added)

2 A2 1 34 34 (the same information for the second individual, A2, a male defined by the code 1 and homozygous for allele 34)

PopB (now the corresponding data for the second subpopulation)

N 2

Ne 2

newN 2 2

3 B1 0 34 38 (note that the first position corresponds to incremental numbering for the whole population)

4 B2 1 38 38

restrictions (indicates the following restrictions, if applicable, for mating—optional)

1 2 3 5 (in this case, female number 1, following the incremental code, should mate male 2 and produce 3 female and 5 male offspring)

3 4 2 2 (all females should be represented in the restriction list)

forbidden (optional section, indicates which males cannot mate, if applicable, with each female)

1 1 4 (female 1 cannot mate with one male, male 4)

3 0 (female 3 can mate with all males)

matrix (provides the ordered genealogical coancestry matrix—optional section, required if no molecular genotypes are available)

0.5 0.125 0 0

0.125 0.5 0 0
0 0 0.5 0.125
0 0 0.125 0.5

After data file loading, there is a chance to edit the data and save that edit in a new file with the corresponding button.

2.3. Option Configuration

When data is ready, the user should press "edit the settings" button to configure the miscellaneous, population genetics, and population management options (Fig. 2).

2.3.1. Miscellaneous Options

- *Data for coancestries.* This option allows for determining the nature of data to use in the population and management analyses. If "Use matrix" is selected, the program will use the optional coancestry matrix provided in the input data (under "matrix" section). If "Use molecular data" is selected, the program builds a coancestry matrix from the molecular data. This molecular matrix is displayed in the tab called "Coancestry matrix" after the analysis.

- *Weighting options.* This option allows for determining the weight given to each subpopulation in order to calculate the averages over subpopulations in the analysis. The three options are weighted by N (the number of individuals available in each subpopulation), weighted by N_e (the effective size of each subpopulation that should be provided in the data; see Fig. 1), or not weighted at all (this assumes that all subpopulations have the same census size).

2.3.2. Population Analysis Options

- *Show results for each locus.* This gives the results for every locus separately as well as the results when averaging over all loci.

- *Calculate confidence intervals by bootstrapping.* This option allows performing a bootstrapping routine in order of obtain confidence intervals for some population parameters. This routine resamples over loci with replacement. The user can choose the level of confidence (α) and the number of bootstrap replicates desired (see Note 3).

- *Option for allelic diversity analysis: Rarefaction.* If marked, this allows performing the rarefaction technique to correct for sample size (21).

- *Ranking populations.* This shows a couple of options to perform a simulated annealing algorithm (22) to obtain the relative contribution of each subpopulation to produce a single pool (a synthetic population or a germplasm bank) of maximal diversity. It allows a weighting factor (λ) balancing the relative importance of the within- and between-population components of gene diversity. It also allows setting a minimum contribution for any subpopulation. Furthermore, the parameters for the simulated annealing algorithm can be changed here (see Note 4).

2.3.3. Population Management Options

- *Management method.* This option allows choosing the method used for management between the following: (1) Dynamic, the method described by Fernández et al. (6) (see Note 5), (2) Isolated subpopulations, which searches for the contributions to minimize the coancestry (and optionally the inbreeding too) in each subpopulation without any migration, and (3) OMPG (One-Migrant-Per-Generation rule), which applies the Isolated subpopulations method and then forces that two randomly chosen subpopulations exchange an individual each generation.

- λ. This is the factor weighting the relative importance of within-subpopulation diversity (W) relative to between-subpopulation (B) diversity as the minimizing criterion for the dynamic method ($B + \lambda W$). Thus, if λ is zero, all weight is given to the between-subpopulation component of coancestry. If $\lambda \gg 1$, more weight is given to the within-subpopulation component. Finally, if $\lambda = 1$, total global coancestry is minimized, and global diversity maximized.

- *Maximum number of migrants.* The user can specify the maximum number of migrations allowed for the whole population. Migrants would be offspring moved to a subpopulation different from that of their parents.

- *Maximum Delta F.* It makes a constraint in the maximum per-generation rate of inbreeding (ΔF) allowed in the population. This option is alternative to the use of the λ parameter in the optimization procedure.

- *Monogamy.* The algorithm searches for solutions where males only mate to one female.

- *Max. offspring per mate.* This option sets a limit for the number of offspring allowed for any mating pair.

- *Force restricted mates.* The program allows for specific mating pairs (provided in the optional "restrictions" section of the data file, as shown at the bottom of Fig. 1) to be forced and the maximum number of contributed progeny from each pair to be set up. This is useful, for example, when the initial optimization has produced a set of mating pairs and contributions, but one or more of them have actually failed or the required number of progeny has not been generated. Thus, a second run of the program can be carried out including these restrictions, so that the optimization is made under the given constraints.

- *Avoid forbidden mates.* The program looks for mating pairs of minimum coancestry for these restrictions, but specific mating pairs (provided in the optional "forbidden" section of the data file) can be avoided with this option.

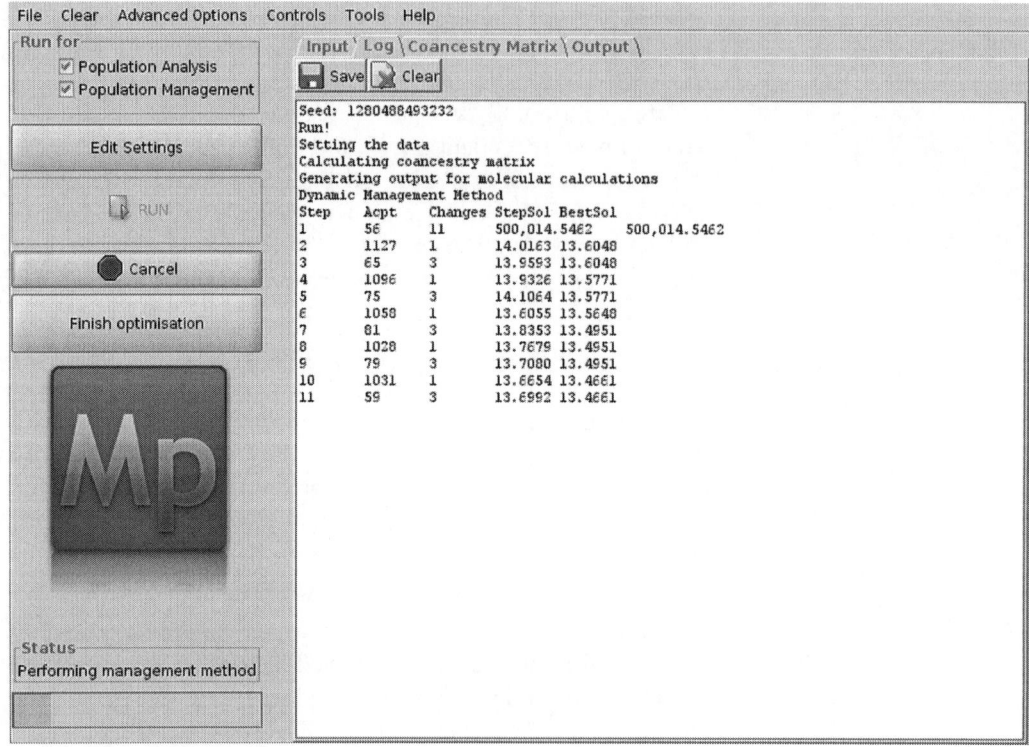

Fig. 3. Screenshot of METAPOP during optimization of management strategies.

- *Simulated annealing parameters.* As well as in ranking populations, the user can choose the parameters used during the optimization performed by the management methods (see Note 4).

2.4. Running the Program

After the options were conveniently set by the user, the analysis and/or management can be launched by the Run button. During the run, the log tab (Fig. 3) shows information about the progress and errors, if applicable. The duration of the run strongly depends on the amount of data (number of subpopulations, individuals, loci and alleles per locus) and the amount of memory available for the Java Virtual Machine (in Note 6, we explain how to increase this). In case you have a large number of loci or alleles per locus, it could be a good idea to customize the output (via Advanced Options in the menu) and unmark the allele frequencies in each subpopulation and in the whole population as this can save a lot of memory. As an illustration, the analysis of a data set of 36 haplotypes for 927 individuals in seven subpopulations (as shown in Subheading 3) with more than 500 alleles per locus runs for 35 min in a computer under GNU/Linux. After the execution, results are displayed in the output tab (Fig. 4), where they can be edited (i.e., by adding some text), saved into an html file (that can be viewed later in any Web browser) or even copied (ctrl + c) and then pasted (ctrl + v) in any spreadsheet or text processor.

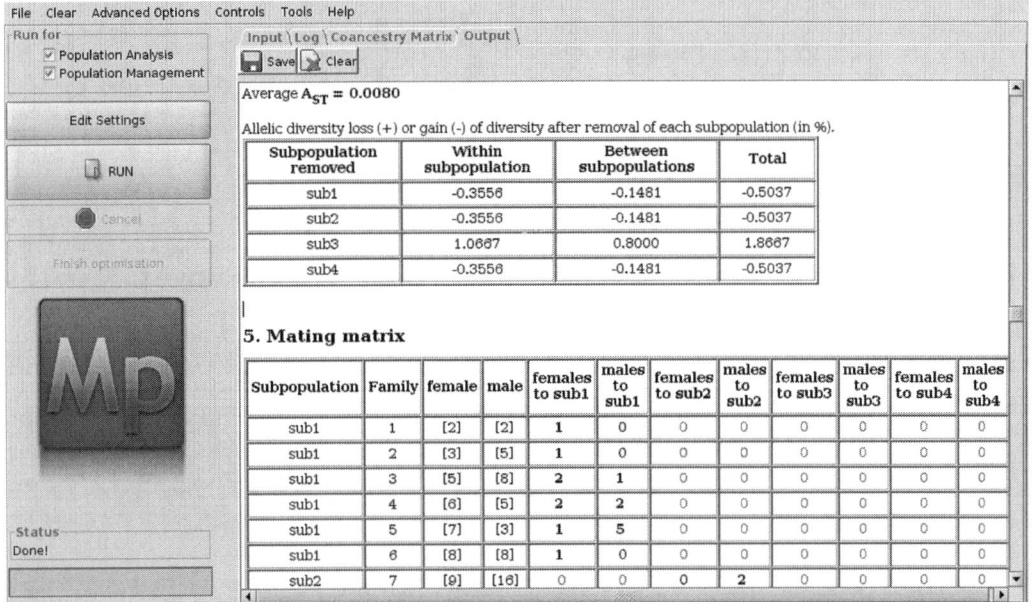

Fig. 4. Screenshot of METAPOP's output showing results for allelic diversity and mating management.

2.5. Results Visualization

Results for the population analysis are obtained as follows. From genotypic data on autosomal codominant molecular markers the coancestry between all pairs of individuals is obtained from the information on the frequencies of the markers, as made explicit by Caballero and Toro (4). Alternatively, a pedigree coancestry matrix can be used as input for the program. The software makes a number of calculations regarding gene diversity for each subpopulation i: the average self-coancestry of individuals (S_i), the average coancestry among individuals (f_{ii}), the average distance between individuals (d_{ii}), the average inbreeding coefficient (F_i), the deviation from Hardy–Weinberg proportions (α_i), and the proportion of gene diversity among individuals (G_i). In addition, for pairs of subpopulations (i, j), it calculates the average coancestry (f_{ij}), the Nei's minimum (DM_{ij}) and standard distances (DS_{ij}), and the Reynolds' distance (DR_{ij}). The overall gene diversity (GD_T) and its partition in within-individual (GD_{WI}), between-individual (GD_{BI}), within-subpopulation (GD_{WS}), and between-subpopulation (GD_{BS}) components are also made explicit. Wright's statistics (F_{IS}, F_{ST}, F_{IT}; (11)) are calculated following Nei (7). Regarding allelic diversity, the software makes a number of calculations following Caballero & Rodríguez-Ramilo (5) for each subpopulation i (as seen in Fig. 4): within-population allelic diversity (A_{Si}), allelic distance between subpopulations (Da_{ij}), the contribution of each subpopulation to the total allelic diversity of the population, and the coefficient of allelic differentiation between subpopulation, A_{ST}. The impact of the removal of a given subpopulation i on the global gene or allelic diversity (as a percentage of loss or gain in diversity) and its components is also calculated, as illustrated in Subheading 3.

If management is applied, the output shows a table with the mating matrix (Fig. 4). This table provides the mating individuals and their number of offspring per sex and subpopulation. If a mate pair has offspring in a subpopulation different from its own one, then that implies migration or movement of individuals. Finally, the program output gives the coancestry matrix among the expected progeny. This matrix can be directly included in the input window in order to run again the program and find the contributions of individuals and migrations expected in a further generation. This is useful, for example, to compare different alternative methods or different parameters and restrictions (migration rates, restrictions on inbreeding, etc.), what can be extremely useful for the manager in order to take the current decisions on the population.

3. Example

As an illustration of the partition of gene and allelic diversity within different components and of the contribution of each subpopulation to the total diversity, we used the human data from Rosenberg et al. (23) and Conrad et al. (24). Rosenberg et al. (23) analyzed a set of 1,056 humans subdivided into 52 populations genotyped for 377 microsatellite loci. These 52 populations were grouped within 7 geographical regions that we consider as subpopulations, for simplicity: Africa ($N=119$), Middle-East ($N=178$), Europe ($N=161$), Central-South Asia ($N=210$) East-Asia ($N=241$), Oceania ($N=39$) and America ($N=108$). Conrad et al. (24) analyzed a subset of 927 humans from these 52 populations genotyped for 2,834 SNPs. These 52 populations were grouped again within the same geographical regions (subpopulations) with the following sample sizes: Africa ($N=103$), Middle-East ($N=158$), Europe ($N=149$), Central-South Asia ($N=199$) East-Asia ($N=229$), Oceania ($N=27$), and America ($N=62$). The rarefaction correction (21) was applied for the analysis of allelic diversity using the smallest sample size corresponding to the Oceania region. Microsatellite and SNP data were used directly for the analysis. Obviously, allelic diversity may have less sense in the case of SNP data, because only two alleles are usually present for each marker. The calculation however is possible and will be shown. An alternative way to treat SNP variation is through the analysis of haplotypes within genomic regions when the SNP phase is known. Each genomic region can be considered a "locus" with the different haplotype variants regarded as different "alleles," making it possible to run an analysis of allelic variation with SNP data. For this purpose, we considered the 36 genomic regions studied by Conrad et al. (24). These 36 genomic regions include 16 regions scattered across the autosomes, 16 from chromosome 21 and four from the non-pseudoautosomal X chromosome. Each region was studied using 84 SNPs, including a high-density core of 60 SNPs spaced at an average of 1.5 kb apart, flanked by two sets of 12 SNPs at 10 kb average spacing. For our analysis, we excluded the two sets of 12 flanking SNPs in each genomic region. The observed haplotype frequencies were calculated within each genomic region. Therefore, this analysis is equivalent to use 36 "loci" (genomic regions) with a large number of "alleles" (haplotypes).

The average number of alleles for the 377 microsatellite loci was 12.42 (ranging from 4 to 32) and the average number of haplotypes for the 36 regions was 683.1 (ranging from 130 to 1,006). Figure 5 shows the proportional contribution to within, between and total gene (Fig. 5a) or allelic diversity (Fig. 5b) of each of the seven geographic regions (subpopulations). The results obtained with the three types of data give a similar and concordant picture. The African region has the largest contribution to within- and between-subpopulation diversity, and there is a declining overall contribution for regions increasingly distant from Africa, with Oceania and America showing the lowest contributions to overall diversity. The results are in agreement with the Recent African Origin hypothesis and subsequent colonizations characterized by a number of small bottlenecks (25).

4. Notes

1. METAPOP is a program that is still growing and changing. This implies that it could be slightly different when the reader checks the software relative to when we wrote this chapter. Improvements, new features, and suggestions from users are welcome. For example, the current program only allows for autosomal codominant markers. Dominant marker data and X-linked data are expected to be incorporated soon. In addition, the population management option only makes use of gene diversity and an extension to allelic diversity is expected to be developed in the future. However, this chapter was written taking into account that some differences in the interface could still happen. On the software Web site, there is updated information about the current version and changes as well as several ways of subscription to more information.

2. The conversion from Genepop or Structure data files is obviously limited to molecular data. However, as METAPOP's file requires information about sex, this would be included into the converted file. If the original Genepop file has information

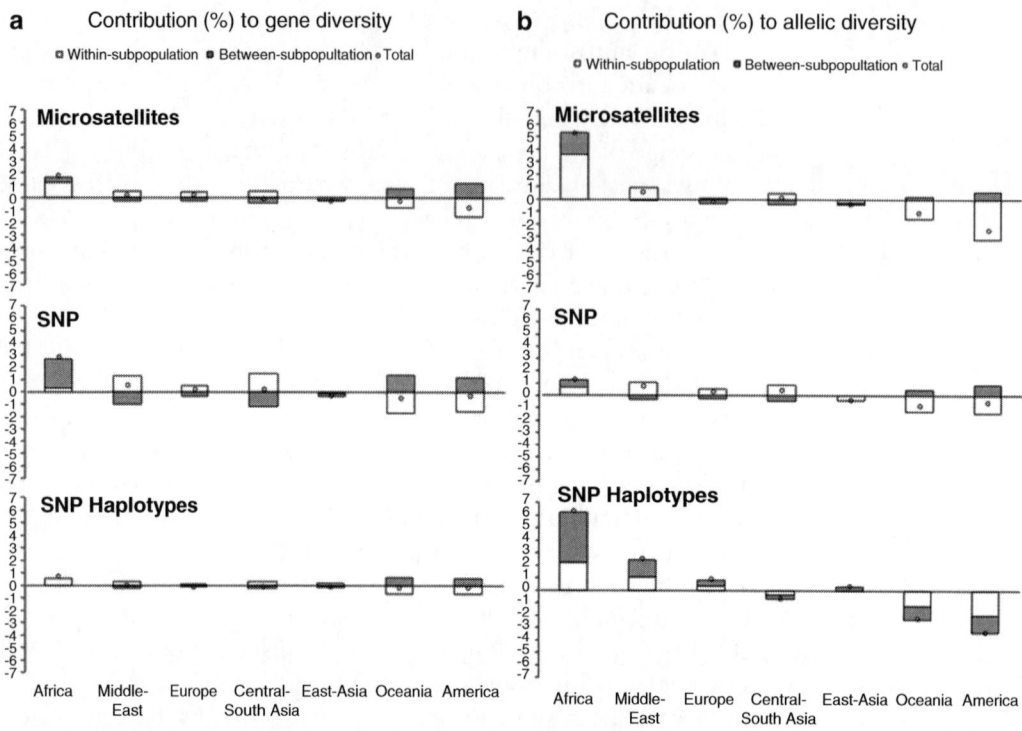

Fig. 5. Proportional contribution to gene diversity (**a**) and allelic diversity (**b**) of each of the seven subpopulations (human regions) using data from microsatellites (23) and SNPs (24). *Positive values* indicate a loss of diversity when the subpopulation is removed from the population, and vice versa. *Circles* indicate the contribution to overall gene/allelic diversity, *white and dark boxes* give the contribution to within- and between-subpopulation diversity, respectively.

about sex coded in the individual label, this should be maintained. If not, the converter will assign female sex to all individuals (allowing for population analysis but not for management). Both file converters are additional features of the program and they are provided, in METAPOP's Web site, as separate python scripts (that requires a python's interpreter with the Biopython library installed in the computer), although in future versions they will be integrated in the METAPOP main application.

3. The calculation of confidence intervals by bootstrapping can be very computationally intensive and it is only provided, for now, for Wright's (11) statistics (F_{IS}, F_{ST}, F_{IT}), so the user should decide if it is worth obtaining them. In general, every replicate of bootstrapping would last the same time as the analysis, so it would be a good idea to run the raw analysis and then consider run the bootstrapping.

4. Simulated annealing (SA) algorithm (22, 26) is a metaheuristic algorithm for optimization of a given objective function, such as global diversity. In METAPOP, SA is used to obtain the

relative contributions of subpopulations and the mating pairs and their offspring for population management. The name and inspiration come from annealing in metallurgy, a technique involving heating and controlled cooling of a material to increase the size of its crystals and reduce their defects. By analogy with this physical process, each step of the SA replaces the current solution by a random solution, changing some of the solution members (for METAPOP, a different mating pair or individual offspring contribution), chosen with a probability that depends on the difference between the objective function values (energy) and on a parameter called *temperature*, which is gradually decreased during the process. The dependency is such that the current energy changes almost randomly when the *temperature* is large, but increasingly reduces as the *temperature* goes to zero. The allowance for wrong moves saves the method from becoming stuck at local optima. In METAPOP, the user can set different parameters affecting the performance of the simulated annealing routine used for the management methods, which can be different from the ones set for the routine calculating the contributions to a pool. Temperature sets the initial value for *temperature*. Each step consists of 5,000 evaluated solutions using the same temperature. From a step to the next, the *temperature* changes by a rate defined by k. The maximum number of steps is also defined by the user.

In order to get accuracy and speed of computation, some previous runs may be performed to find the optimal values for the particular problem to deal with. During the execution, the log tab (see Fig. 3) gives information about the SA, so if the Best Solution column changes very fast, it indicates that parameters would be not the best ones for those data. However, the default parameters would be enough for most of the applications in conservation. It is also worth mentioning that given that this is a heuristic method, different runs will produce different solutions (optimal solution is not ensured with this algorithm). In such cases, the final Energy of the solution is provided, and the user should choose the solution with the lowest energy.

5. When individual pedigrees are known or molecular data are available to infer them, a dynamic method can be used to optimize the individuals' contribution by minimizing the whole population coancestry (6). The method takes into account both the within- and between-subpopulation components of coancestry (with variable weights depending on the importance desired for each one) and, at the same time, provides the optimum scheme of migrations under a possible restriction in the total number of movements if required.

6. When dealing with a large amount of data, there are a couple of tips to improve the performance of METAPOP and avoid a "lack of memory" error. It is hard to determine the data limit that METAPOP can handle as it depends on the distribution of data and on the computer used. We tested up to 10,000 SNPs (biallelic) in 1,000 individuals without crashing. In general, the memory allocated by the Java Virtual Machine is quite limited and not suitable for more than 200 loci with 5–6 alleles per locus. To increase the amount of memory available for the program, it can be executed from the command-line/DOS console with "java -jar Metapop.jar -Xms255M -Xmx1500M" indicating that it will start (-Xms) with 255 Mb of memory and it can increase up to a maximum (-Xmx) of 1,500 Mb. These values can be customized depending on the computer used. We recommend setting the maximum at half of the RAM available in the system if running under Windows, or at ¾ if running under GNU/linux or Mac OS. Other ways to improve the performance are to display only the needed sections in the output (via the Advanced Options menu), to avoid bootstrapping or population ranking as well as to remove all monomorphic or useless loci.

Acknowledgments

We thank Miguel Toro and Jesús Fernández for helpful discussions. This work was funded by the Ministerio de Ciencia e Innovación and Fondos Feder (CGL2009-13278-C02), and a grant for Consolidación e estruturación de unidades de investigación competitivas do sistema universitario de Galicia, Consellería de Educación e Ordenación Universitaria, Xunta de Galicia.

References

1. Frankham R, Ballou JD, Briscoe DA (2002) Introduction to conservation genetics. Cambridge University Press, Cambridge

2. Allendorf FW, Luikart G (2007) Conservation and the genetics of populations. Blackwell Publishing, Malden, MA

3. Pérez-Figueroa A, Saura M, Fernández J, Toro MA, Caballero A (2009) METAPOP – a software for the management and analysis of subdivided populations in conservation programs. Conserv Genet 10:1097–1099

4. Caballero A, Toro MA (2002) Analysis of genetic diversity for the management of conserved subdivided populations. Conserv Genet 3:289–299

5. Caballero A, Rodríguez-Ramilo ST (2010) A new method for the partition of allelic differentiation within and between subpopulations and its application in conservation. Conserv Genet 11:2219–2229. doi:10.1007/s10592-010-0107-7

6. Fernández J, Toro MA, Caballero A (2008) Management of subdivided populations in conservation programs: development of a novel dynamic system. Genetics 179:683–692

7. Nei M (1973) Analysis of gene diversity in subdivided populations. Proc Natl Acad Sci USA 70:3321–3323

8. Allendorf FW (1986) Genetic drift and the loss of alleles versus heterozygosity. Zoo Biol 5:181–190

9. Luikart G, Allendorf F, Cornuet JM, Sherwin W (1998) Distortion of allele frequency distributions provides a test for recent population bottlenecks. J Hered 89:238–247

10. Santiago E, Caballero A (1998) Effective size and polymorphism of linked neutral loci in populations under selection. Genetics 149:2105–2117

11. Wright S (1969) Evolution and the genetics of populations. The theory of gene frequencies, vol 2. University of Chicago Press, Chicago

12. Petit RJ, El Mousadik A, Pons O (1998) Identifying populations for conservation on the basis of genetic markers. Conserv Biol 12:844–855

13. Foulley JL, Ollivier L (2006) Estimating allelic richness and its diversity. Livest Sci 101:150–158

14. Toro MA, Fernández J, Caballero A (2009) Molecular characterization of breeds and its use in conservation. Livest Sci 120: 174–195

15. Eding H, Crooijmans PMA, Groenne MAM, Meuwissen THE (2002) Assessing the contribution of breeds to genetic diversity in conservation schemes. Genet Sel Evol 34:613–633

16. Ollivier L, Foulley JL (2005) Aggregate diversity: new approach combining within- and between-breed genetic diversity. Liv Prod Sci 95:247–254

17. Ballou J, Lacy R (1995) Identifying genetically important individuals for management of genetic variation in pedigreed populations. In: Ballou JD, Gilpin M, Foose TJ (eds) Population management for survival and recovery. Columbia University Press, New York, pp 76–111

18. Meuwissen THE (2007) Operation of conservation schemes. In: Oldenbroek K (ed) Utilisation and conservation of farm animal genetic resources. Wageningen Academic Publishers, Wageningen, The Netherlands, pp 167–193

19. Fernández J, Toro MA, Caballero A (2001) Practical implementations of optimal management strategies in conservation programmes: a mate selection method. Anim Biodivers Conserv 24:17–24

20. Ávila V, Fernández J, Quesada H, Caballero A (2010) An experimental evaluation with *Drosophila melanogaster* of a novel dynamic system for the management of subdivided populations in conservation programs. Heredity 106: 765–774

21. El Mousadik A, Petit RJ (1996) High level of genetic differentiation for allelic richness among populations of the argan tree [*Argania spinosa* (L.) Skeels] endemic to Morocco. Theor Appl Genet 92:832–839

22. Kirkpatrick S, Gelatt CD Jr, Vecchi MP (1983) Optimization by simulated annealing. Science 220:671–680

23. Rosenberg NA, Pritchard JK, Weber JL et al (2002) Genetic structure of human populations. Science 298:2381–2385

24. Conrad DF, Jakobsson M, Coop G et al (2006) A worldwide survey of haplotype variation and linkage disequilibrium in the human genome. Nature Genet 38:1251–1260

25. Handley LJ, Manica A, Goudet J, Balloux F (2007) Going the distance: human population genetics in a clinal world. Trends Genet 23:432–439

26. Press WH, Flannery BP, Teukolsky SA, Vetterling WT (1989) Numerical recipes. Cambridge University Press, Cambridge

Chapter 16

DetSel: An R-Package to Detect Marker Loci Responding to Selection

Renaud Vitalis

Abstract

In the new era of population genomics, surveys of genetic polymorphism ("genome scans") offer the opportunity to distinguish locus-specific from genome-wide effects at many loci. Identifying presumably neutral regions of the genome that are assumed to be influenced by genome-wide effects only, and excluding presumably selected regions, is therefore critical to infer population demography and phylogenetic history reliably. Conversely, detecting locus-specific effects may help identify those genes that have been, or still are, targeted by natural selection. The software package DetSel has been developed to identify markers that show deviation from neutral expectation in pairwise comparisons of diverging populations. Recently, two major improvements have been made: the analysis of dominant markers is now supported, and the estimation of empirical P-values has been implemented. These features, which are described below, have been incorporated into an R package, which replaces the stand-alone DetSel software package.

Key words: Adaptive divergence, AFLP, DetSel, Population genomics, Selection

1. Introduction

Population genomics (1, 2) focuses on population-based studies of genome-wide genetic variation. This approach aims at detecting the footprints of selective pressures on specific genomic regions that are under positive selection. Population genomics may naturally come as the first step in an integrated analysis because it does not require a fine knowledge of the nature of the traits involved in the adaptive response of organisms to local environmental conditions (3). This approach consists in a genome scan of locus-specific signatures of adaptive population divergence, revealed by unusually high level of population differentiation at specific marker loci (1–3). Those loci that are involved in adaptation to local environmental

François Pompanon and Aurélie Bonin (eds.), *Data Production and Analysis in Population Genomics: Methods and Protocols*, Methods in Molecular Biology, vol. 888, DOI 10.1007/978-1-61779-870-2_16, © Springer Science+Business Media New York 2012

conditions are indeed expected to exhibit increased differentiation among populations (along with a decreased diversity within-population). This is so because divergent selection favors different alleles in different populations. Increased differentiation may also result from hitchhiking with locally adapted variants at linked sites (4), at a rate that depends upon the relative strength of selection and recombination. Because they explore such hitchhiking effects along the genome, genome scans require a high density of markers. Yet, as a by-product benefit of this genotyping effort, the marker loci identified as responding to selection should presumably be tightly linked to the locus of interest (3). Genome-wide information on the relative distribution of the genetic signatures of ongoing selection offers the unique opportunity to understand the genetic basis of adaptive population divergence (5) and speciation (6).

With the availability of high throughput molecular methods, several genome scans have been performed in humans (7, 8), in model organisms such as *Drosophila* (9) and *Arabidopsis* (10), and in a growing number of nonmodel organisms, such as the periwinkle *Littorina saxatilis* (11), the larch budmoth *Zeiraphera diniana* (12), the salmon *Salmo salar* (13), the common frog *Rana temporaria* (14), among others. Although the available methods for data analysis are still in their infancy and require further development, these studies have proved that genome scans are an effective tool for the identification of functionally important genomic regions. Population-based genome scans have indeed several advantages (3): (1) as opposed to QTL mapping that requires controlled crossings in the laboratory, they can be applied on any species, provided that a high-density genome coverage is obtained; (2) they should be capable of identifying loci that have undergone a history of weak selection, over long periods of time; (3) they may lead to the identification of selected loci that underlie phenotypic traits whose adaptive significance was previously unforeseen.

When a very large number of markers are genotyped across multiple populations (typically, hundreds of thousands), it has been advocated that signatures of natural selection can simply be identified in the extreme tails of the empirical distribution of F_{ST} estimates (15). These model-free approaches have been applied to both the Perlegen (7) and the HapMap (8) Single Nucleotide Polymorphism (SNP) datasets (16–19). Such methods are intended to be immune to arbitrary assumptions about the (unknown) demographic history of the sample, when the number of markers is large. Dependence upon the unknown demography (including the geographic and historical relationship among populations) was indeed a severe criticism of the Lewontin–Krakauer's tests of selective neutrality (20), based on the sampling distribution of the parameter F_{ST} (21, 22). Yet, recent refinements of this controversial test showed that the distribution of F_{ST} estimates should be relatively robust to demographic effects, which prevents the need to model the demography explicitly (23, 24). This robustness to the effects of demography stems from the properties of gene genealogies in structured populations (25).

With a lower number of markers (typically, several thousands) model-based approaches have been developed to approximate the variation of F_{ST} estimates among loci. One of such approaches is FDIST, developed by Beaumont and colleagues (23), and another is DETSEL, developed by Vitalis and colleagues (24, 26). Other methods have been more recently developed, which are based on a Bayesian approach: BAYESFST (27) and BayeScan (28). Both FDIST and DETSEL rely on summary statistics to describe the variation of F_{ST} (or F_{ST}-like) estimates among loci expected under the null hypothesis of neutrality. More precisely, they are based on the estimation of the joint distribution of some summary statistics, by means of stochastic simulations of neutral gene genealogies in a simple population model. The simulations are constructed so that the average values of the summary statistics over thousands of simulations closely match those measured on the dataset of interest. The expected distribution under the null hypothesis of neutrality is then compared to the observed distribution of the statistics estimated from the dataset. Those loci that depart significantly from the simulated distribution (outliers) are considered as targeted by selection.

The main difference between FDIST and DETSEL lies in the underlying demographic model. While FDIST considers an island model of population structure, i.e., a set of populations with constant and equal deme sizes that are connected by gene flow (29), DETSEL considers a pure divergence model, in which an ancestral population splits into two daughter populations. Hence, the choice between FDIST and DETSEL should primarily be made according to the supposed demographic model of the species under scrutiny.

2. Program Usage

2.1. The Underlying Demographic Model

Going forward in time, DETSEL considers an ancestral population at mutation-drift equilibrium with constant size N_e. Generations do not overlap. Then, this population may go through a bottleneck of stationary size N_0 during t_0 generations. Last, this population splits into two daughter populations of constant sizes N_1 and N_2 (so that the daughter populations may have different sizes), which evolve independently from each other for t generations. By independent evolution, it is meant that the populations do not exchange any migrants between the time of the split and the present. This model of population divergence is illustrated in Fig. 1.

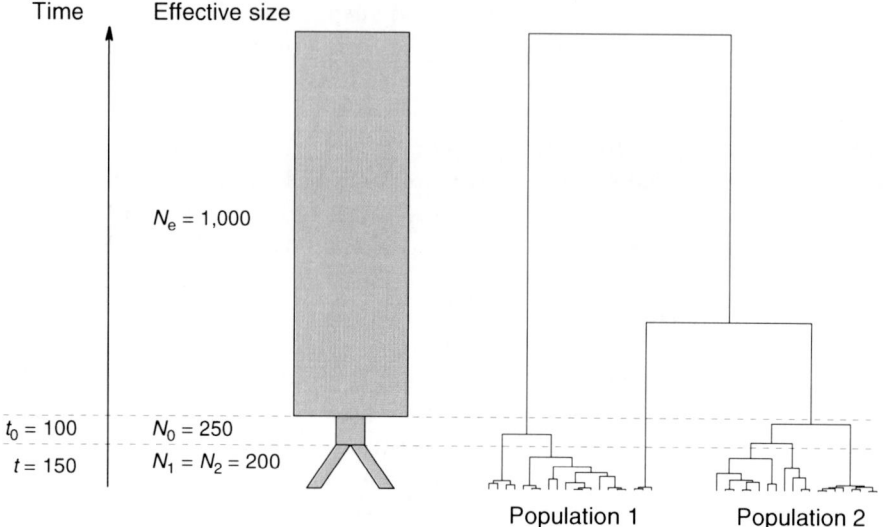

Fig. 1. DᴇᴛSᴇʟ underlying demographic model and illustrated example of a simulated genealogy. The parameter values of this example are provided along the time scale on the left-hand side of the graph, for the different periods of the demographic model. The demographic model is schematized in *gray*, in the middle of the graph. On the right-hand side of the graph, the genealogy of a sample of 2 × 20 genes is shown.

2.2. The Parameters of Interest

DᴇᴛSᴇʟ calculates the joint distribution of the parameters F_1 and F_2, which represent the population-specific differentiation of daughter populations 1 and 2, respectively (24, 30). These parameters are defined as:

$$F_i = \frac{Q_{w,i} - Q_b}{1 - Q_b}, \quad \text{for } i \in \{1, 2\}$$

where $Q_{w,i}$ is the probability of identity in state for pairs of genes sampled within population i, and Q_b is the probability of identity in state for pairs of genes sampled between populations 1 and 2 (24). Hence, F_i has the form of an intraclass correlation for the probability of identity in state ("IIS correlation") for genes within population i, relatively to genes between populations (31, 32). It is worth noting that the average of F_i over the two populations, weighted by the number of pairs of sampled genes in each population, gives the intraclass correlation for the probability of identity in state for genes within populations, relatively to genes between populations, i.e., F_{ST}. If new mutations arising during the divergence process are neglected, it can be shown that:

$$F_i \approx 1 - \exp(-T_i)$$

where $T_i \equiv t / N_i$ is generally referred to as the "branch length" for population i (24). This is a well-known result when the two daughter populations have the same size ($N_1 = N_2 = N$), so that $F_1 = F_2 \approx 1 - \exp(-t/N)$ (33). An important result is that, in the low-mutation limit, the new parameters F_1 and F_2 do not depend on the nuisance parameters of the model, i.e., t_0, N_0, N_c and the average mutation rate $\bar{\mu}$. This suggests that a simple moment-based estimator of the branch length parameter T_i may be derived as

$$\hat{T}_i = -\log(1 - \hat{F}_i)$$

where $\hat{F}_i$ is an estimator of the parameter F_i. The $\hat{F}_i$ estimate of population-specific differentiation of population i is calculated from the estimated allele frequencies, using Eq. 8 in ref. (30).

2.3. Coalescent-Based Simulations

For each set of parameter values, a sequence of artificial datasets is generated using standard coalescent simulations as described by, e.g., Hudson (34) or Hein et al. (35). The simulations are performed as follows (see Fig. 1 for an illustrated example of a simulated genealogy). For each population, the genealogy of a sample of n_i genes is generated for a period of time ranging from present to t generations in the past. During this period, all the coalescent events are separated by exponentially distributed time-intervals, with mean $N_i / \binom{n_i}{2}$ in the ith population. At time t, the number n_0 of lineages that remain represents the ancestors of all the genes sampled in populations 1 and 2. The genealogy of these lineages is generated backward for the time-period $[t; t + t_0)$, and all the coalescence events are separated by exponentially distributed time-intervals, with mean $N_0 / \binom{n_0}{2}$. At time $t + t_0$, the lineages that remain are the ancestors of all the genes sampled in populations 1 and 2. The genealogy of these n_c genes is generated for the period $[t + t_0; +\infty)$, with all coalescent events separated by exponentially distributed time-intervals with mean $N_c / \binom{n_c}{2}$. Once the complete genealogy is obtained, the mutation events are superimposed on the coalescent tree.

The divergence model underlying DetSel has a number of nuisance parameters, i.e., parameters that are largely unknown, but that nonetheless must be accounted for in the divergence model (e.g., ancestral population size, divergence time, mutation rate). Because of the uncertainty in the nuisance parameter values, it is recommended to perform simulations using different combinations of values for the ancestral population size, divergence time, and bottleneck parameters. It is worth mentioning that under the model assumptions, F_i depends upon divergence time t only through the ratio $T_i \equiv t / N_i$. Hence, absolute values of t are unimportant, since the T_i ratios relevant for the simulations are calculated from the $\hat{F}_1$ and $\hat{F}_2$ multilocus estimates. On average, mutations arise at rate $\bar{\mu}$ per generation. Locus-specific mutation rates are drawn from a Gamma distribution with shape parameter 2 and rate parameter $\bar{\mu} / 2$. Most importantly, the expected joint conditional distribution of the $\hat{F}_1$ and $\hat{F}_2$ estimates conditioned upon the number of alleles in the pooled sample (i.e., the total number of alleles across populations 1 and 2) is almost independent on the nuisance parameters (24). This result provides the justification for using the conditional distributions to analyze the homogeneity in the patterns of genetic differentiation revealed by a (large) set of markers.

As compared to the previous version of DETSEL (26), two major improvements have been made in the latest version: the analysis of dominant markers is now supported, and the estimation of empirical P-values has been implemented. These features, which are described below, have been incorporated into an R package (36) which replaces the stand-alone DETSEL software package. The stable version is available from CRAN: http://cran.r-project.org/. The development version is available at http://r-forge.r-project.org/projects/detsel. Both versions can be installed directly from R.

2.4. Biallelic Dominant Markers

While on model organisms (whose genome sequence is known) genome scans have been performed using hundreds of thousands of SNPs (19), most of the genome scans on nonmodel organisms have been carried out using hundreds of Amplified Fragment Length Polymorphisms (AFLPs) (11, 14). AFLPs have indeed been successfully developed on a wide range of organisms and were found to be highly efficient in gathering genomic data from unknown genomes (37). AFLPs are biallelic dominant markers, which means that heterozygotes and homozygote individuals for the [+] allele cannot be distinguished. It is therefore challenging to estimate the underlying allele frequencies, and preliminary assumptions (e.g., about deviations from Hardy–Weinberg equilibrium) are needed (38). Since the allele frequencies are estimated from the proportion of homozygous genotypes for the [−] allele, the sampling variance is large when the [−] allele is rare, i.e., when the number of homozygous genotypes for the [−] allele is low. As a result, the variance of F_{ST}-like estimates is larger when the [−] allele is rare than when it is frequent, and this property must be accounted for when estimating the joint distributions of summary statistics.

Specifically, the latest version of DETSEL simulates biallelic genotypes from which one allele is randomly assigned to be recessive. This procedure results in "band absence" and "band presence" phenotypes that resemble the polymorphism observed at AFLP markers. For both the observed and the simulated data, the underlying allele frequencies are calculated following Zhivotovsky's (39) Bayesian method with nonuniform prior distribution of allele frequencies as implemented, e.g., in the AFLP-SURV 1.0 software package (40). This method has been shown to provide reliable [–] allele frequency estimates, even with moderate departure from Hardy–Weinberg equilibrium (38). Then, the $\hat{F}_1$ and $\hat{F}_2$ estimates of population-specific differentiation of populations 1 and 2, respectively, are calculated from the estimated allele frequencies, using Weir and Hill's (30) Eq. 8. This approach ensures that the inflated sampling variance of $\hat{F}_1$ and $\hat{F}_2$ (when the recessive allele is rare) is comparable between observed and simulated data. Along with the estimation of the joint distribution of the $\hat{F}_1$ and $\hat{F}_2$ estimates, the updated version of DETSEL also calculates the expected joint distribution of Weir and Cockerham's F_{ST} estimate (41) for each population pair and Nei's heterozygosity H_e of the pooled sample, under the null hypothesis of neutrality.

2.5. Identification of Outliers

As a last improvement to the original DETSEL software package, I developed a new algorithm to calculate the empirical P-values associated with each observation in the dataset. To that end, the cumulative distribution function (CDF) is evaluated empirically from the joint distribution of all the pairwise observations ($\hat{F}_1$, $\hat{F}_2$) within the simulated dataset. Then, the empirical P-value for a given marker locus i is calculated as one minus the CDF evaluated at locus i, i.e., P-value(locus i) = $1 - \text{CDF}(\hat{F}_{1,i}, \hat{F}_{2,i})$, where $\hat{F}_{1,i}$ and $\hat{F}_{2,i}$ are the observed values of $\hat{F}_1$ and $\hat{F}_2$ at locus i. For multiallelic markers, the joint distribution of all the pairwise observations ($\hat{F}_1$, $\hat{F}_2$) within the simulated dataset is computed from a two-dimensional array, where the ($\hat{F}_1$, $\hat{F}_2$) pairs are binned, and then smoothed using the Average Shifted Histogram (ASH) algorithm (42) as implemented in the "ash" R package. Because the distribution of ($\hat{F}_1$, $\hat{F}_2$) estimates for biallelic markers is discontinuous with many ties, the CDF is computed instead by enumerating all ($\hat{F}_1$, $\hat{F}_2$) pairs in the simulated data.

3. Examples

The rationale of DETSEL is to estimate the population parameters F_1 and F_2 from the real data, and then to simulate artificial data to compute the expected distribution of these population-specific estimates of differentiation under selective neutrality, conditionally on the total number of alleles in the pooled sample (24). Then, the empirical P-values and the density of the statistics are computed. Below, I provide a step-by-step tutorial to run the sequence of analyses.

3.1. Running DETSEL Step-by-Step

In R, load the DETSEL package, using the following command:

```
> library(DetSel)
```

Before running the following analyses, the user shall copy the input data file into the R working directory. DETSEL takes the same input file format as Beaumont and Nichols' FDIST2 or Dfdist (23), as detailed in Fig. 2. For dominant data, it is important to note that the frequency of the homozygote individuals for the recessive allele appears first in either the rows or columns of the data matrix (see data file format in Fig. 2). For codominant data, it is possible to convert a data file formatted in GENEPOP format into the DETSEL format by running, e.g., the command line:

```
>  genepop.to.detsel(infile = "mygenepopfile",
outfile = "mydata.dat")
```

This command converts a data file in GENEPOP format (see refs. 43, 44), which name can be specified using the *infile* argument into a data file in DETSEL format, which name can be specified

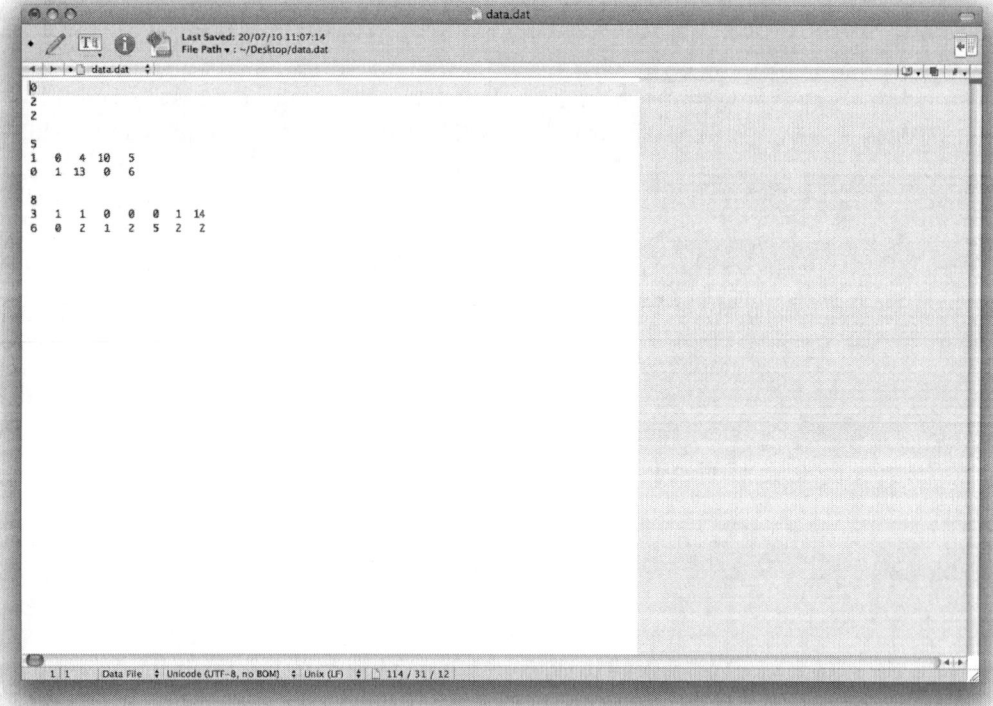

Fig. 2. Example input file in DETSEL format. The first line is a 0/1 indicator. "0" indicates that the data matrix for each locus is a populations × alleles matrix; "1" indicates that the data matrix for each locus is an alleles × populations matrix. The *second line* contains the number of populations. The *third line* contains the number of loci. Then, the data for each locus consists in the number of alleles at that locus, followed by the data matrix at that locus, with each row corresponding to the same allele (if the indicator variable is 1) or to the same population (if the indicator variable is 0). For dominant data, the data consists in the number of genotypes, not the number of alleles. For dominant data, it is important to note that the frequency of the homozygote individuals for the recessive allele appear first in either the rows or columns of the data matrix. In the illustrated example, the data consists in two populations and two loci, with five alleles at the first locus and eight alleles at the second locus.

using the *outfile* argument. Then, to read the data file, run the command line in R:

```
> read.data(infile,dominance,maf,a,b)
```

where *infile* is the input file name, which must stand in the R working directory; *dominance* is a logical variable, which is FALSE if codominant data are considered (e.g., microsatellite markers, SNPs, etc.), or TRUE, if biallelic dominant data are considered (e.g., AFLPs); *a* and *b* are the parameters for the beta prior distribution, used in Zhivotovsky's (39) Bayesian method to compute the underlying allele frequencies; the default values are $a = b = 0.25$, as suggested by Mark A. Beaumont in the DFDIST manual, yet the user may alternatively chose to use Zhivotovsky's Equation (Eq. 13) to compute estimates of *a* and *b* from the data (39) [note that neither *a* nor *b* parameters are needed if *dominance* = FALSE]; *maf* is the maximum allele frequency (the frequency of the most frequent allele over the full sample) to be considered in both the input file and the simulated data. Hence, the command `read.data(infile = "mydata.dat", dominance = FALSE, maf = 0.99)` will read an input file named *data.dat* that contains codominant markers, and a maximum allele frequency of 0.99 will be applied (i.e., by removing marker loci in the observed and simulated datasets that have an allele with frequency larger than 0.99). The command `read.data(infile = "mydata.dat", dominance = TRUE, maf = 0.99, a = 0.25, b = 0.25)` will read an input file named *data.dat* that contains dominant biallelic markers; a maximum allele frequency of 0.99 will be applied and the *a* and *b* parameters for the beta prior distribution (39) will be set to 0.25.

The command line `read.data()` creates a file named *infile. dat*, a file named *sample_sizes.dat* and a set of files named *plot_i_j. dat* where *i* and *j* correspond to population numbers, so that each file *plot_i_j.dat* corresponds to the pairwise analysis of populations *i* and *j*. In the file *infile.dat*, each line corresponds to the pairwise analysis of populations *i* and *j*. Each line contains (in that order): the name of the output simulation file, the numbers *i* and *j*, the multilocus estimates of F_1 and F_2, and Weir and Cockerham's estimate of F_{ST} (41). The file *sample_sizes.dat* contains sample sizes information, for internal use only. In the files *plot_i_j.dat*, each line corresponds to one locus observed in the dataset. Each line contains (in that order): the locus-specific estimates of F_1 and F_2, Weir and Cockerham's estimate of F_{ST} (41), Nei's heterozygosity (H_e), the number of alleles at that locus in the pooled sample, and the rank of the locus in the dataset.

Once the `read.data()` command line has been executed, the following command line executes the simulations:

```
> run.detsel()
```

The user is first asked to provide the total number of simulations for the entire set of parameter values (default: 500,000). For biallelic data, I recommend to run no less than 1,000,000 simulations to estimate correctly the P-values for each empirical locus. With less than a million simulations, indeed, simulation tests have shown that the P-values may be biased (data not shown). The user is then asked to provide the average mutation rate $\bar{\mu}$, and the mutation model: type "0" for the infinite allele model, where each mutation creates a new allelic state in the population; type "1" for the stepwise mutation model, where each mutation consists in increasing or decreasing by one step, the size of the current allele; and type any integer k (with $k > 1$) for a k-allele model, where each mutation consists in drawing randomly one allele among k possible states, provided it is different from the current state. For example, for SNP data, type "2". Finally, the user is asked to provide the number of distinct sets of nuisance parameters (the default is a single set of parameters). Because of the uncertainty in the nuisance parameter values, it is recommended to perform simulations using different combinations of values for the ancestral population size, divergence time, and bottleneck parameters. Then, the user is asked to provide as many sets of parameters as he/she indicated. Each set comprises four parameters that should be given in the following order: t, N_0, t_0 and N_c. Here, the user must chose parameter values, including the mutation model, that correspond to his/her knowledge of the biological model. Most importantly, he/she must check a posteriori that for each pairwise comparison, the $\hat{F}_1$ and $\hat{F}_2$ estimates averaged over the simulated data do not deviate much from the observed values in the real dataset (see below).

The command line run.detsel() creates a list of files named *Pair_i_j_n_i_n_j.dat*, where i and j are the indices of populations pairs, and n_i and n_j are the sample sizes of populations i and j, respectively. Because some marker loci may have missing data, several *Pair_i_j_n_i_n_j.dat* files may be created for a given pair of populations. Simulating the exact sample size for each locus is required to precisely calculate the empirical P-values, especially for biallelic markers (see Note 1). Note that if negative multilocus F_i estimates are observed for a pairwise comparison, then the simulations will not be run for that pair (see Note 2). Each line of the *Pair_i_j_n_i_n_j.dat* files contains the locus-specific estimates of F_1 and F_2, Weir and Cockerham's estimate of F_{ST} (41), Nei's heterozygosity (H_e), and the number of alleles at that locus in the pooled sample. The command line run.detsel() also creates a file named *out.dat* that contains the estimates of the above statistics averaged over all the simulated data. In the file *out.dat*, each line corresponds to the pairwise analysis of populations i and j with sample sizes and n_i and n_j. Each line contains (in that order): the name of the output simulation file (*Pair_i_j_n_i_n_j.dat*), the multilocus estimates of F_1 and

F_2, and Weir and Cockerham's estimate of F_{ST} (41). An important point to consider is to make sure that for each pairwise comparison, the $\hat{F}_1$ and $\hat{F}_2$ estimates averaged over the simulated data (in the file *out.dat*) closely match to the observed values in the real dataset (in the file *infile.dat*). If not, this suggests that the simulated datasets do not fit to the observed data, which urges to choose other parameter values for the nuisance parameters.

Last, the command line `compute.p.values(x.range,y.range,n.bins,m)` produces an output file, named *P-values_i_j.dat*, with the *P*-value associated with each observation. The arguments *x.range* and *y.range* are the range of values in the *x*- and *y*-axis, respectively, which both take the default values *x.range* = *y.range* = c(-1,1); *n.bins* is the size of the two-dimensional array of $n \times n$ square cells used to bin the $\hat{F}_1$ and $\hat{F}_2$ estimates, which takes the default value *n.bins* = c(100,100); *m* gives the smoothing parameters of the ASH algorithm, which takes the default value *m* = c(2,2). Because a large number of markers are usually considered, it may be advisable to control the error rate due to multiple testing (see Note 3). Combining *P*-values across pairwise comparisons may be achieved by several means (see Note 4).

Once the `run.detsel()` and `compute.p.values()` command lines have been executed, the following command line can be used to plot graphs with an estimation of the density of $\hat{F}_1$ and $\hat{F}_2$ estimates, as detailed in the appendix in Vitalis et al. (24):

```
> draw.detsel.graphs(i,j,x.range,y.range,n.bins,m,alpha)
```

where *i* and *j* are the population indices, *x.range*, *y.range*, *n.bins*, *m* have the same meaning as in the function `compute.p.values()` and *alpha* is the α-level (hence 1 – *alpha* is the proportion of the distribution within the plotted envelope), which takes the default value *alpha* = 0.05. Hence, the command `draw.detsel.graphs(i=1,j=2,n.bins=c(50,50),alpha=0.01)` plots the 99% confidence regions corresponding to the pair of populations 1 and 2, using a 50×50 two-dimensional array. Note that if the arguments *i* and *j* are missing, then all the population pairs are plotted. It is noteworthy that our estimation of the density of the $\hat{F}_1$ and $\hat{F}_2$ estimates might be discontinuous, because of the discrete nature of the data (the allele counts). This is particularly true when the number of alleles upon which the distribution is conditioned is small. The command line `draw.detsel.graphs()` produces as many conditional distributions per population pair as there are different allele numbers in the pooled sample. All the observed data points are plotted in each graph. The outlier loci, for which the empirical *P*-value is below the threshold α-level, are plotted with a star symbol. For the latter, the locus number (i.e., its rank in the data file) is provided on the graph.

3.2. Example on a Simulated Dataset

In this example, I simulated a dataset of 100 marker loci. The ancestral population size was set to $N_c = 20{,}000$, and I considered that no bottleneck occurred between the ancestral equilibrium state and the population split (hence, $t_0 = N_0 = 0$). The divergence time was set to $t = 100$ generations. All the markers were simulated using a K-allele model with five possible allelic states and the average mutation rate was set to $\bar{\mu} = 0.0001$. I considered that among the 100 markers, 95 were "neutral". These were simulated with $N_1 = N_2 = 2{,}000$. Among the 100 markers, five were simulated as "selected". To simulate those selected markers, I simply considered that their effective size was reduced to $N_1 = 100$ in population 1 (4, 45, 46). 50 genes (25 diploids) were sampled in each population, and a maximum allele frequency of 0.99 was fixed. The simulated data were formatted as in Fig. 2 and are available by typing the command line:

```
> make.example.file(file = "data.dat")
```

This will copy the example file, which will be named *data.dat*, into the user's working directory. Then, running the command line:

```
> read.data(infile = "data.dat", dominance = FALSE, maf = 0.99)
```

produces a file named *infile.dat* that contains the multilocus estimates of the population-specific differentiation parameters ($\hat{F}_1 = 0.085$ and $\hat{F}_2 = 0.054$) and a file named *plot_1_2.dat* that contains the locus-specific estimates of F_1 and F_2, Weir and Cockerham's estimate of F_{ST} (41), Nei's heterozygosity (H_e), the number of alleles at that locus in the pooled sample, and the rank of the locus in the dataset. Then, running the command line:

```
> run.detsel(example = TRUE)
```

produces a file named *Pair_1_2_50_50.dat* that contains the simulated data, and a file named *out.dat* that contains the multilocus estimates of the population-specific differentiation parameters over all simulations ($\hat{F}_1 = 0.092$ and $\hat{F}_2 = 0.058$). Note that these values are very close to the values observed in the dataset. The command line `compute.p.values()` then produces a file named *P-values_1_2.dat*, were one can identify five significant P-values (out of 100) at the $\alpha = 0.01$ level. All these five P-values (0.003, 0.0002, 0.007, 0.004, 0.0009) correspond to the selected loci (ranked 96 to 100 in the dataset). Last, running the command line:

```
> draw.detsel.graphs(i = 1, i = 2, alpha = 0.01)
```

produces the four outputs in Fig. 3. All marker loci with P-values < 0.01 are plotted with star symbols, with the locus number indicated nearby.

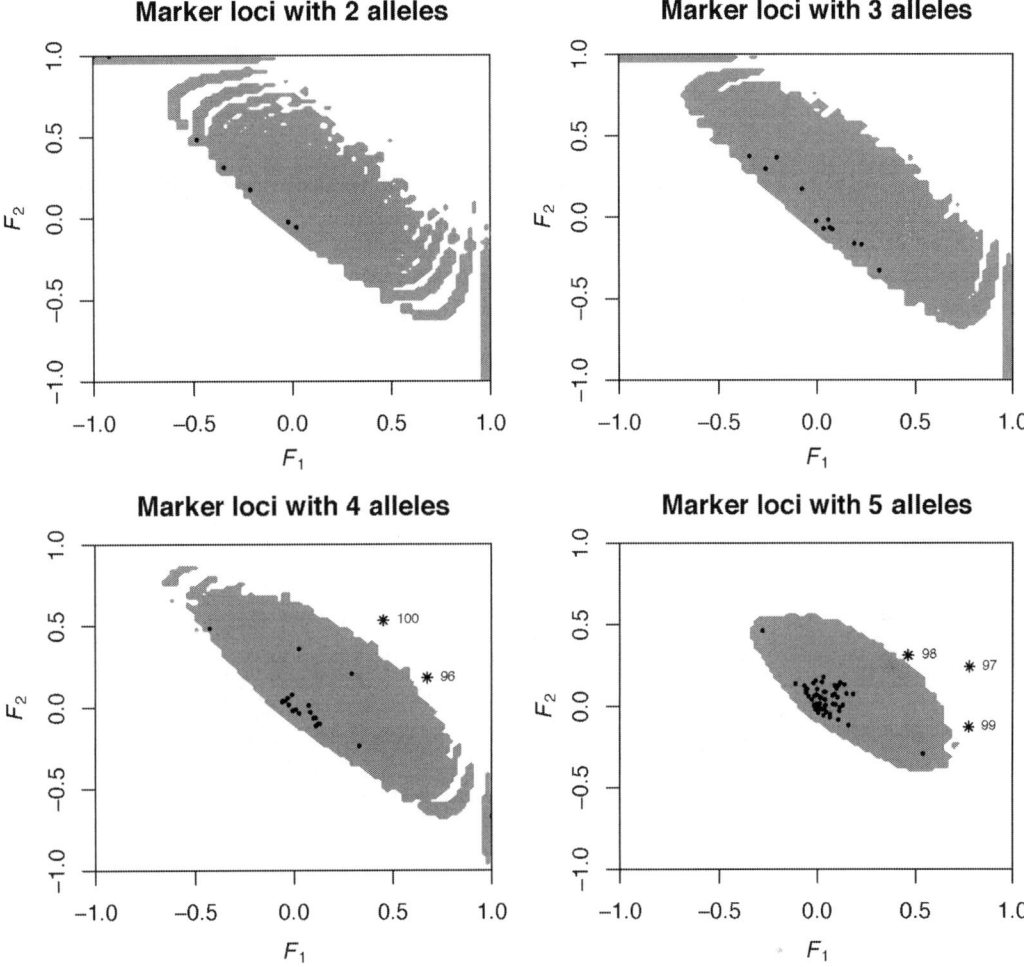

Fig. 3. Example output of the DETSEL analysis of data simulated with $N_e = 20{,}000$, $t_0 = N_0 = 0$, $t = 100$ generations, $\bar{\mu} = 0.0001$ and 95 "neutral" markers with $N_1 = N_2 = 2{,}000$ and five selected markers with $N_1 = 100$ and $N_2 = 2{,}000$. The four graphs represent the four 99% envelopes for the distributions of ($\hat{F}_1$, $\hat{F}_2$) estimates, conditioned upon the presence of 2, 3, 4, and 5 alleles in the pooled sample, respectively. All marker loci with P-values < 0.01 are plotted with an *asterisk* symbol, with the locus number indicated nearby. All other marker loci are plotted in smaller size.

4. Notes

1. Evaluation of type I error for biallelic dominant markers (i.e., AFLPs). In order to estimate the false-positive rate at any given nominal type α-level (type I error), I ran DETSEL analyses on artificially generated datasets under selective neutrality. I performed the analyses on 100 simulated datasets of 1,000 loci each, generated using the DETSEL coalescent-based algorithm with the following simulation parameters: $N_e = 20{,}000$,

$t_0 = N_0 = 0$ (hence, no bottleneck between the ancestral equilibrium state and the population split), $t = 100$ generations. All the markers were simulated as dominant, biallelic markers, with $\bar{\mu} = 0.00001$. The target value of the differentiation parameters was set to $F_1 = F_2 = 0.05$. DetSel detected on average 1.21 (SD = 1.09), 10.15 (SD = 3.23), and 50.22 (SD = 8.03) outliers at the 0.1, 1, and 5% α-level, respectively. The medians were equal to 1, 10, and 50 outliers, respectively, which perfectly matches the expected number of outliers expected in a 1,000-locus dataset, for $\alpha = 0.1$, 1, and 5%, respectively. I checked that the P-values in each dataset were uniformly distributed, as expected, by means of Kolmogorov–Smirnov tests (no significant test among the 100 datasets, at the 0.1% level).

2. Negative multilocus F_i estimates. When the collected samples are barely differentiated, DetSel multilocus estimates of F_1 and F_2 may be slightly negative. Since the F_i parameters, which cannot take negative values, are used to parameterize the coalescent-based simulations, DetSel analyses cannot be performed in such cases. Very low genetic differentiation between populations thus prevents the use of DetSel.

3. Multiple testing. Because a large number of markers are usually considered, it may be advisable to control the error rate due to multiple testing. Because classical methods used to account for multiple testing (e.g., Bonferroni correction) are generally considered as too conservative, an alternative is to compute the q-values, using the observed distribution of P-values (47). The Q-value at each locus provides the minimum false discovery rate that is incurred when considering this locus as an outlier. In brief, considering a locus with $p = 0.01$ as significant implies that 1% of all loci are also expected to be detected as significant, be they real or false positives. On the other hand, considering $q = 0.01$ as a threshold for significance ensures that, among the significant loci, 1% are expected to be false positive. Computing q-values may be achieved using the QVALUE package for R or the standalone software package QVALUE available from http://www.bioconductor.org/packages/release/bioc/html/qvalue.html. In that case, QVALUE should be used to compute the q-values from the P-values calculated with DetSel, which are recorded in the files *P-values_i_j.dat*.

4. Combining information across pairwise comparisons. When considering a set of populations, DetSel provides a probability for each locus to be an outlier of the neutral distribution in each population pair. Hence, it is desirable to combine the information brought by all the analyses. In particular, when the populations have been sampled across two contrasting

environments, it may be desirable to identify the loci that have a high probability to be under selection in a specific environment. One possibility is then to combine the P-values at each locus across all the pairs of populations that represent different environments using the Z-transform test by Whitlock (48), as in Meyer et al. (49). An alternative is to combine the P-values for each marker across the pairwise analyses using the Fisher's method (50). The rationale for this approach is to combine the genetic information from different replicates and hence to give more support to the outliers detected in more than one pair of populations.

Combining P-values has the advantage to take all the P-values from the relevant pairwise comparisons into account, instead of considering only the P-values above an arbitrary significance threshold. Yet, the assumption of statistical independence of the P-values is violated because (1) some of the pairwise comparisons are not independent from each other, and (2) the populations may have a shared history and/or may be connected by gene flow.

Nosil et al. (51) have suggested an alternative approach to minimize the false-positive rate. This approach is aimed at giving more support to outliers repeated in independent analyses of population pairs taken from contrasting environments, and to tentatively avoid sources of selection independent from the environmental conditions considered. This approach, which has been used, e.g., in Midamegbe et al. (52), consists in (1) retaining all marker loci that depart significantly from neutrality in all pairwise comparisons that involve samples collected from different environments; (2) from this set of outliers, discarding all marker loci that depart significantly from neutrality in one pairwise comparison at least that involves samples collected from the same environment. Although this approach is appealing, it necessarily depends upon an arbitrarily chosen significance threshold.

Acknowledgments

I am grateful to Hélène Fréville and François Pompanon for useful comments on a previous version of this chapter. I also sincerely acknowledge Eric Bazin for providing me some hints and tips during the packaging of DetSel into R. Part of this work was carried out by using the resources of the Computational Biology Service Unit from the Museum national d'Histoire naturelle (MNHN—CNRS UMS 2700). This work was funded by the ANR grants "EMILE" (09-BLAN-0145) and "NUTGENEVOL" (07-BLAN-0064).

References

1. Black WCT, Baer CF, Antolin MF, DuTeau NM (2001) Population genomics: genome-wide sampling of insect populations. Annu Rev Entomol 46:441–469

2. Luikart G, England PR, Tallmon D et al (2003) The power and promise of population genomics: from genotyping to genome typing. Nat Rev Genet 4:981–994

3. Storz JF (2005) Using genome scans of DNA polymorphism to infer adaptive population divergence. Mol Ecol 14:671–688

4. Maynard Smith JM, Haigh J (1974) Hitchhiking effect of a favorable gene. Genet Res 23:23–55

5. Schlötterer C (2003) Hitchhiking mapping – functional genomics from the population genetics perspective. Trends Genet 19:32–38

6. Wu C-I (2001) Genes and speciation. J Evol Biol 14:889–891

7. Hinds DA, Stuve LL, Nilsen GB et al (2005) Whole-genome patterns of common DNA variation in three human populations. Science 307:1072–1079

8. International HapMap Consortium (2003) The International HapMap Project. Nature 426:789–796

9. Andolfatto P, Przeworski M (2000) A genome-wide departure from the standard neutral model in natural populations of *Drosophila*. Genetics 156:257–268

10. Nordborg M, Hu TT, Ishino Y et al (2005) The pattern of polymorphism in *Arabidopsis thaliana*. PLoS Biol 3:e196

11. Wilding CS, Butlin RK, Grahame J (2001) Differential gene exchange between parapatric morphs of *Littorina saxatilis* detected using AFLP markers. J Evol Biol 14:611–619

12. Emelianov I, Marec F, Mallet J (2004) Genomic evidence for divergence with gene flow in host races of the larch budmoth. Proc Roy Soc B 271:97–105

13. Vasemägi A, Nilsson J, Primmer CR (2005) Expressed sequence tag-linked microsatellites as a source of gene-associated polymorphisms for detecting signatures of divergent selection in atlantic salmon (*Salmo salar* L.). Mol Biol Evol 22:1067–1076

14. Bonin A, Taberlet P, Miaud C, Pompanon F (2006) Explorative genome scan to detect candidate loci for adaptation along a gradient of altitude in the common frog (*Rana temporaria*). Mol Biol Evol 23:773–783

15. Goldstein DB, Chikhi L (2002) Human migrations and population structure: what we know and why it matters. Annu Rev Genomics Hum Genet 3:129–152

16. Akey JM, Zhang G, Zhang K et al (2002) Interrogating a high-density SNP map for signatures of natural selection. Genome Res 12:1805–1814

17. Barreiro LB, Laval G, Quach H et al (2008) Natural selection has driven population differentiation in modern humans. Nat Genet 40:340–345

18. Deng LB, Tang XL, Kang JA et al (2007) Scanning for signatures of geographically restricted selection based on population genomics analysis. Chin Sci Bull 52:2649–2656

19. Weir BS, Cardon LR, Anderson AD et al (2005) Measures of human population structure show heterogeneity among genomic regions. Genome Res 15:1468–1476

20. Lewontin RC, Krakauer J (1973) Distribution of gene frequency as a test of the theory of the selective neutrality of polymorphisms. Genetics 74:175–195

21. Nei M, Maruyama T (1975) Lewontin-Krakauer test for neutral genes. Genetics 80:395

22. Robertson A (1975) Remarks on the Lewontin-Krakauer test. Genetics 80:396

23. Beaumont M, Nichols RA (1996) Evaluating loci for use in the genetic analysis of population structure. Proc Roy Soc B 263:1619–1626

24. Vitalis R, Dawson K, Boursot P (2001) Interpretation of variation across marker loci as evidence of selection. Genetics 158:1811–1823

25. Beaumont MA (2005) Adaptation and speciation: what can F_{ST} tell us? Trends Ecol Evol 20:435–440

26. Vitalis R, Dawson K, Boursot P, Belkhir K (2003) DetSel 1.0: a computer program to detect markers responding to selection. J Hered 94:429–431

27. Beaumont MA, Balding DJ (2004) Identifying adaptive genetic divergence among populations from genome scans. Mol Ecol 13:969–980

28. Foll M, Gaggiotti O (2008) A genome-scan method to identify selected loci appropriate for both dominant and codominant markers: a Bayesian perspective. Genetics 180:977–993

29. Wright S (1931) Evolution in Mendelian populations. Genetics 16:97–159

30. Weir BS, Hill WG (2002) Estimating F-statistics. Annu Rev Genet 36:721–750

31. Cockerham CC, Weir BS (1987) Correlations, descent measures: drift with

migration and mutation. Proc Natl Acad Sci USA 84:8512–8514

32. Rousset F (1996) Equilibrium values of measures of population subdivision for stepwise mutation processes. Genetics 142:1357–1362

33. Reynolds J, Weir BS, Cockerham CC (1983) Estimation of the coancestry coefficient: basis for a short-term genetic distance. Genetics 105:767–779

34. Hudson RR (1990) Gene genealogies and the coalescent process. In: Futuyma RJ, Antonovics J (eds) Oxford survey in evolutionary biology. Oxford University Press, Oxford

35. Hein J, Schierup MH, Wiuf C (2005) Gene genealogies, variation and evolution. A primer in coalescent theory. Oxford University Press, Oxford

36. R Development Core Team (2009) R: a language and environment for statistical computing. R Foundation for Statistical Computing, Vienna

37. Bensch S, Akesson M (2005) Ten years of AFLP in ecology and evolution: why so few animals? Mol Ecol 14:2899–2914

38. Bonin A, Ehrich D, Manel S (2007) Statistical analysis of amplified fragment length polymorphism data: a toolbox for molecular ecologists and evolutionists. Mol Ecol 16:3737–3758

39. Zhivotovsky LA (1999) Estimating population structure in diploids with multilocus dominant DNA markers. Mol Ecol 8:907–913

40. Vekemans X, Beauwens T, Lemaire M, Roldan-Ruiz I (2002) Data from amplified fragment length polymorphism (AFLP) markers show indication of size homoplasy and of a relationship between degree of homoplasy and fragment size. Mol Ecol 11:139–151

41. Weir BS, Cockerham CC (1984) Estimating F-statistics for the analysis of population structure. Evolution 38:1358–1370

42. Scott DW (1992) Multivariate density estimation: theory, practice, and visualization. Wiley, New York

43. Raymond M, Rousset F (1995) Genepop (version 1.2): population genetics software for exact tests and ecumenicism. J Hered 86:248–249

44. Rousset F (2007) Genepop'007: a complete re-implementation of the genepop software for Windows and Linux. Mol Ecol Notes 8:103–106

45. Barton NH (1995) Linkage and the limits to natural selection. Genetics 140:821–841

46. Kaplan NL, Hudson RR, Langley CH (1989) The "hitchhiking effect" revisited. Genetics 123:887–899

47. Storey JD, Tibshirani R (2003) Statistical significance for genomewide studies. Proc Natl Acad Sci USA 100:9440–9445

48. Whitlock MC (2005) Combining probability from independent tests: the weighted Z-method is superior to Fisher's approach. J Evol Biol 18:1368–1373

49. Meyer CL, Vitalis R, Saumitou-Laprade P, Castric V (2009) Genomic pattern of adaptive divergence in *Arabidopsis halleri*, a model species for tolerance to heavy metal. Mol Ecol 18:2050–2062

50. Fisher RA (1932) Statistical methods for research workers. Oliver and Boyd, Edinburgh

51. Nosil P, Egan SP, Funk DJ (2008) Heterogeneous genomic differentiation between walking-stick ecotypes: "isolation by adaptation" and multiple roles for divergent selection. Evolution 62:316–336

52. Midamegbe A, Vitalis R, Malausa T et al (2011) Scanning the European corn borer (*Ostrinia* spp.) genome for adaptive divergence between host-affiliated sibling species. Mol Ecol 20:1414–1430

Chapter 17

Use of Qualitative Environmental and Phenotypic Variables in the Context of Allele Distribution Models: Detecting Signatures of Selection in the Genome of Lake Victoria Cichlids

Stéphane Joost, Michael Kalbermatten, Etienne Bezault, and Ole Seehausen

Abstract

When searching for loci possibly under selection in the genome, an alternative to population genetics theoretical models is to establish allele distribution models (ADM) for each locus to directly correlate allelic frequencies and environmental variables such as precipitation, temperature, or sun radiation. Such an approach implementing multiple logistic regression models in parallel was implemented within a computing program named Matsam. Recently, this application was improved in order to support qualitative environmental predictors as well as to permit the identification of associations between genomic variation and individual phenotypes, allowing the detection of loci involved in the genetic architecture of polymorphic characters. Here, we present the corresponding methodological developments and compare the results produced by software implementing population genetics theoretical models (Dfdist and BayeScan) and ADM (Matsam) in an empirical context to detect signatures of genomic divergence associated with speciation in Lake Victoria cichlid fishes.

Key words: Genome scans, Signature of selection, Genotype × phenotype association, Environmental variables, Logistic regression, Cichlid fishes, Seascape genetics

1. Introduction

On the basis of data produced by genome scans, the main approach to identify loci under directional selection – or likely to be linked to genomic regions under directional selection – is to use population genetics theoretical models to detect outlier molecular markers showing a larger genetic differentiation than expected under the neutral hypothesis (1–3).

François Pompanon and Aurélie Bonin (eds.), *Data Production and Analysis in Population Genomics: Methods and Protocols*, Methods in Molecular Biology, vol. 888, DOI 10.1007/978-1-61779-870-2_17, © Springer Science+Business Media New York 2012

An alternative is to establish allele distribution models (ADM) for each examined locus to directly correlate allelic frequencies with the variation of explanatory variables of interest (e.g., environmental predictors such as precipitation, temperature, sun radiation, etc.) (4, 5). In this case, the geographic coordinates (spatial variables) of sampled individuals are used to link molecular data (presence or absence of a given allele at a genetic marker) with existing environmental variables (value of an environmental variable at the location where individuals were sampled). Of course, environmental variables can also be directly recorded with sensors in the field.

Mitton et al. (6) first had the idea to correlate the frequency of alleles with an environmental variable (elevation) to look for a signature of selection in ponderosa pine. They detected significant association between gene frequencies and slopes of different aspects. In another paper also dedicated to ponderosa pine, Mitton et al. (7) discovered that excess of heterozygosity was associated with xeric habitats. Then, Stutz and Mitton (8) applied the same approach to Engelmann spruce and showed that natural selection was varying with soil moisture. At the beginning of the 2000s, Joshi et al. (9) and Skøt et al. (10) implemented such association studies on a broad scale to study adaptation in common plant species. But until then, the number of loci considered remained very low, for instance six AFLP loci analyzed together with temperature data in Skøt et al. in 2002 (10). A few years later, Joost (11) contrasted a higher number of loci (and alleles) with ecoclimatic variables in goat, frog, and brown bear. For the purpose of running many simultaneous univariate logistic regressions, an application named MATSAM was developed with MATLAB (The MathWorks Inc.) (12). This software was successfully used to study adaptation in pine weevil and sheep (4), in common frog (13), in goat breeds (14), in fish (15), and in plants (16, 17). The results produced by MATSAM were compared and/or validated by the application of theoretical population genetics approaches to the same data sets in all publications mentioned above.

In this chapter, we describe the principles of MATSAM, its limits, and additional features implemented in a new version released in summer 2010. Then a case study applied to Lake Victoria cichlids illustrates the novel functions.

2. MATSAM

In its first version, MATSAM computes multiple simultaneous univariate logistic regressions to test for association between allelic frequencies at marker loci and quantitative ecoclimatic variables (12). To ensure the robustness of the method, two statistical tests (likelihood ratio G and Wald) assess the significance of coefficients calculated by the logistic regression function.

The molecular data sets used for analysis are in the form of matrices; each row of the matrix corresponds to a sampled individual, while columns are organized according to the sampled individual's geographic coordinates and contain binary information (1 or 0) related to the genetic information observed at each genetic marker. Dominant biallelic markers (e.g., AFLPs) can be used directly as they provide binomial information. Codominant multiallelic markers (e.g., microsatellites) need to be encoded as described in (4), and this is also the case for codominant biallelic markers (e.g., SNPs) (used in ref. 14).

The initial Matsam stand-alone application comes with two Excel macros developed in Visual Basic able to (a) automatically process the large amount of results provided and highlight the most significant associations and (b) draw graphs of the logistic functions (sigmoids) corresponding to any pair of genetic markers vs. environmental variables constituting the models.

The second version of Matsam released in 2010 also includes an upgrade allowing qualitative predictors to be correlated with the presence/absence of alleles. Indeed, many environmental databases available contain nominal or ordinal data that cannot be processed as quantitative variables by Matsam (e.g., CORINE land cover or FAO soil map).

2.1. Design Variables

For continuous quantitative predictors, logistic regression models contain parameters (β_i) which represent the change in the response (y) according to a change of the predictor (x). For categorical predictors, parameters represent the different categories of a predictor. The predictor x is chosen to exclude or include a parameter for each observation. Hence, it is called a *design* or *dummy* variable. At least design variables need to be defined for categories, groups, or classes. Consequently, the model will in any case be multivariate with at least $m - 1$ parameters (to process m categories).

There are multiple ways to define design variables (18). Presently, Matsam implements three of them: the *reference*, the *symmetrical*, and the *independent* parametrization. The difference between these different types of parametrization is highly dependent on the conceptual interpretation that is made of the categories of predictors.

2.1.1. Reference Design Variables

In this case, a group (or a category) always has to be set as the reference group. For groups, one is defined as the reference and the other groups are simply an increase of the expected value compared to the reference one. Each defined combination of design parameters will reflect the expected value of y. Thus, the type of design variable has to be carefully chosen regarding what the predictor conceptually represents (e.g., low–medium–high shrub density; see ref. 18).

If the first group is used as a reference, then the practical implementation regarding the expected value gives:

$$E(\Upsilon_1) = \mu,$$

$$E(\Upsilon_2) = \mu + \alpha_1,$$

...

$$E(\Upsilon_m) = \mu + \alpha_1 + \cdots + \alpha_{m-1}.$$

Thus, the first group will represent the mean value of the groups, and all other groups will represent the mean ± a certain variation. Moreover, the predictor will have to be recoded according to the following scheme:

$$\text{Group 1: } \begin{bmatrix} 1 & 0 & \ldots & 0 \end{bmatrix}$$

$$\text{Group 2: } \begin{bmatrix} 1 & 1 & 0 & \ldots & 0 \end{bmatrix}$$

...

$$\text{Group } m: \begin{bmatrix} 1 & \ldots & 1 \end{bmatrix}$$

Finally, all predictor values affiliated to group m will be recoded into a multivariate regression having an intercept and β_i ($i = 1,\ldots,m-1$) (see an example in Subheading 2.2).

2.1.2. Symmetrical Design Variables

Here, the groups are treated symmetrically. That is to say, it is necessary to define a central group around which the symmetry is distributed:

$$E(\Upsilon_1) = \mu,$$

$$E(\Upsilon_2) = \mu + \alpha_1,$$

...

$$E(\Upsilon_m) = \mu - \alpha_1 - \cdots - \alpha_u.$$

We need $[m/2]$ variables to express the relationships between the groups. The recoding scheme is of size $m \times [m/2]$:

$$\text{Group 1: } \begin{bmatrix} 1 & 0 & \ldots & 0 \end{bmatrix}$$

$$\text{Group 2: } \begin{bmatrix} 1 & 1 & 0 & \ldots & 0 \end{bmatrix}$$

...

$$\text{Group } m: \begin{bmatrix} 1 & -1 & \ldots & -1 \end{bmatrix}$$

For example, let us imagine a categorical predictor composed of five groups with the following ordinal values: "very bad," "bad," "average," "good," and "very good." The "average" group is the central group in this case and the other groups are distributed around it:

$$E\left(\Upsilon_{very\,bad}\right) = \mu - \alpha_1 - \alpha_2,$$

$$E\left(\Upsilon_{bad}\right) = \mu - \alpha_1,$$

$$E\left(\Upsilon_{average}\right) = \mu,$$

$$E\left(\Upsilon_{good}\right) = \mu + \alpha_1,$$

$$E\left(\Upsilon_{very\,good}\right) = \mu + \alpha_1 + \alpha_2.$$

And the corresponding recoding scheme is:

Group "very bad" $\begin{bmatrix} 1 & -1 & -1 \end{bmatrix}$

Group "bad" $\begin{bmatrix} 1 & -1 & 0 \end{bmatrix}$

Group "average" $\begin{bmatrix} 1 & 0 & 0 \end{bmatrix}$

Group "good" $\begin{bmatrix} 1 & 1 & 0 \end{bmatrix}$

Group "very good" $\begin{bmatrix} 1 & 1 & 1 \end{bmatrix}$

Conceptually, μ represents the overall average effect and α_i the group differences. Moreover, the sum of expected values has to be null (18):

$$\left[E\left(\Upsilon_{very\,bad}\right) - \mu\right] + \left[E\left(\Upsilon_{bad}\right) - \mu\right] + \left[E\left(\Upsilon_{average}\right) - \mu\right]$$
$$+ \left[E\left(\Upsilon_{good}\right) - \mu\right] + \left[E\left(\Upsilon_{very\,good}\right) - \mu\right]$$
$$= -\alpha_2 - \alpha_1 - \alpha_1 + \alpha_1 + \alpha_1 + \alpha_2 = 0.$$

2.1.3. Independent Design Variables

In this last parametrization scheme, each group is independent of the other m groups (corresponding to nominal qualitative variables). It means that there is no intercept in the regression model:

$$E\left(\Upsilon_1\right) = \alpha_1,$$

$$E\left(\Upsilon_2\right) = \alpha_2,$$

$$\ldots$$

$$E\left(\Upsilon_m\right) = \alpha_m.$$

Table 1
Example of the recoding of the location predictor

Location	Location$_1$	Location$_2$	Location$_3$
Central Europe	1	0	0
Southern Europe	1	1	0
Alps	1	1	1

Design variables enable such an implementation by defining recoding values as follows:

$$\text{Group 1:} \begin{bmatrix} 1 & 0 & \dots & 0 \end{bmatrix}$$

$$\text{Group 2:} \begin{bmatrix} 0 & 1 & 0 & \dots & 0 \end{bmatrix}$$

$$\dots$$

$$\text{Group } m: \begin{bmatrix} 0 & \dots & 0 & 1 \end{bmatrix}$$

This recoding scheme is indeed similar to an identity matrix of size $m \times m$. Furthermore, as there is no intercept, the convergence of the model may become problematical. Indeed, a regression without an intercept becomes less stable (increase of the number of degrees of freedom of the model) and thus the maximum log-likelihood might not converge.

2.2. Example

We introduce here a small example to illustrate how design variables might be used. Let us consider a data set including a categorical predictor called "Location." It is a nominal predictor characterizing the habitat location of animals. It is made of three classes: "Alps," "Central Europe," and "Southern Europe." These location values are purely nominal and cannot be recoded into quantitative values or intervals. Furthermore, the class "Central Europe" will be used as a reference value. This consideration is completely conceptual, but it forces this predictor in a specific way. It implies that all location values will be recoded into three new location predictors (see Table 1) computed automatically by MATSAM. This clearly implies the computation of a multivariate regression:

$$y = \beta_1 \cdot \text{Location}_1 + \beta_2 \cdot \text{Location}_2 + \beta_3 \cdot \text{Location}_3 + \varepsilon.$$

As "Location$_1$" shows a value 1 everywhere, it substitutes and becomes the model intercept.

2.3. Limitations

Some limitations must be expressed regarding the use of design variables, in particular, the fact that if a predictor has many categories, the number of parameters in the model may be too high. This issue increases the number of degrees of freedom of the model and might result in overfitting problems. Moreover, this overfitting produces unstable estimated standard error (19), which is the consequence of an almost singular variance matrix. This happens most of the time when the maximum log-likelihood does not converge.

To limit the effect caused by the problem mentioned above, one should always try to reduce as much as possible the number of required design variables. To this end, the type of recoding scheme has to be carefully chosen, and the type of recoding should always be the one requiring the lowest number of design variables as possible. As a result, one should first choose the *symmetrical* recoding. If not possible, the second choice should be the *reference* recoding scheme, and at last the *independent* one.

2.4. Additional Improvements

In addition to its capacity to process qualitative variables, Matsam stand-alone application is able to:

(a) Generate a graph for each association model (without any complementary Excel Macro)

(b) Produce histograms to show the allelic frequency at each molecular marker for different values of environmental variables under investigation

(c) Produce the file containing the results with the names of genetic markers and of environmental variables defined by the user to create the input matrix

(d) Characterize the different types of errors that can be generated during the processing of the models (e.g., the model does not converge)

(e) Produce a matrix containing pseudo-R^2 (Efron, MacFadden, Cox & Snell, and Nagelkerke/Cragg & Uhler), Akaike information criterion (AIC) and Bayesian information criterion (BIC) goodness-of-fit indicators for each model

Another important change is that the different parameters to configure the application have to be indicated in a parameter file. The main parameters define the type of qualitative environmental variables (nominal, ordinal) in order to generate the adequate design variables.

Several improvements of Matsam are in progress and will mainly address spatial autocorrelation issues. They include, in particular, the processing and the mapping of Moran's I, of local indicators of spatial autocorrelation (LISA), and of geographically weighted regression (GWR). Moran's I and LISA are classical tools to measure spatial autocorrelation (see http://geoplan.asu.edu/anselin), while GWR is a family of regression models recently developed in which the β coefficients are allowed to vary spatially and therefore permit to reduce residuals (see http://ncg.nuim.ie/ncg/GWR/). The software will also permit to process multivariate models.

The logistic regression-based method developed here can be accurately used to detect statistical associations between genotype at any individual genomic locus with variations of environmental variables, in order to identify loci potentially playing a role in adaptation.

But the same approach can also be used to identify associations between genomic variation and individual phenotypes, to ultimately help reveal genomic regions involved in the genetic architecture of polymorphic characters. In the case of phenotypic traits known to be subject to divergent selection among study populations, this method can then be used to discover genomic signatures of selection associated with each of these traits specifically. The following case study applied to cichlid fishes will provide examples of both cases.

3. Signature of Genomic Divergence Associated with Speciation in Lake Victoria Cichlids

The Lake Victoria cichlid flock is one of the most explosive examples of adaptive radiation, with more than 500 species having evolved during the last 15,000 years. The repetitive occurrence of the same adaptively important traits in unrelated taxa makes the Lake Victoria flock an ideal model system for studying adaptive radiation in shape, ecology, and behavior.

Within the Lake Victoria cichlid radiation, *Pundamilia pundamilia* and *Pundamilia nyererei* are two sympatric sister species, inhabiting the shores of rocky islands and widely distributed in the lake (20). They differ not only in male nuptial coloration but also in other ecological characters, as feeding ecology, depth distribution, photic environment, visual pigment, and female mating preference for male nuptial coloration (21). However, such divergences appear only in near islands with high water transparency, whereas in near islands with low water transparency, genetic differentiation is reduced or absent and intermediate color phenotypes are common or even dominate (22–24) (Fig. 1).

Along the Mwanza Gulf in the Southern part of the lake, the rocky islands show a continuous gradient of water clarity, from turbid in the South to clear in the North, associated with an increased heterogeneity of the light environment. Following this gradient, populations of *Pundamilia* exhibit different stages of speciation, from a single polymorphic panmictic population to well-differentiated sibling species, constituting a "speciation transect" (21). Furthermore, pieces of evidence have been uncovered for divergent/disruptive selection acting on male breeding color and opsin gene variants, as well as on eco-morphological traits (23, 24). Finally, the global pattern of genetic differentiation among populations suggested a parallel divergence between divergent eco-morphs off the shore of each island along the Manza Gulf (23). The example presented here is taken from a larger population genomic study aiming at identifying the dynamics of genomic differentiation along the gradient of speciation in *Pundamilia* (25).

3.1. Method

We studied four replicate pairs of divergent *Pundamilia* populations along this speciation transect using an AFLP genome-scan

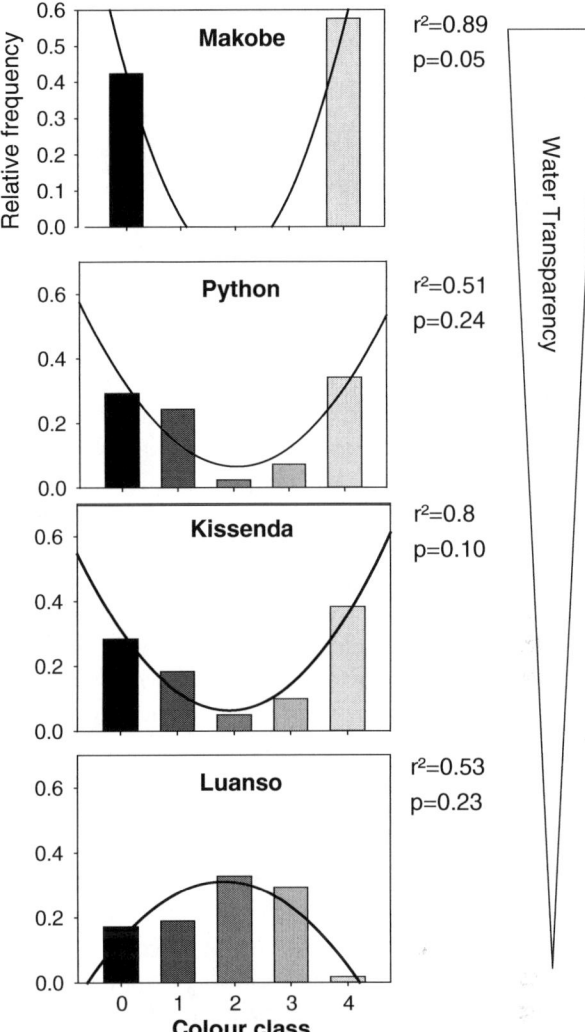

Fig. 1. Distribution of the color phenotypes within the analyzed samples of the two divergent *Pundamilia* species at each of the island studied along the speciation transect. *Fitted curves* are polynomial quadratic fits. The increase in relative frequency of class 2 (intermediate color phenotypes) with the decrease of water transparency and the collapse of populations into unimodal intermediate phenotypic distribution is notable from the quadratic fits.

based on 234 individuals and 520 loci to identify (a) signatures of divergent selection as well as (b) associations between genomic loci and eco-morphological traits under selection. For this purpose, we used a logistic regression approach combined with F_{ST}-outlier-based methods (see Note 1).

The investigation of genotype×phenotype association using logistic regression was conducted with the method implemented in the second version of MATSAM v2 (http://www.econogene.eu/software/sam). In parallel, we ran two F_{ST}-outliers detection methods, DFDIST (2, 3) and BAYESCAN (26). To allow a comparison between the three methods, we conducted a similar analysis independently with each of these software programs, and considered two distinct levels of detection for the divergent loci: one stringent corresponding to "significant" loci and the second–less conservative–corresponding to "marginally significant" loci (see Notes 2 and 3). Here, it is important to stress the fact that the determination of the chosen significance thresholds used to identify "significant" and "marginally significant" loci for each method is not rigorous but based on what is generally admitted in the literature. These significance thresholds are summarized in Table 2 (see Note 3).

Pundamilia males were randomly sampled off the shore of each island. Then individual male nuptial coloration was assessed on the basis of a 5-point color-scale system (Figs. 1 and 2), reflecting the increase of redness of the dorsal color pattern of the individual, ranging from "0" for completely blue phenotype to "4" for completely red-dorsum phenotype (23, 27). Analyses were then carried out based on either (a) the individual male "color-score" or (b) the color-score grouping reflecting "species" morphotypes. First, the analysis based on individual male color-score allows testing the relationship between precise morphotype and genetic variation (with individual-based approach, i.e., "color" variable). Second, the definition of divergent "species" categories based on color-categories (i.e., individuals scored as 0 or 1 included in a "blue group" (*P. pundamilia*), individuals scored as 3 or 4 included in a "red group" (*P. nyererei*), and individuals scored as 2 included in an "intermediate group") allows testing association between morphotype and genotype simultaneously with individual-based and population-based approaches (i.e., "species" variable, Fig. 2). In this second case, only the two extreme phenotype groups, representing divergent species, were used in population-based approaches to quantify the differentiation between divergent eco-morph populations, whereas the "intermediate group" was also considered for the individual-based approach (see Note 4).

We focused on the identification of signatures of divergent selection between the two *Pundamilia* species or eco-morphs. Analyses were then conducted (a) independently within each replicate pair of divergent sympatric populations (i.e., at the island level), to detect outlier loci within each of the four study islands; as well as (b) across all island populations grouped by color-morph (i.e., blue *P. pundamillia* vs. red *P. nyererei*), to detect global outliers over the entire study area. This led to five comparison tests in total. Such pattern of independent replicate divergences across closely related population-pairs with a very low level of hierarchical genetic structure appears particularly suitable for the detection of signatures of selection within and across populations (28, 29) (see Note 5).

3.2. D*fdist* and B*aye*S*can* Results

Over all five comparison tests, the combination of the two F_{ST}-outlier-based approaches allowed the identification of 49 loci potentially under selection with at least one method (Table 2). Among them, all loci detected with BayeScan were also detected with Dfdist (i.e., 15 loci, representing 31% of the detected outlier loci, Fig. 3),

Table 2
For the three analysis methods used in the *Pundamilia* genome-scan, comparison of (a) analysis parameters, (b) sample sets and grouping/tested variables, and (c) outlier loci detection results presented within and among methods, as well as the level of congruence between all pairs of methods

Methods	F_{ST}-outlier approach		Logistic regression
Software	DFDIST	BAYESCAN	MATSAM
Analysis parameters			
Detection thresholds			
Significant	$P<0.01$	$\log_{10}(BF)\geq 1$ (equivalent $P<0.24$)	$P<0.05$
Marginally significant	$P<0.05$	$\log_{10}(BF)\geq 0.5$ (equivalent $P<0.09$)	$P<0.1$
Additional detection parameters	Sequential background F_{ST} estimate	F_{IS} was estimated from microsatellites	Detection with both Wald and G-tests
Sample set and test variables			
- Comparison tests	Separately within each island and across all islands populations ($n=5$ tests in total)		
- Sample sets	The two groups of extreme morphotypes (excluding intermediate phenotypes)		All individuals (including intermediate)
- Analyzed variables	Species (based on color phenotype)		Species and color (habitat, depth, morphometrics)
Results			
Loci detection			
- per method (signif. + marg. signif.)	49 (17+32)	15 (8+7)	21 (11+10)
- with outlier methods	49		
- with the three methods	55		
Repeated detection			
- between pairs of populations	2	1	1
- across populations	11	5	5
- between DFDIST and BAYESCAN	15 (31%)		
- between DFDIST and MATSAM		15 (27%)	
- between BAYESCAN and MATSAM			11 (44%)

Fig. 2. Representation of the different male nuptial color phenotypes occurring in the two divergent *Pundamilia* species along the transect of speciation of the Mwanza Gulf; five discrete color-categories have been described relative to the increase of redness of the pattern; then the five color-scores correspond to (0) totally blue phenotype (absence of yellow and red colors), (1) yellow coloration on the flank (absence of red), (2) yellow flank with the presence of red along the lateral line, (3) yellow flank with a partially red dorsum, and (4) totally red dorsum. The color of the anal fin is not taken into account for the attribution of the color-score (for more details, see refs. 21 and 27).

Fig. 3. Relationship between heterozygosity (He) and locus-specific F_{ST} between the divergent *Pundamilia* populations computed using Dfdist, conducted independently overall four islands with divergent populations grouped by color-morphs across islands (**a**) as well as separately within each island (**b**–**e**). The loci detected by each of the three analysis methods are indicated: for Dfdist outliers at $P > 0.99$ and $P > 0.95$ are represented in *gray* and *black*, respectively; outlier loci detected by BayeScan with $\log_{10}(BF)$ 0.05 are marked by a *circle*; and loci for which an association with a phenotypic variable has been detected using MATSAM are marked by a *cross* (for Kissenda Island, association with depth and morphometric variables are represented separately).

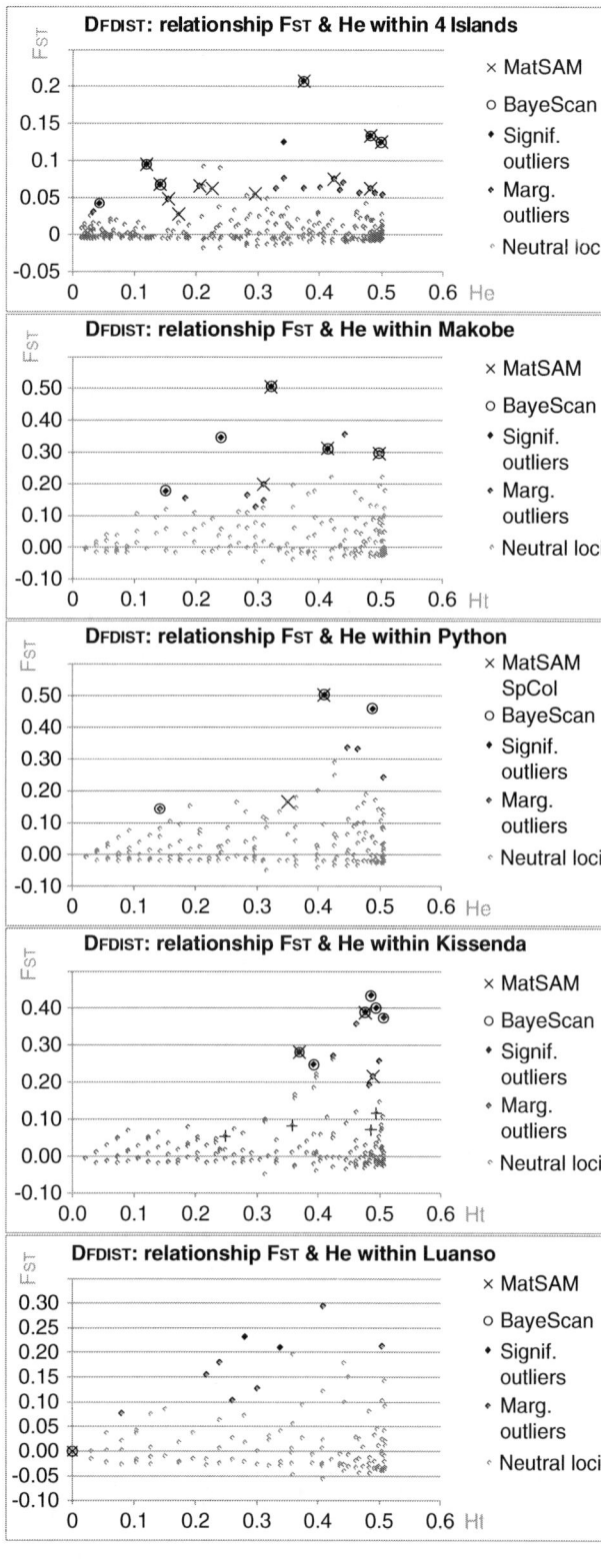

reflecting the relatively lower stringency of Dfdist, especially when the background genomic differentiation between population is low (26, 30, 31). Then 7–12 loci were detected with either one or both methods at the island level and 21 loci across all islands (Table 3 and Fig. 3). Furthermore, 11 outlier loci were detected repeatedly within a specific island as well as across all islands, and only two loci (i.e., 2G225 and 1G117) were detected repeatedly in two different islands, representing, respectively, 22% and 4% of the detected outlier loci (see Table 2).

3.3. Matsam Results

Logistic regression was conducted over the four study islands to identify associations between genomic loci and male nuptial coloration or species belonging. Additionally, within Kissenda Island, logistic regression was also conducted to test for genotypic association with habitat depth and 12 morphological variables.

Morphometric characters were considered as strictly continuous variables, color and species variables were encoded and analyzed as categorical ordinal variables (symmetrical parametrization, i.e., distribution of categories around the intermediate phenotype). For the purpose of methodological testing, habitat depth was analyzed either as a quantitative continuous or as a categorical variable (in the latter case with the reference set to 0 m). This allowed comparing the statistical power of multivariate logistic regression models - implied when using categorical variables (see Subheading 2.1) - with univariate models commonly used for continuous variables and theoretically expected to provide a higher detection power. Among all variables and comparison tests, the observation of a high proportion of detected associations with both univariate and multivariate models (61%) - while in 30% of the cases association was only detected in the context of univariate models and 9% in the context of multivariate models only - is in accordance with the slight reduction of detection power when using multivariate models.

Over the five comparison tests, associations were detected with 21 loci at significant or marginally significant levels for at least one model (Table 2): 17 loci with species or color variables, two with depth, and five with morphometric characters (see Fig. 4).

Furthermore, even if the majority of the loci showed an association with only one category of variable, three loci simultaneously exhibited an association with the color phenotype and a morphometric trait, probably due to the statistic association of these characters in the populations. The absence of co-association with any other type of character for the two loci associated with habitat depth (and their lack of detection as outlier loci, see Fig. 3) suggests a relatively wider independence of individual habitat adaptation from species belonging, compared to other phenotypic characters.

Zero to six loci were detected within each of the island-specific analyses, and 12 loci were detected across islands (Table 3 and Fig. 3). Furthermore, five loci were detected repeatedly at the

Table 3

Summary of environmental characteristics at the four islands along the Mwanza gulf transect, number of sampled individuals for each island, estimators of genetic diversity and differentiation, and number of potentially divergent loci detected by each method as well as by all three methods, and finally estimate of the fraction of genomic loci under selection

Locality	Makobe	Python	Kissenda	Luanso	All_Islands	Total
Code	Ma	Py	Ks	Lu	2Col	Loci
Islands environmental characteristics						
Water transparency[a]	225 ± 67	96 ± 21	78 ± 24	50 ± 10	—	—
Light slope[b]	8×10^{-3}	7.6×10^{-2}	7.9×10^{-2}	9.6×10^{-2}	—	—
Sample sets (number of individuals)						
Pundamilia nyererei	34	30	29	26	119	
Pundamilia pundamilia	25	26	28	14	93	
Intermediate	0	1	3	18	22	
Island community	59	57	60	58	234	
Genetic diversity and differentiation						
Number of polymorphic loci ($P<0.99$)	382	394	334	308	369	520
Detected divergent loci						
Dfdist	12	7	12	10	21	49
BayeScan	6	3	6	0	6	15
MATSAM	5	2	8 (+4)	0	12	21
Across all methods	12	8	13 (+4)	10	24	55
Percentage of divergent loci	2.31%	1.54%	2.5% (3.27%)	1.92%	4.62%	11%

[a]Secchi depth in centimeters

[b]The light slope is the steepness of the light gradient. It is calculated by regressing the transmittance orange ratio against the mean distance (in meters) from the shore, measured along the lake floor in transects

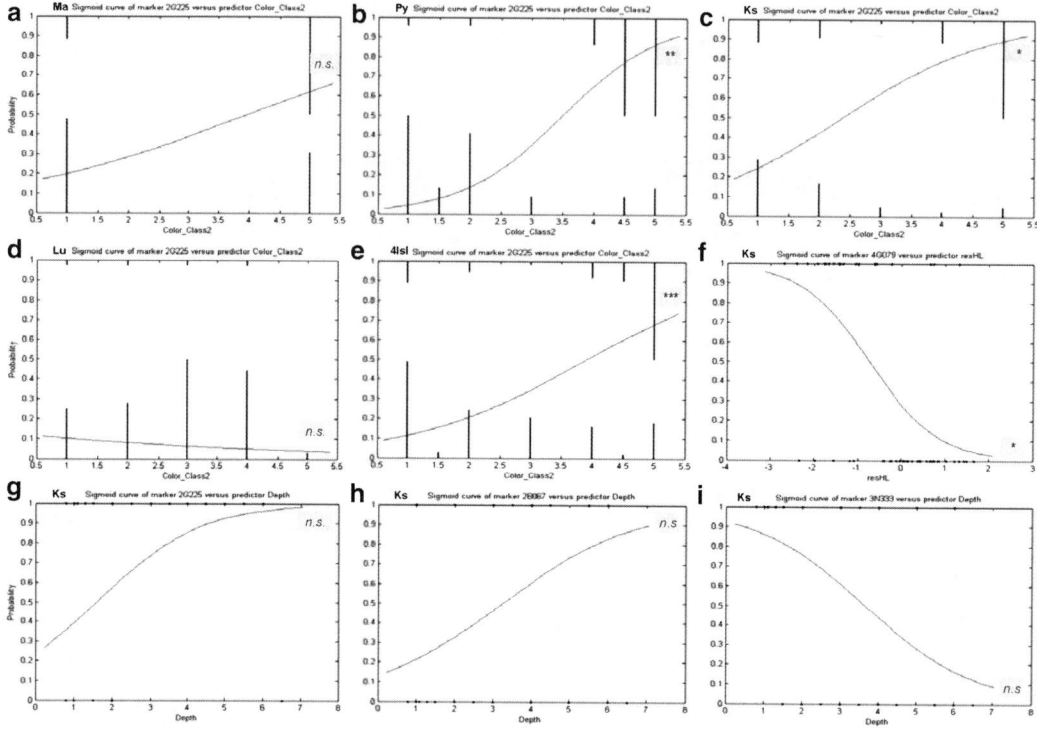

Fig. 4. Representation of different models of logistic regression to investigate association between locus genotype and phenotypic variables conducted with MATSAM; (**a–e**) test of association between genotype at locus 2G225 and individual color-score (considered as ordinal variable, ranging from 1 to 5) estimated independently within each island and across all islands; then within Kissenda Island populations, (**f**) test of association between genotype at locus 2G225 and a morphometric variable divergent between eco-morphs, the head length (HL), and (**g–i**) test of association between genotype at three loci (2G225, 2B067, and 3N333) and habitat depth. Respective levels of significance of the association are indicated for each model (***$P<0.001$, **$P<0.01$, *$P<0.05$, applying Bonferroni correction; *n.s.* nonsignificant).

island level as well as across island comparison tests, and only one locus (i.e., 2G225) was detected repeatedly in two different island comparison tests, representing, respectively, 24% and 5% of the detected outlier loci.

3.4. Discussion

When comparing the F_{ST}-outlier and logistic regression approaches, a very high proportion of loci potentially under selection was also significantly associated with at least one phenotypic variable studied, representing 78.6% of the loci detected by both F_{ST}-outlier methods. All genotype×phenotype associations detected here strictly involved phenotypic variables previously identified to be under divergent selection between the two *Pundamilia* species (24). This supports the capacity of the logistic regression approach to (a) identify genomic loci under selection by association with characters targeted by selection, and then to (b) identify loci involved in the genetic architecture of these traits.

This high proportion of genomic loci exhibiting both a signature of selection and association with divergent color-morph characters (73% of significant and marginally significant loci) suggests a predominant action of selection on male nuptial coloration in the divergence between *P. pundamilia* and *P.nyererei*, as further demonstrated by detailed population genomics study (25).

4. Notes

1. Aside from F_{ST}-outlier-based methods, logistic regression can also be used to identify genomic signature of selection, when testing associations between genomic variation and traits identified previously as subjected to selection (e.g., by F_{ST}/Q_{ST} analysis (32, 33)). Rather than identifying the signature of selection based on the comparison of local (locus-specific) differentiation with background genomic differentiation, logistic regressions can identify genomic regions associated with the different targets of selection.

   On condition that traits under selection were identified among study populations or species, this provides two sets of methods based on different assumptions, which can be used complementarily to identify signatures of selection. The congruent detection with both methods can then be taken as a strong support factor (i.e., low probability to detect false positives by both methods).

2. The difference in detection power between F_{ST}-outlier methods and ADM (i.e., logistic regression) is primarily due to the fact that the former are population-based approaches, while the later are individual-based approaches. In F_{ST}-outlier methods, the number of discrete populations sampled is particularly important to depict the global pattern of differentiation within the system, while the number of individuals (total or per population) will mostly affect the accuracy (i.e., confidence interval) of the differentiation estimators. On the other hand, in logistic regression, the detection power is mostly affected by the total number of individuals analyzed and their continuous distribution among the range of the values of the tested variables. Indeed, sampled individuals have to be continuously and homogeneously distributed over the study area, in order to (a) ensure a maximum environmental representativeness and maximize the chance to encompass contrasted environments and (b) avoid superimposition of similar associations (see Note 5). Consequently, more accuracy and statistical power is expected in ADM when carrying out analysis on a global scale than on a regional or local one.

3. As significance is estimated from very different statistical tests in each of the three approaches used, it appears difficult to determine objectively detection thresholds a priori for each method to allow similar stringency among them. In the present study, we selected empirical significant thresholds generally used in similar studies.

   The method implemented in MATSAM implies the realization of numerous logistic regression tests between all possible loci and explanatory variables. Such a design then requires correcting the raw significance threshold for multiple testing. The Bonferroni correction was chosen due to its conservativeness, expected to minimize the false discovery rate. However, this correction stringency is also likely to reduce the true discovery rate, especially if numerous loci are analyzed. To compensate these two antagonist parameters, it is necessary to explore the results obtained for a large and systematic number of different significance thresholds and observe the number of significant associations well beyond the classic 95% and 99% significance thresholds, and also with lower significance thresholds (correction for multiple comparisons always included). Alternatively, other correction methods for multiple testing are available as referenced in (13) and could be tested in this context.

4. F_{ST}-outlier methods and logistic regressions allow us to detect signatures of selection according to different types of "grouping variables." On the one hand, F_{ST}-outlier methods are based on interpopulation divergence measurements and require the analysis of "grouping variables" allowing a strict assignment of individuals to produce discrete populations (i.e., absence of intermediate individuals). On the other hand, logistic regression allows the analysis of "grouping variables" (i.e., predictors), which could also have a continuous or even overlapping distribution among populations. This is reflected by two variables in our case study. First, considering the "species" variable, individuals were assigned to three categories based on their nuptial coloration, the two extreme ones representing the strictly alternative morph/species groups, whereas the third (minority) one represented intermediate morphs. All three categories were then included in the analysis with logistic regression, while only the two extreme ones were included in the F_{ST}-based analyses (i.e., the intermediate one being excluded). Second, the analysis of the "habitat depth" was only possible with the logistic regression approach, due to the fact that individuals show a fully continuous and a widely overlapping distribution between the two species under study (i.e., it is impossible to cluster individuals into unambiguous discrete categories).

5. The existence of a hierarchical genetic structure among studied populations can generate an increase of the false discovery

rate if it is not taken into account, as previously shown in the case of F_{ST}-based approaches (34). Similar patterns are expected with logistic regression when the variability along explanatory variables cannot be disentangled from a strong (hierarchical or not) population genetic structure. In our case study, we detected a lower genetic divergence between sympatric populations than between allopatric conspecific ones. This genetic pattern of differentiation suggests independent replicated island-specific divergence. In such a case, adaptive divergence and background genetic structure are not confounded even at the global level, and consequently permits to obtain reliable results.

Acknowledgments

We would like to thank Geoffrey Dheyongera, Salome Mwaiko, and Isabel Magalhaes for their participation to the project. We are also grateful to Martine Maan, Alan Hudson, Kay Lucek, and Mathieu Foll for their contribution to fruitful scientific discussions about the issues presented in this chapter.

References

1. Lewontin RC, Krakauer J (1973) Distribution of gene frequency as a test of the theory of the selective neutrality of polymorphisms. Genetics 74:175–195

2. Beaumont MA, Nichols RA (1996) Evaluating loci for use in the genetic analysis of population structure. Proc R Soc Lond B 263: 1619–1626

3. Beaumont MA, Balding DJ (2004) Identifying adaptive genetic divergence among populations from genome scans. Mol Ecol 13:969–980

4. Joost S, Bonin A, Bruford MW et al (2007) A spatial analysis method (SAM) to detect candidate loci for selection: towards a landscape genomics approach to adaptation. Mol Ecol 16:3955–3969

5. Poncet BN, Herrmann D, Gugerli F et al (2010) Tracking genes of ecological relevance using a genome scan in two independent regional population samples of *Arabis alpina*. Mol Ecol 19:2896–2907

6. Mitton JB, Linhart YB, Hamrick JL, Beckman JS (1977) Observations on genetic structure and mating system of ponderosa pine in Colorado Front Range. Theor Appl Genet 51:5–13

7. Mitton JB, Sturgeon KB, Davis ML (1980) Genetic differentiation in ponderosa pine along a steep elevational gradient. Silvae Genet 29:100–103

8. Stutz HP, Mitton JB (1988) Genetic variation in *Engelmann spruce* associated with variation in soil moisture. Arctic Alpine Res 20: 461–465

9. Joshi J, Schmid B, Caldeira MC et al (2001) Local adaptation enhances performance of common plant species. Ecol Lett 4:536–544

10. Skøt L, Hamilton NRS, Mizen S, Chorlton KH, Thomas ID (2002) Molecular genecology of temperature response in *Lolium perenne*: 2. association of AFLP markers with ecogeography. Mol Ecol 11:1865–1876

11. Joost S (2006) The geographic dimension of genetic diversity: a GIScience contribution for the conservation of animal genetic resources. In: School of Civil and Environmental Engineering (ENAC), p. 178. No 3454, Ecole Polytechnique Fédérale de Lausanne (EPFL), Lausanne

12. Joost S, Kalbermatten M, Bonin A (2008) Spatial analysis method (SAM): a software tool combining molecular and environmental data

to identify candidate loci for selection. Mol Ecol Resour 8:957–960

13. Joost S, Bonin A (2007) Quantitative geography and genomics: spatial analysis to detect signatures of selection along a gradient of altitude in the common frog (*Rana temporaria*). In: Institute of Geography (ed) 15th European colloquium on theoretical and quantitative geography. University of Lausanne, Montreux, Switzerland

14. Pariset L, Joost S, Marsan PA, Valentini A, Consortium E (2009) Landscape genomics and biased F_{ST} approaches reveal single nucleotide polymorphisms under selection in goat breeds of North-East Mediterranean. BMC Genet 10

15. Tonteri A, Vasemagi A, Lumme J, Primmer CR (2010) Beyond MHC: signals of elevated selection pressure on Atlantic salmon (*Salmo salar*) immune-relevant loci. Mol Ecol 19:1273–1282

16. Parisod C, Joost S (2010) Divergent selection in trailing- versus leading-edge populations of *Biscutella laevigata*. Ann Bot 105:655–660

17. Freeland JR, Biss P, Conrad KF, Silvertown J (2010) Selection pressures have caused genome-wide population differentiation of *Anthoxanthum odoratum* despite the potential for high gene flow. J Evol Biol 23:776–782

18. Dobson AJ, Barnett AG (2008) An introduction to generalized linear models, 3rd edn. CRC. 307p. Boca Raton, Florida 307p

19. Hosmer DW, Lemeshow S (2000) Applied logistic regression, 2nd edn. Wiley Series in Probability and Statistics. 375p. Hoboken, New Jersey

20. Seehausen O, van Alphen JM (1999) Can sympatric speciation by disruptive sexual selection explain rapid evolution of cichlid diversity in Lake Victoria? Ecol Lett 2:262–271

21. Seehausen O (2009) Progressive levels of trait divergence along a "speciation transect" in the Lake Victoria cichlid fish Pundamilia. In: Butlin RK, Schluter D, Bridle J (eds) Speciation and patterns of diversity. Cambridge University Press. Cambridge, UK

22. Seehausen O, van Alphen JJM, Witte F (1997) Cichlid fish diversity threatened by eutrophication that curbs sexual selection. Science 277:1808–1811

23. Seehausen O, Terai Y, Magalhaes IS et al (2008) Speciation through sensory drive in cichlid fish. Nature 455:620–623

24. Magalhaes IS, Mwaiko S, Schneider MV, Seehausen O (2009) Divergent selection and phenotypic plasticity during incipient speciation in Lake Victoria cichlid fish. J Evol Biol 22:260–274

25. Bezault E, Dheyongera G, Mwaiko S, Magalhaes I, Seehausen O. (in prep.) Genomic signature of divergent adaptation along replicated environmental gradients in Lake Victoria cichlid fish

26. Foll M, Gaggiotti O (2008) A genome-scan method to identify selected loci appropriate for both dominant and codominant markers: a Bayesian perspective. Genetics 180:977–993

27. Seehausen O (1997) Distribution of and reproductive isolation among color morphs of a rock-dwelling Lake Victoria cichlid (*Haplochromis nyererei*). Ecol Freshw Fish 6:57

28. Wilding CS, Butlin RK, Grahame J (2001) Differential gene exchange between parapatric morphs of *Littorina saxatilis* detected using AFLP markers. J Evol Biol 14:611–619

29. Campbell D, Bernatchez L (2004) Generic scan using AFLP markers as a means to assess the role of directional selection in the divergence of sympatric whitefish ecotypes. Mol Biol Evol 21:945–956

30. Caballero A, Quesada H, Rolan-Alvarez E (2008) Impact of amplified fragment length polymorphism size homoplasy on the estimation of population genetic diversity and the detection of selective loci. Genetics 179:539–554

31. Nielsen EE, Hemmer-Hansen J, Poulsen NA et al (2009) Genomic signatures of local directional selection in a high gene flow marine organism; the Atlantic cod (*Gadus morhua*). BMC Evol Biol 9 , 276p. doi:10.1186/1471-2148-9-276

32. Spitze K (1993) Population-structure in *Daphnia obtusa* – quantitative genetic and allozymic variation. Genetics 135:367–374

33. Whitlock MC (2008) Evolutionary inference from QST. Mol Ecol 17:1885–1896

34. Excoffier L, Hofer T, Foll M (2009) Detecting loci under selection in a hierarchically structured population. Heredity 103:285–298

Chapter 18

Genomic Scan as a Tool for Assessing the Genetic Component of Phenotypic Variance in Wild Populations

Carlos M. Herrera

Abstract

Methods for estimating quantitative trait heritability in wild populations have been developed in recent years which take advantage of the increased availability of genetic markers to reconstruct pedigrees or estimate relatedness between individuals, but their application to real-world data is not exempt from difficulties. This chapter describes a recent marker-based technique which, by adopting a genomic scan approach and focusing on the relationship between phenotypes and genotypes at the individual level, avoids the problems inherent to marker-based estimators of relatedness. This method allows the quantification of the genetic component of phenotypic variance ("degree of genetic determination" or "heritability in the broad sense") in wild populations and is applicable whenever phenotypic trait values and multilocus data for a large number of genetic markers (e.g., amplified fragment length polymorphisms, AFLPs) are simultaneously available for a sample of individuals from the same population. The method proceeds by first identifying those markers whose variation across individuals is significantly correlated with individual phenotypic differences ("adaptive loci"). The proportion of phenotypic variance in the sample that is statistically accounted for by individual differences in adaptive loci is then estimated by fitting a linear model to the data, with trait value as the dependent variable and scores of adaptive loci as independent ones. The method can be easily extended to accommodate quantitative or qualitative information on biologically relevant features of the environment experienced by each sampled individual, in which case estimates of the environmental and genotype × environment components of phenotypic variance can also be obtained.

Key words: Amplified fragment length polymorphism, Genetic determination, Genomic scan, Heritability, Multiple linear regression, Phenotypic variance components

1. Introduction

The study of contemporary evolution in wild populations poses two major challenges, namely, quantifying selection on the phenotypic traits of interest (phenotypic selection) and evaluating the extent to which individual variation in these traits is due to genetic differences (heritability) (1). Phenotypic selection occurs when individuals with different phenotypic characteristics differ

François Pompanon and Aurélie Bonin (eds.), *Data Production and Analysis in Population Genomics: Methods and Protocols*, Methods in Molecular Biology, vol. 888, DOI 10.1007/978-1-61779-870-2_18, © Springer Science+Business Media New York 2012

in their fitness because of differences in viability and/or fecundity. Following the development of a theoretical framework and associated analytical toolbox for estimating the shape and strength of selection on quantitative traits (1–5), studies of phenotypic selection on wild populations have proliferated in the past 30 years, and have shown that phenotypic selection is widespread in nature (reviews in refs. 6–8). Nevertheless, "natural selection is not evolution" ((9), p. vii), and changes in phenotypic value distributions arising from selection can be constrained or altered by the pattern of inheritance of the traits of interest. Phenotypic selection studies on wild populations usually fall short of demonstrating ongoing evolutionary change because of the absence of information on the genetic basis of variation in the phenotypic traits of interest ((7); but see, e.g., refs. 10–12 for some exceptions), a limitation that stems from the problems involved in estimating heritability and genetic correlations under natural field conditions (13).

Traditional quantitative genetics techniques to estimate heritability rely on information about the relationships among individuals, generally in the form of a pedigree structure or a matrix of pairwise individual relatedness (14, 15). Except for some special cases, as when pedigrees can be inferred from detailed observations of matings (16), neither pedigrees nor information on individual relatedness are available for natural populations. This difficulty has been customarily circumvented by artificially breeding individuals of known pedigree through controlled crosses between field-collected progenitors of known phenotypes, and keeping the experimental populations in controlled artificial environments. Many organisms, however, are not amenable to these methods. Conventional quantitative genetics methods based on experimental crossing schemes are impractical or impossible with long-lived species that cannot be artificially crossed, propagated vegetatively or kept in captivity, or with those that reproduce irregularly or have long periods until first reproduction. In addition, the expression of quantitative traits usually differ between environments, and heritability estimates may differ widely between wild and artificial populations of the same species (17), hence field-based estimates are essential to investigate the possibility for evolution in the settings where the organisms naturally occur (18).

Alternative methods for estimating trait heritability in wild populations have been developed in recent years which take advantage of the increased availability of genetic markers to reconstruct pedigrees or to estimate relatedness between individuals (reviewed in refs. 19–21). These methods, however, are not a panacea, as their application also faces a number of difficulties. Most of these procedures either make assumptions on the structure of relatedness in the population (e.g., assuming known, discrete classes of relatedness) or require additional information on parentage of individuals, which greatly limits their application to complex natural populations where individuals may differ widely in age and relatedness, as in uneven-aged populations of long-lived organisms. Ritland regression procedure (18, 22, 23), which is based on assessing the strength of the relationship between pairwise phenotypic similarity and relatedness for the individuals in a population, stands apart from the rest of marker-based methods in that it allows direct estimation of quantitative genetic parameters without requiring specification of an explicit pedigree or prior knowledge of population structure (19).

At first view, Ritland method appears widely applicable, since it is based on the assessment of the simple relationship between phenotype and genotype, and it has been used to estimate genetic parameters for wild, unpedigreed populations (24–27). Nevertheless, this method has two intrinsic limitations (see ref. 19 for a detailed discussion). First, since it depends heavily on the efficiency of the estimation of pairwise relatedness, the large statistical errors associated with these estimators will produce broad standard errors around genetic parameter estimates. Second, the method requires the existence of significant variance in relatedness in the sample analyzed (18, 22, 23). The limited number of applications of Ritland method and their modest success are perhaps attributable to the joint influence of these intrinsic limitations, and despite its appeal the utility of the method may be problematic (19, 25, 28–30). In the same pessimistic vein, it has been also suggested that, despite the increased availability of genetic markers, marker-based methods in general might not be as useful for genetic parameters estimation as previously thought (13, 19).

A new marker-based technique has been recently proposed by Herrera and Bazaga (31) which, by focusing on individual genotypes and adopting a genomic scan approach, avoids the problems associated with estimating pedigrees or individual relatedness. Although this method is not necessarily aimed at the direct estimation of quantitative genetic parameters, it does provides an alternative, relatively easy way for addressing one of the key aspects in evolutionary studies, namely, the quantitative assessment of the genetic component of phenotypic variance in wild populations. As stressed by van Kleunen and Ritland (26), "even the possibility of determining the presence of heritable genetic variation is a big improvement for studies testing for the potential for evolution in natural populations." The motivation and a step-by-step description of the genomic scan-based method of Herrera and Bazaga are presented in this chapter, followed by an example of application and a comparison with results from the Ritland procedure for the same data.

2. Overview and Motivation

A trait can be "hereditary" in the sense of being transmitted from parent to offspring or in the sense of being determined by the genotype. Under the first meaning, the heritability of a character is "the relative importance of heredity in determining phenotypic values," and is known as heritability in the narrow sense (14). Marker-based methods for estimating heritability mentioned in the preceding section are framed in this specific context, hence their reliance on pedigree reconstruction or relatedness estimation. Under the second meaning, a trait's heritability is simply "a quantity defined as the proportion of phenotypic variance that is genetically determined" (32). This is the extent to which individual phenotypes are determined by the genotypes, and is known as heritability in the broad sense or degree of genetic determination (14). The method described in this chapter is framed in this second context, as it focuses on quantifying the fraction of population-wide phenotypic variance that is attributable to genetic differences between individuals. These genetic differences are evaluated directly through the application of a genomic scan approach at the within-population scale. Rather than targeting the narrow sense heritability of the phenotypic trait of interest by inferring the relatedness of individuals and then relating it to their phenotypic similarity, as done with the Ritland method, the present method directly proceeds to estimate the proportion of observed phenotypic variance that can be statistically accounted for by adaptive genetic differences between individuals, elucidated through a variant of explorative genomic scan tailored for within-population use.

Consider a population for which a sample of individuals are simultaneously characterized with regard to some quantitative phenotypic trait and a large number of anonymous, putatively neutral genetic markers. If the number of markers is large and are randomly distributed across the whole genome, a certain proportion of the markers is expected to be in linkage disequilibrium with non-neutral, adaptive loci having some causative effect on the phenotypic value of the trait of interest. It should therefore be possible to single out these non-neutral markers by searching for significant statistical associations between markers and phenotypic values across individuals, an approach usually known as association mapping or linkage disequilibrium mapping (33–35). This method has been used, for example, to infer associations between quantitative trait loci (QTLs) and phenotypic traits, and the same underlying principles can motivate genome screenings aimed at identifying functional associations between anonymous markers and phenotypic traits. Once a set of non-neutral (or linked to non-neutral) markers has been identified which are significantly correlated with phenotypic values, statistical models can be fitted to the data which estimate the proportion of phenotypic variance in the sample statistically accounted for by the multilocus differences between individuals in these non-neutral loci.

The success of the method will depend on whether non-neutral genomic regions that are causatively associated with the phenotypic trait of interest, or adjacent regions in linkage disequilibrium, are "hit" by some of the anonymous markers chosen, so that genotype–phenotype correlations across individuals can be revealed. Selection of markers is thus critical. The larger the number of markers and the more thorough its distribution across the genome, the higher the expected probability of identifying non-neutral genomic regions causatively associated with the phenotypic trait of interest. The amplified fragment length polymorphism (AFLP) technique (36) allows for a virtually unlimited number of markers without prior sequence knowledge. Detailed linkage maps have also revealed that AFLP markers are thoroughly distributed over the whole genome (37–40). These features make AFLP markers particularly suited to genomic scan and linkage disequilibrium mapping approaches with non-model organisms (35, 41, 42), and also render them the markers of choice for the method described in this chapter. In the immediate future, large single-nucleotide polymorphism (SNP) datasets made available for non-model species by next-generation sequencing will provide additional opportunities for the application of the method described here, with the added advantage that SNPs could be easily selected in or near coding regions of the genome.

The method described here rests on the assumption that the set of loci found to be statistically related to individual variation in phenotypic traits are functionally, causatively related to such variation, or linked to loci functionally affecting such variation, and can therefore be used as valid descriptors of individual genotypes. In the case of AFLP markers, and here lies yet another reason supporting their use, three lines of evidence show that this assumption will probably hold true in most instances: (a) high-density linkage maps often reveal a close proximity of AFLP markers to QTLs or genes with known functionality (43–46); (b) AFLP allelic frequencies are responsive to artificial selection on quantitative traits (47–49); and (c) studies on populations of the same species combining phenotypic information with population-genomic scans often reveal concordant phenotypic and AFLP variation (50, 51).

3. Methods

In the following, it is assumed that the analysis focuses on a single quantitative phenotypic trait that has been measured in a randomly chosen sample of individuals from a single study population.

The same individuals are also fingerprinted using AFLP markers, so that phenotypic trait values and multilocus AFLP data are simultaneously available.

3.1. AFLP Fingerprinting

As noted above, using a large number of markers is essential to increase the chances of hitting at or near (linked) genomic regions causatively associated with the phenotypic trait under consideration. Proportions of non-neutral loci reported by previous AFLP-based genomic scans of natural populations of non-model organisms mostly fall between 1 and 5% of the polymorphic markers assayed ((35, 51) and references therein), which suggests the rule of thumb that individuals should be fingerprinted using a minimum of several hundred polymorphic AFLP markers. The probability of success will be enhanced if this large number of markers is achieved by selecting a variety of primer pairs for each of several combinations of restriction enzymes, rather than by selecting a single enzyme combination and a few primer pairs each yielding many markers. This recommendation is motivated by the observation that non-neutral AFLP markers often are nonrandomly distributed and clustered on particular primer pairs (51–53) or restriction enzyme combinations (31). AFLP markers obtained using $EcoRI$ and $PstI$ as rare cutter enzymes often differ widely in their distribution across genomes (38, 45, 54, 55), thus using a combination of $EcoRI$–$MseI$ and $PstI$–$MseI$ primer pairs will generally be a better starting off strategy than using primer pairs based on only one rare cutter. This is illustrated by results of the study by Herrera and Bazaga (31) on the wild violet $Viola\ cazorlensis$, where AFLP markers significantly associated with individual fecundity were about three times more frequent among $PstI$-based (4.4% of polymorphic loci) than among $EcoRI$-based combinations (1.6% of polymorphic loci).

3.2. Checking for Cryptic Population Substructuring

In natural populations, marker–phenotype associations may not reflect an underlying causal relationship. Significant marker–phenotype associations can be caused by admixture between sub-populations that have different allele frequencies at marker loci and different values of the trait (33, 56). Such situation might occur, for example, if the study population comprises several genetically distinct subgroups that originated from independent colonization events and differ in average phenotypes. The possibility of spurious marker–phenotype associations due to such cryptic population substructuring should therefore be confidently ruled out prior to conducting any further analyses. This can be accomplished with any of the model-based Bayesian assignment methods currently available for the identification of admixture events using multilocus molecular markers, as implemented, for example, in the programs structure (57–59) or baps (60, 61).

If results of these preliminary analyses do not suggest any cryptic substructuring, then it would be justified to proceed to the analytical step in Subheading 3.3. Alternatively, if there are reasons to suspect that the study population is cryptically substructured, one should further check whether genetic subgroups arising from the Bayesian analysis are phenotypically homogeneous (e.g., by comparing trait means across subgroups). If subgroups are phenotypically similar, despite genetic distinctiveness, it would be safe to proceed to the next step. If subgroups were found to differ phenotypically, it would still be possible to proceed by splitting the sample and conducting separate analyses on each subgroup. This would allow for testing the intriguing possibility that genetically distinct subgroups differ in the degree of genetic determination of the phenotypic trait under study.

3.3. Identifying Phenotype-Related Markers

To identify the AFLP markers associated with the phenotypic trait under consideration, separate logistic regressions will be run for each polymorphic marker, using band presence as the dependent binary variable and the phenotypic trait as the continuous, independent one. The statistical significance of each marker–trait relationship will be assessed by consideration of the p-value obtained from a likelihood ratio test. Although several tests are available for assessing the significance of logistic regressions, likelihood ratio tests are more powerful and reliable for sample sizes often used in practice (62). Given the large number of logistic regressions involved (as many as polymorphic AFLP markers), precautions must be taken to account for the possibility of obtaining by chance alone an unknown number of false significant regressions (see Note 1).

3.4. Estimating the Genetic Component of Variance

In this step, the combined phenotypic effects of all the AFLP markers found in Subheading 3.3 to be significantly related to the trait under consideration ("adaptive loci" hereafter) is estimated by fitting a multiple linear regression to the individual data, using trait value as the dependent variable (y) and all adaptive loci as independent ones (x_i; $i=1$ to k significant markers), coded as binary scores. In multiple linear regression, a response variable is predicted on the basis of an assumed linear relationship with several independent variables. The combined explanatory value of the x_i variables for predicting y is assessed with the multiple correlation coefficient R, and R^2 is known as the coefficient of determination or the squared multiple correlation (63). The R^2 from the fitted multiple regression linking the phenotypic trait with the set of adaptive loci can therefore be properly interpreted as an estimate of the proportion of phenotypic variance in the trait that is jointly explained by the additive, linear effects of adaptive loci scores. In other words, the regression R^2 provides an estimate of the degree of genetic determination, heritability in the broad sense, or genetic component of variance, of the focal phenotypic trait. The overall statistical significance of the multiple linear regression can be determined using an ordinary F test.

Three aspects must be noted at this point. First, it is extremely unlikely that most or even a sizable fraction of the genomic regions influencing a given phenotypic trait (or closely linked to such regions) are hit in a genomic screening involving only several hundred markers. Estimates of the degree of genetic determination obtained with this method are thus expected to be biased downwards, because there will always be an unknown number of genomic regions whose effects on the phenotype are left out from the regression, and because adding independent variables x_i to a linear model always increases, never decreases, the value of R^2 (63).

Second, it follows from linear models' properties that, unless all the x's are mutually orthogonal, the R^2 of the multiple regression cannot be partitioned into k components each uniquely attributable to an x_i (63). In the present context, this would imply that the allelic states of adaptive loci entering the model are uncorrelated (i.e., completely unlinked). If this condition is proven for the particular set of individuals sampled, then the contributions of individual loci to phenotypic variance could be inferred from the multiple regression, for example, by using some stepwise fitting procedure. Alternatively, if the x's are statistically non-independent (i.e., in linkage disequilibrium), this property of linear models leads one to predict that the regression R^2 will be robust to the multicollinearity of the independent variables.

And third, if the number of individuals sampled is about the same order of magnitude than the number k of adaptive loci included as independent variables in the regression, then it is possible to obtain a large value of R^2 that is not meaningful. In this case, some x's that do not contribute independently to predicting y may appear to do so, and the regression may provide a spurious estimate of the degree of genetic determination. To correct at least in part for this tendency, the adjusted R^2 should be used instead of the ordinary uncorrected R^2 (63). An additional way to circumvent the problem of a large number of adaptive loci relative to sample size is to reduce the dimensionality of the (adaptive) multilocus genotype of individuals. This can be achieved by obtaining individual scores on the first few axes of a principal coordinates analysis of the pairwise genetic distance matrix (e.g., using the program genalex; (64)), and then using these scores as independent variables in the multiple regression rather than the original binary scores of adaptive loci. Reducing the dimensionality of the multilocus genotypes of individuals based on adaptive loci can also be useful for other purposes, as shown in the next section.

3.5. Possible Enhancements: Environmental and Genotype × Environment Components of Phenotypic Variance

The multiple regression method described in Subheading 3.4 will estimate the genetic component of phenotypic variance using exclusively information on the phenotypes and marker-defined genotypes of sampled individuals. But some study systems allow gathering information on features of the environment naturally experienced by the individuals sampled. In these cases, some enhancements of the method are possible which permit to evaluate also the environmental and genotype × environment components of phenotypic variance. Consider, for example, a wild-growing plant population where individuals differ in the microclimate (e.g., temperature, humidity) and soil physicochemical properties of growing sites. Because of the individuals' phenotypic plasticity (the ability of a given genotype to produce different phenotypes when exposed to different environments; (65)), environmental heterogeneity of the individuals' growing sites will generate a certain amount of phenotypic variance, which clearly corresponds to an environmental variance component. In addition, a genotype × environment variance component will also arise whenever individual plants (genotypes) respond differently to variations in environmental factors, such as temperature or soil fertility, because of differences in reaction norms (65).

The proportions of genetic, environmental, and genotype × environment components of phenotypic variance can be simultaneously assessed if information on the environment experienced by individuals is available in addition to their phenotypes and AFLP fingerprints. This can be achieved by fitting a random effect linear model to the individual data, using the focal phenotypic trait as the response variable, and genotype (as described by the set of adaptive loci), environment (as described by biologically relevant parameters), and genotype × environment as independent variables, all treated as random effects. By using a restricted maximum likelihood method (REML), estimates of the genetic, environmental, and genotype × environment components of phenotypic variance of sampled individuals under natural field conditions can be obtained from this model. Computations can be performed using the procedure MIXED in the SAS statistical package, specifying a VC or "variance components" type of covariance structure, which assumes a different variance component for each random effect in the model.

Unless the number of individuals sampled is really large, which is an unlikely situation in most field studies, the dimensionality of the genetic and environmental data should be reduced prior to fitting the mixed model, since otherwise a large number of variance components would be non-estimable. This would result from the large number of model parameters that should be estimated if separate genotype × environment effects are modeled for each adaptive locus. Dimensionality of the multilocus genotype of individuals can be reduced by obtaining individual scores on the first few axes of a principal coordinates analysis on the genetic distance matrix, as described in Subheading 3.4 above. The method for reducing the dimensionality of environmental variables will depend on the number and nature (discrete vs. continuous variables) of the environmental parameters associated with each individual. For sets of quantitative environmental data (e.g., microclimate parameters), scores on the first axes from a principal component analysis could be used as descriptors of individual environments. For sets of qualitative environmental data (e.g., compass orientation, substrate type), individual scores on the axes from a nonmetric multidimensional scaling ordination could provide reduced-dimensionality descriptors of individual environments (see ref. 66 for a review of multivariate applications to genetic marker data).

4. Application to a Case Study

The methods described above were applied by Herrera and Bazaga (31) to investigate the genetic basis of individual variation in long-term cumulative fecundity in a sample of 52 wild-growing individuals of the long-lived Spanish violet *Viola cazorlensis* monitored for 20 years. Plants were fingerprinted using eight *Eco*RI + 3 / *Mse*I + 3 and eight *Pst*I + 2 / *Mse*I + 3 primer combinations, which yielded a total of 365 polymorphic AFLP markers. The same plants were also characterized for quantitative floral traits including length of the floral peduncle, floral spur, and upper, middle, and lower petal blades, and an earlier study revealed significant phenotypic selection on most of these floral features (67). Application of the genomic scan method to evaluate the genetic basis of individual variation in these traits will provide insights on the possibility of evolutionary change in floral characteristics in that population.

Results are shown in Table 1, where the genetic determination estimate for 20-year cumulative fecundity for the same individuals is also included (31). Between 2 and 4 AFLP loci were significantly associated with individual variation in floral traits, and multiple regressions between floral traits and their associated AFLP loci were all statistically significant, which reveals that individual variation in all floral traits had a significant genetic basis. The number of statistically significant AFLP loci was lower for floral traits than for fecundity. This difference may reflect a real disparity between the two types of phenotypic traits in the number of genomic regions underlying their variation, but could also be due to a lower statistical power for detecting significant loci in the case of floral traits, e.g., because of lower phenotypic variance. The greater phenotypic variability of individual fecundity (coefficient of variation, CV = 138%) in comparison to floral traits (CV range = 8–23%) is

Table 1
Application of the genomic scan and Ritland methods for evaluating the genetic component of phenotypic trait variance in a wild population of the violet *Viola cazorlensis*

Phenotypic trait	Significant AFLP loci	R^2	F-value	p-Value	Ritland method (h^2)[a]
	Genomic scan method (regression between trait and significant loci)				
Floral peduncle	2	0.383	16.85	<0.0001	−0.042 (−0.416)
Floral spur	4	0.353	7.96	<0.0001	0.064 (−0.204)
Upper petal	3	0.279	7.59	0.0003	0.178 (−0.380)
Middle petal	2	0.190	6.99	0.0021	0.102 (−0.477)
Lower petal	2	0.199	7.33	0.0016	0.110 (−0.470)
20-year fecundity	11	0.631	7.00	< 0.0001	0.471 (−0.076)

[a]The lower 5% quantile of the bootstrap distribution is given in parentheses. For the heritability value to be significant at $p < 0.05$, this value must be larger than zero (26)

compatible with the latter interpretation. Anyway, these results for *V. cazorlensis* suggest that even modest numbers of significant AFLP loci can be sufficient to estimate the genetic component of phenotypic variance. Furthermore, the demonstration that individual variation in floral traits has a genetic basic provides the necessary complement to earlier results showing phenotypic selection on these traits (67), and suggests that ongoing selection is likely to promote evolutionary change of the average floral phenotype at the study population.

Estimates of the genetic component of phenotypic variation obtained with the genomic scan and Ritland methods are not strictly comparable, because they correspond to two different concepts of heritability as noted earlier. Keeping this caveat in mind, it is still interesting to compare Ritland h^2 and genomic scan-based R^2 figures for the same dataset and traits, to see whether the general conclusions reached by the two methods are similar. Ritland h^2 estimates for the six traits studied in the *Viola cazorlensis* dataset obtained with the program mark (68) are shown in Table 1. Estimates vary widely among traits, but none of them reaches statistical significance (i.e., $h^2 > 0$) at the 0.05 level. Ritland h^2 and genomic-scan R^2 values are correlated across traits, although the relationship does not reach statistical significance ($r = 0.655$, $N = 6$, $p = 0.16$). For this dataset, therefore, Ritland and genomic scan methods provide contrasting conclusions regarding the genetic component of phenotypic variation. While Ritland estimates would point to the absence of a significant additive genetic component underlying the variation of floral traits, the genomic scan method does provide evidence of a significant genetic component in all cases, as generally found in other species (7). In addition, genomic-scan R^2 values for *Viola cazorlensis* in Table 1 are comparable to broad-sense heritability estimates obtained for morphological and life history traits in other species of *Viola* using quantitative genetics methods (69).

5. Note

1. The possibility of obtaining by chance alone an unknown number of false significant phenotype-markers logistic regressions (false positives or type I errors) could be controlled by using some simple correction to the *p*-value threshold required for significance, such as Bonferroni-style corrections (70, 71). Nevertheless, Bonferroni-style corrections can be overly conservative, and the more-sophisticated approach proposed by Storey and Tibshirani (72) is better suited to genome-wide studies, where the number of simultaneous tests can be very large. In this method, a measure of statistical significance called the *q*-value is associated with each separate test. The *q*-value is similar to the well-known *p*-value, except it is a measure of significance in terms of the false-discovery rate rather than the false-positive rate. In the present context, the *q*-value obtained for a given marker is the expected proportion of false positives incurred for the whole set of regressions when the logistic regression for that particular marker is considered as statistically significant. Using the qvalue package (72) for the *R* environment (73), *q*-values for all the marker–trait regressions can be computed, ranked, and the largest *q*-value leading to an expectation of less than one falsely significant regression [i.e., *q*-value×(number of regressions accepted as significant)<1] used as the threshold for determining the statistical significance of marker–trait associations. In this way, the statistical power for detecting as many significant regressions as possible is enhanced while keeping with the conservative constraint of avoiding any false positive.

Acknowledgments

I am grateful to Pilar Bazaga for her invaluable contribution to the development of the method described here; Antonio Castilla for discussion and comments on the manuscript; and Aurélie Bonin and François Pompanon for the opportunity to contribute to this volume. Work supported by grant CGL2006-01355 (Ministerio de Educación y Ciencia, Gobierno de España).

References

1. Endler JA (1986) Natural selection in the wild. Princeton University Press, Princeton

2. Lande R, Arnold SJ (1983) The measurement of selection on correlated characters. Evolution 37:1210–1226

3. Arnold SJ, Wade MJ (1984) On the measurement of natural and sexual selection: theory. Evolution 38:709–719

4. Arnold SJ, Wade MJ (1984) On the measurement of natural and sexual selection: applications. Evolution 38:720–734

5. Manly BF (1985) The statistics of natural selection. Chapman and Hall, London

6. Kingsolver JG, Hoekstra HE, Hoekstra JM et al (2001) The strength of phenotypic selection in natural populations. Am Nat 157:245–261

7. Geber MA, Griffen LR (2003) Inheritance and natural selection on functional traits. Int J Plant Sci 164:S21–S42

8. Kingsolver JG, Pfennig DW (2007) Patterns and power of phenotypic selection in nature. Bioscience 57:561–572

9. Fisher RA (1958) The genetical theory of natural selection, 2nd edn. Dover, New York

10. Andersson S (1996) Floral variation in *Saxifraga granulata*: phenotypic selection, quantitative genetics and predicted response to selection. Heredity 77:217–223

11. Campbell DR (1996) Evolution of floral traits in a hermaphroditic plant: field measurements of heritabilities and genetic correlations. Evolution 50:1442–1453

12. Conner JK (1997) Floral evolution in wild radish: the roles of pollinators, natural selection, and genetic correlations among traits. Int J Plant Sci 158:S108–S120

13. Moore AJ, Kukuk PF (2002) Quantitative genetic analysis of natural populations. Nat Rev Genet 3:971–978

14. Falconer DS, MacKay TFC (1996) Introduction to quantitative genetics, 4th edn. Longman, Essex

15. Lynch M, Walsh B (1998) Genetics and analysis of quantitative traits. Sinauer, Sunderland

16. Garant D, Kruuk LEB, McCleery RH et al (2004) Evolution in a changing environment: a case study with great tit fledging mass. Am Nat 164:E115–E129

17. Roff DA (1997) Evolutionary quantitative genetics. Chapman and Hall, New York

18. Ritland K (1996) A marker-based method for inferences about quantitative inheritance in natural populations. Evolution 50:1062–1073

19. Garant D, Kruuk LEB (2005) How to use molecular marker data to measure evolutionary parameters in wild populations. Mol Ecol 14:1843–1859

20. Fernández J, Toro MA (2006) A new method to estimate relatedness from molecular markers. Mol Ecol 15:1657–1667

21. Pemberton JM (2008) Wild pedigrees: the way forward. Proc R Soc B 275:613–621

22. Ritland K (2000) Marker-inferred relatedness as a tool for detecting heritability in nature. Mol Ecol 9:1195–1204

23. Ritland K (2000) Detecting inheritance with inferred relatedness in nature. In: Mousseau TA, Sinervo B, Endler J (eds) Adaptive genetic variation in the wild. Oxford University Press, Oxford

24. Ritland K, Ritland C (1996) Inferences about quantitative inheritance based on natural population structure in the yellow monkeyflower, *Mimulus guttatus*. Evolution 50: 1074–1082

25. Klaper R, Ritland K, Mousseau TA, Hunter MD (2001) Heritability of phenolics in *Quercus laevis* inferred using molecular markers. J Hered 92:421–426

26. van Kleunen M, Ritland K (2005) Estimating heritabilities and genetic correlations with marker-based methods: an experimental test in *Mimulus guttatus*. J Hered 96:368–375

27. Andrew RL, Peakall R, Wallis IR et al (2005) Marker-based quantitative genetics in the wild?: the heritability and genetic correlation of chemical defenses in eucalyptus. Genetics 171:1989–1998

28. Thomas SC, Coltman DW, Pemberton JM (2002) The use of marker-based relationship information to estimate the heritability of body weight in a natural population: a cautionary tale. J Evol Biol 15:92–99

29. Wilson AJ, McDonald G, Moghadam HK et al (2003) Marker-assisted estimation of quantitative genetic parameters in rainbow trout, *Oncorhynchus mykiss*. Genet Res 81:145–156

30. Rodríguez-Ramilo ST, Toro MA, Caballero A, Fernández J (2007) The accuracy of a heritabil-

ity estimator using molecular information. Conserv Genet 8:1189–1198

31. Herrera CM, Bazaga P (2009) Quantifying the genetic component of phenotypic variation in unpedigreed wild plants: tailoring genomic scan for within-population use. Mol Ecol 18: 2602–2614

32. Hedrick PW (2005) Genetics of populations, 3rd edn. Jones and Bartlett, London

33. Mackay TFC (2001) The genetic architecture of quantitative traits. Ann Rev Gen 35:303–339

34. Ellegren H, Sheldon BC (2008) Genetic basis of fitness differences in natural populations. Nature 452:169–175

35. Stinchcombe JR, Hoekstra HE (2008) Combining population genomics and quantitative genetics: finding the genes underlying ecologically important traits. Heredity 100: 158–170

36. Vos P, Hogers R, Bleeker M et al (1995) AFLP: a new technique for DNA fingerprinting. Nucleic Acids Res 23:4407–4414

37. Van Eck HJ, Van der Voort JR, Draaistra J et al (1995) The inheritance and chromosomal localization of AFLP markers in a non-inbred potato offspring. Mol Breeding 1:397–410

38. Castiglioni P, Ajmone-Marsan P, van Wijk R et al (1999) AFLP markers in a molecular linkage map of maize: codominant scoring and linkage group distribution. Theor Appl Genet 99:425–431

39. Peng J, Korol AB, Fahima T et al (2000) Molecular genetic maps in wild emmer wheat, *Triticum dicoccoides*: genome-wide coverage, massive negative interference, and putative quasi-linkage. Genome Res 10:1509–1531

40. Rogers SM, Isabel N, Bernatchez L (2007) Linkage maps of the *dwarf* and normal lake whitefish (*Coregonus clupeaformis*) species complex and their hybrids reveal the genetic architecture of population divergence. Genetics 175:375–398

41. Bonin A, Ehrich D, Manel S (2007) Statistical analysis of amplified fragment length polymorphism data: a toolbox for molecular ecologists and evolutionists. Mol Ecol 16:3737–3758

42. Meudt HM, Clarke AC (2007) Almost forgotten or latest practice? AFLP applications, analyses and advances. Trends Plant Sci 12:106–117

43. Raman H, Moroni JS, Sato K et al (2002) Identification of AFLP and microsatellite markers linked with an aluminium tolerance gene in barley (*Hordeum vulgare* L.). Theor Appl Genet 105:458–464

44. Herselman L, Thwaites R, Kimmins FM et al (2004) Identification and mapping of AFLP markers linked to peanut (*Arachis hypogaea* L.) resis-

tance to the aphid vector of groundnut rosette disease. Theor Appl Genet 109:1426–1433

45. Yuan L, Dussle CM, Muminovic J et al (2004) Targeted BSA mapping of *Scmv1* and *Scmv2* conferring resistance to SCMV using *Pst*I/*Mse*I compared with *Eco*RI/*Mse*I AFLP markers. Plant Breeding 123:434–437

46. Papa R, Bellucci E, Rossi M et al (2007) Tagging the signatures of domestication in common bean (*Phaseolus vulgaris*) by means of pooled DNA samples. Ann Bot 100:1039–1051

47. Cameron ND, van Eijk MJT, Brugmans B et al (2003) Discrimination between selected lines of pigs using AFLP markers. Heredity 91:494–501

48. Jump AS, Peñuelas J, Rico L et al (2008) Simulated climate change provokes rapid genetic change in the Mediterranean shrub *Fumana thymifolia*. Global Change Biol 14:637–643

49. Freeland JR, Biss P, Conrad KF, Silvertown J (2010) Selection pressures have caused genome-wide population differentiation of *Anthoxanthum odoratum* despite the potential for high gene flow. J Evol Biol 23:776–782

50. Klappert K, Butlin RK, Reinhold K (2007) The attractiveness fragment – AFLP analysis of local adaptation and sexual selection in a caeliferan grasshopper, *Chorthippus biguttulus*. Naturwissenschaften 94:667–674

51. Herrera CM, Bazaga P (2008) Population-genomic approach reveals adaptive floral divergence in discrete populations of a hawk moth-pollinated violet. Mol Ecol 17:5378–5390

52. Campbell D, Bernatchez L (2004) Genomic scan using AFLP markers as a means to assess the role of directional selection in the divergence of sympatric whitefish ecotypes. Mol Biol Evol 21:945–956

53. Scotti-Saintagne C, Mariette S, Porth I et al (2004) Genome scanning for interspecific differentiation between two closely related oak species [*Quercus robur* L. and *Q. petraea* (Matt.) Liebl.]. Genetics 168:1615–1626

54. Vuylsteke M, Mank R, Antonise R et al (1999) Two high-density AFLP linkage maps of *Zea mays* L.: analysis of distribution of AFLP markers. Theor Appl Genet 99:921–935

55. Ojha BR (2005) Comparison of Eco-Mse and Pst-Mse primer combinations in generating AFLP map of tomato. J Inst Agric Anim Sci 26:27–35

56. Cardon LR, Palmer LJ (2003) Population stratification and spurious allelic association. Lancet 361:598–604

57. Pritchard JK, Stephens M, Donnelly P (2000) Inference of population structure using multilocus genotype data. Genetics 155: 945–959

58. Evanno G, Regnaut S, Goudet J (2005) Detecting the number of clusters of individuals

using the software STRUCTURE: a simulation study. Mol Ecol 14:2611–2620

59. Falush D, Stephens M, Pritchard JK (2007) Inference of population structure using multilocus genotype data: dominant markers and null alleles. Mol Ecol Notes 7:574–578

60. Corander J, Waldmann P, Sillanpää MJ (2003) Bayesian analysis of genetic differentiation between populations. Genetics 163: 367–374

61. Corander J, Marttinen P (2006) Bayesian identification of admixture events using multilocus molecular markers. Mol Ecol 15: 2833–2843

62. Agresti A (2007) An introduction to categorical data analysis, 2nd edn. Wiley, Hoboken

63. Rencher AC, Schaalje GB (2008) Linear models in statistics, 2nd edn. Wiley, Hoboken

64. Peakall R, Smouse PE (2006) GENALEX 6: genetic analysis in Excel. Population genetic software for teaching and research. Mol Ecol Notes 6:288–295

65. Pigliucci M (2001) Phenotypic plasticity: beyond nature and nurture. Johns Hopkins University Press, London

66. Jombart T, Pontier D, Dufour AB (2009) Genetic markers in the playground of multivariate analysis. Heredity 102:330–341

67. Herrera CM (1993) Selection on floral morphology and environmental determinants of fecundity in a hawk moth-pollinated violet. Ecol Monogr 63:251–275

68. Ritland K (2005) Multilocus estimation of pairwise relatedness with dominant markers. Mol Ecol 14:3157–3165

69. Antlfinger AE, Curtis WF, Solbrig OT (1985) Environmental and genetic determinants of plant size in *Viola sororia*. Evolution 39:1053–1064

70. Rice WR (1989) Analyzing tables of statistical tests. Evolution 43:223–225

71. Narum SR (2006) Beyond Bonferroni: less conservative analyses for conservation genetics. Conserv Genet 7:783–787

72. Storey JD, Tibshirani R (2003) Statistical significance for genomewide studies. Proc Natl Acad Sci USA 100:9440–9445

73. R Development Core Team (2009) R: a language and environment for statistical computing. R Foundation for Statistical Computing, Vienna

INDEX

François Pompanon and Aurélie Bonin (eds.), *Data Production and Analysis in Population Genomics: Methods and Protocols*,
Methods in Molecular Biology, vol. 888, DOI 10.1007/978-1-61779-870-2, © Springer Science+Business Media New York 2012

Printed by Publishers' Graphics LLC
SO20130301.19.18.14